Science
and Sustainability

Science Education for Public Understanding Program
SEPUP

University of California at Berkeley

Lawrence Hall of Science

This project was supported, in part, by the
National Science Foundation
Opinions expressed are those of the authors
and not necessarily those of the Foundation.

Lab-aids
Ronkonkoma, New York

SEPUP Staff

Dr. Barbara Nagle, Director
Manisha Hariani, Associate Director
Lee Amosslee, Instructional Materials Developer
Janet Bellantoni, Instructional Materials Developer
Asher Davison, Instructional Materials Developer
Kate Haber, Instructional Materials Developer
Daniel Seaver, Instructional Materials Developer
Miriam Shein, Publications Coordinator
Roberta Smith, Administrative Coordinator
Ezequiel Gonzalez, Administrative Assistant

Dr. Herbert D. Thier, Founding Director

Contributors / Developers

Daniel Seaver	Barbara Nagle
Herbert D. Thier	Manisha Hariani
Laura Baumgartner	Mike Reeske

2005 update: Asher Davison

Teacher contributors

Victoria Deneroff	Mark Klawiter
Nancy Dibble	Angela Olivares
Dan Jeung	Veronica Peterson

Production

Book design and layout: Miriam Shein
Cover design: Maryann Ohki

Developmental editing: Devi Mathieu
Copyediting and permissions: Naomi Leite
Index: Laura Baumgartner

Content and scientific review

Dr. Mark Christensen, Professor Emeritus, Energy and Resources Group, University of California at Berkeley
Dr. Carolyn Cover-Griffith, Postdoctoral Fellow, Molecular and Cell Biology, University of California at Berkeley
Dr. Robert E. Horvat, Professor of Science Education, Buffalo State College

Research assistance

Marcelle Siegel, Dieter Wilk

Cover photographs © Material World. All are from *Material World: A Global Family Portrait* (Sierra Club Books) unless otherwise noted. Reproduced with permission. Top left (Mali): Peter Menzel. Top right (Great Britain): David Reed. Market scene (India): Peter Ginter. Children running (Japan): Peter Menzel. Bottom left (India): Peter Menzel. Bottom center (Albania, from *Women in the Material World* [Sierra Club Books]): Catherine Karnow. Bottom right (Mexico): Peter Ginter. Flask sphere, top left (Mongolia): Leong Ka Tai. Flask sphere, bottom left (Iceland): Miguel Fairbanks.

Material World appears in this course by special arrangement with Sierra Club Books and Material World, Inc. SEPUP is grateful to both for their assistance throughout this project. Special thanks to Peter Menzel for his encouragement.

4 5 6 7 8 9   09 08 07 06

The preferred citation format for this book is:
SEPUP. (2005). *Science and Sustainability*. Lawrence Hall of Science, University of California at Berkeley. Published by Lab-Aids®, Inc., Ronkonkoma, NY.

SEPUP
Lawrence Hall of Science
University of California at Berkeley
Berkeley, CA 94720-5200
e-mail: sepup@berkeley.edu
Website: www.sepuplhs.org

Published by:

LaB-aiDS

17 Colt Court
Ronkonkoma, NY 11779
Website: www.lab-aids.com

Field Test Centers

The classroom is SEPUP's laboratory for development. We are extremely appreciative of the following center directors and teachers who taught the program during the 1997–98 and 1998–99 school years. These teachers and their students contributed significantly to improving the course.

Anchorage, AK

Donna L. York, Center Director

Jocelyn Friedman, Andy Holleman, Shannon Keegan, Mark Lyke, Tim Pritchett, Gail Raymond, Lori Sheppard-Gillam, Deborah J. Soltis, Leesa Wingo

Brooklyn, NY

Veronica Peterson, Center Director

Bonnie L. Boateng, Regina Chimici, Jennifer Cordova, Geraldine Curulli, Sherry Dellas, Ian Harding, Edward Kosarin, Jeff Levinson, Len Monchick, Howard Paul, Lisa Piccarillo, Juan Urena, Kathy Wingate

Fresno, CA

Jerry D. Valadez and Bill C. Von Felten, Center Co-Directors

Staci Black, Nancy Dibble, Phyllis R. Emparan, Deborah L. Henell, Joel Janzen, Daniel Jeung, Scott Kruse, Dave Peters, Steve Wilson

Jefferson County, KY

Pamela T. Boykin, Center Director

Shannon Conlon, Craig A. DaRif, Nancy J. Esarey, Sherry Fox, Scott Schneider, Kevin Sharon, Melissa Spaulding, Megan Williams

Los Angeles, CA

Irene C. Swanson, Center Director

Justin Albert, Tammy Bird, Tom Canny, Will Carney, Roberto Corea, Victoria K. Deneroff, Elizabeth Garcia, Gaby Glatzer, Ray A. Harner, Barbara Hougardy, Finis Irvin, Eric Jackson, Karen Jin, Ronald Kimura, Christine Lee, William Mocnik, Ephran Moreira, Evelyn Okafor, Glenda Pepin, Jonathon Perez, Dennis Popp, Brian Smith, Tamara Spivak, Keith Sy, Lisa Trebasky, Muriel Waugh, Cynthia Williams

Riverside, CA

Dr. Karen Johnson, Center Director

Trish Digenan, Russell Ellis, Bonnie Ice-Williams, Patrick A. McCarthy, Deanna Smith-Turnage, Jay Van Meter

Winston-Salem, NC

Dr. Stan Hill, Center Director

Alison Cotarelo, Janet Crigler, Aubry Felder, Shelley Johnson, Cindy Kleinlein, Claire Kull, Pamela Lindner, Elizabeth Morgan, Vicki Nicholson, Jane Richards, Valerie Snell, Beverly Triebert

Wisconsin

Julie Stafford, Center Director

Mark Klawiter, Marian Schraufnagel

Independent

Angie Olivares, Irvine, CA; Mike Reeske, San Diego, CA; John L. Roeder, Princeton, NJ

Contents

Science and Sustainability is a different kind of science course. It not only covers many of the scientific concepts usually included in biology, chemistry, and physics classes, but also relates those concepts to issues of sustainability. Most likely, you have already explored many scientific concepts, but you may not be familiar with the term **sustainability**. Sustainability refers to the ability of populations of living organisms to continue, or sustain, a healthy existence in a healthy environment "forever." Many of the activities of today's human societies are not carried out in ways that promote sustainability. The scientific topics introduced in this course were chosen because they relate to sustainable development—that is, the use of environmental resources in a responsible way to ensure that they will continue to be available for use by future generations.

During this course you will participate in a wide range of activities, including many hands-on labs, current and historical readings, role-plays, and debates. You will also make frequent use of the book *Material World*, which provides a pictorial view of life in dozens of countries around the world. These activities will help you become more confident, competent, and independent in the design, analysis, and communication of issue-oriented science activities. You will be challenged to connect the various components of the course as you analyze risks, assess trade-offs, and make decisions that are based on scientific data.

Issues of science and sustainability impact your personal life, your local community, and the world as a whole. By considering some of these important issues in this course, you will gain the skills necessary for making decisions critical to your future and to the future of other living things on Earth. Some of the unsustainable practices of today's societies have resulted in overcrowding in urban areas, decreased air and water quality, reduced biodiversity, increased starvation, and other global problems. This course will present you with choices that might help solve some of these problems. After evaluating the scientific evidence, you will determine which options are appropriate for the problem in question. In some cases, you will even be given a chance to develop your own solutions for a problem.

This course is divided into four parts: "Living on Earth," "Feeding the World," "Using Earth's Resources," and "Moving the World." Each part focuses on the nature and implications of one theme related to sustainability.

Part 1

Living on Earth

Sustaining the existence of Earth's inhabitants requires that populations of each species survive long enough to produce and nurture the next generation. To survive, all organisms rely on other organisms and on Earth's non-living resources, including air, water, and soil. In Part 1 of *Science and Sustainability*, you will begin investigating a wide range of questions about life on Earth:

- What do humans and all other living organisms require to survive?

- Why, scientifically speaking, do organisms have these needs?

- How do organisms, including humans, fulfill those needs?

- How have science and technology contributed to the survival of modern humans?

- How have human attempts to survive and prosper affected the survival of other organisms?

- How have human attempts to survive and prosper affected the quality of the air, water, or soil?

- What effect does population growth have on the ability of humans to survive?

- What is the outlook for human survival in the future?

- What can today's human population do to increase the ability of future generations to survive?

Science and technology will continue to play an extraordinarily important role in our everyday lives. Part 1 of this course will introduce you to some fundamental scientific principles and processes. This information should help you better understand past scientific discoveries and their impact on society and the environment. It will also prepare you to evaluate the potential impact of future discoveries and inventions.

Sustainable Living

1.1   What Do We Want to Sustain?

Purpose **C**ompare the possessions of families in four countries and consider which are essential for survival.

Introduction

Sustainability, at its most basic level, means continued survival. People in different parts of the world own and use different types of materials and energy. Some of these materials are needed for survival; others are not. The book *Material World* contains photographs, statistics, and written descriptions that provide evidence of what life is like for average families in 30 countries around the world. At the beginning of the section on each country is a "big picture" showing a typical family outside their home, surrounded by all their possessions.

The material goods that belong to an individual or a family serve many purposes. In this activity, you will examine the material possessions of families from four different countries, then decide which possessions are essential for survival and which are not. You will also consider how non-essential possessions affect the sustainability of a society.

Prediction

Work with your partner to come up with an estimate for the percentage of items found in a typical household that are essential for survival.

Materials

■■ **For each team of two students**

1   copy of *Material World*

Procedure

1. Take a few moments to look through *Material World* and get a sense of the information it provides.

2. Look at the locator map in Figure 1 and on pages 4–5 in *Material World* to see where each of the following countries is located: Thailand, Iceland, Ethiopia, and Guatemala. Based only on information from the maps, rank these four countries in order from "easiest to survive in" to "most difficult to survive in."

3. Now, look carefully at the photographs in *Material World* of the families from Thailand, Iceland, Ethiopia, and Guatemala. Make a data table similar to the one on the next page to record your observations.

4. For each family, predict the purpose of one of the possessions you do not recognize.

5. Typically, essential natural resources are not considered personal possessions, but are instead shared by all members of the community. Describe any evidence in *Material World* that might indicate whether or not each family has an adequate supply of clean air, water, and soil.

6. Based on the information you now have, again rank the four countries in order from "easiest to survive in" to "most difficult to survive in."

Figure 1   Locator Map

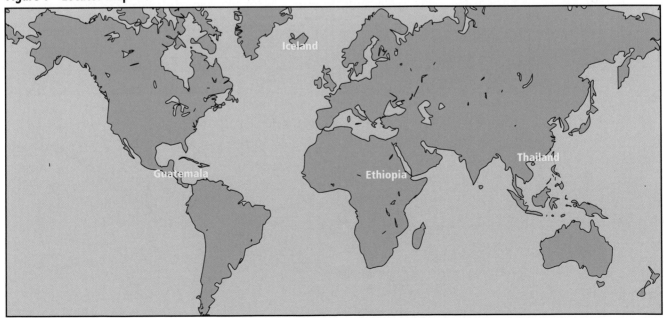

Procedure
(cont.)

Table 1   Family Possessions

	Thailand	Iceland	Ethiopia	Guatemala
Name of family				
5 possessions most essential for survival				
Estimated % of possessions that are essential				
Essential resources not shown in photographs				
Family's 5 most valued possessions				
5 possessions similar to something in your home				
5 possessions you do not recognize				

Analysis

?

Group Analysis

1. Does your family have possessions that meet the same needs as things owned by the families pictured in *Material World*? Explain.

2. Which possessions do you recognize as similar to something your family owns? Do these items tend to be essential or non-essential?

3. In general, are a family's most valued possessions the same as those they need for survival? How do you think people determine the value of their possessions? Explain.

Individual Analysis

4. Do you think the estimate you made in the Prediction is still a good one? Explain why or why not.

5. Think about your own family's possessions. List the five you consider to be the most valuable and the five you consider most essential. Is your list similar to any of the lists you made for families in *Material World*? Are your family's most valued possessions the same as those you need for survival? Explain.

6. Does the number of material goods a family owns have any relationship to soil, air, or water quality? Explain.

7. Using your knowledge of your own community and your observations about the four families from *Material World*, describe two practices that promote sustainability for future generations and two that do not.

1.2 Survival Needs

Purpose ▶ **I**dentify items that are absolutely necessary for human survival.

Introduction

The survival of any organism requires the maintenance of a certain level of activity within its cells. To do so requires access to essential resources that are determined, in part, by the environment in which the organism lives. No matter what the environment may be like, certain resources are needed to meet the basic needs of the human organism.

Scenario

Because of your excellent qualifications, you and your group members have been chosen to go on a 14-day scientific expedition to explore a remote area of planet Earth. Prior to this expedition, the area has never been visited by humans. You will be going to one of the following environments:

Equatorial Desert	Equatorial Jungle
Polar Ice Cap	Mountain Forest

Because of the nature of your expedition, no motorized vehicles can be used. You and your group members will be hiking about 10 miles a day, carrying all your supplies with you.

Materials **For each group of four students**

1 large piece of paper

1 marking pen

Procedure

1. List any essential resources that you can assume will be supplied by the natural environment.

2. Prepare a list of items that your group members agree would have to be brought with you on your expedition.

3. Underline those items that you would need no matter what environment you were in.

Analysis

Group Analysis

1. Decide upon three or four categories (such as protection) into which the supplies on your list can be classified.

2. What basic human needs do the supplies you underlined provide for?

3. What basic human needs do the supplies you did not underline provide for?

4. What types of scientific knowledge could help you explain why humans have these basic needs?

Individual Analysis

5. Imagine what life might have been like for the very first humans on Earth. Describe what you think are the top five ways in which science has made your life easier than it was for them.

1.3 Burn a Nut

Purpose ▶ **W**eigh the advantages and disadvantages of using different energy sources.

Introduction

To survive, our bodies need a constant supply of energy, much of which is provided by the sun. Other commonly used **energy sources** fall into two categories—food and fuel. The energy in foods and fuels is stored within the chemicals that make up these materials. This **stored energy** can then be released during chemical reactions. When the energy is released, it is transferred from the energy source to something else. Foods are used by all organisms to supply the energy and nutrients needed for life. Fuels are used by people and their societies to provide energy for tasks they consider important, such as staying warm, cooking food, or generating electricity. Fuels can be classified as **renewable** or **non-renewable resources**. Renewable fuels, such as firewood and other plant products, can be replenished within a single human lifetime. Non-renewable fuels, like coal and gasoline, cannot be replenished during a single human lifetime.

Why does our society tend to use non-renewable fuels rather than renewable ones? Whenever you make a choice, there are **trade-offs** that must be considered. What are the advantages of using non-renewable fuels? What are the disadvantages?

Nuts are an example of a renewable resource that can be used as a food or a fuel. In this activity, you will actually burn a nut and measure the amount of energy stored in it. During the combustion process, the energy stored in the nut will be transferred from the nut to some water. The amount of **energy transfer** can be evaluated by measuring the change in the temperature of the water. You will also compare the amount of energy stored in a nut to the amount stored in kerosene, which is a non-renewable, non-edible fuel.

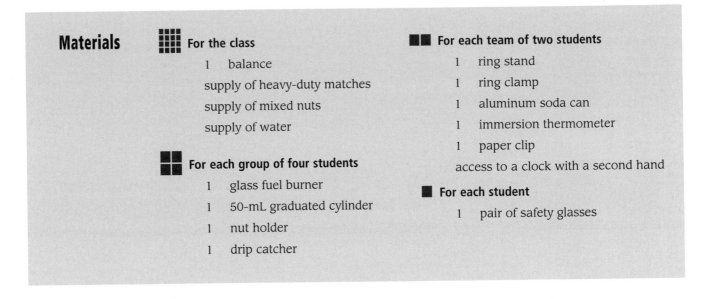

Materials

▦ For the class

1	balance
	supply of heavy-duty matches
	supply of mixed nuts
	supply of water

◰ For each group of four students

1	glass fuel burner
1	50-mL graduated cylinder
1	nut holder
1	drip catcher

◫ For each team of two students

1	ring stand
1	ring clamp
1	aluminum soda can
1	immersion thermometer
1	paper clip
	access to a clock with a second hand

◼ For each student

1	pair of safety glasses

Safety Note

You will be working with open flame, so be very careful. Wear safety glasses and be sure to protect your hands and clothes from hot laboratory equipment. If you have long hair, it should be pulled back to avoid exposure to the flame. You will also be working with glassware, including a glass thermometer. If any glass breaks, notify your teacher immediately. Do not clean it up without your teacher's approval.

Procedure

Part A   Measuring the Energy in a Nut

1. Prepare a data table similar to the one below.

2. Determine the combined mass of your nut, nut holder, and drip catcher.

3. Arrange your nut on the nut holder as shown in Figure 2.

4. Pour 50 mL of room-temperature water into your can, record the initial water temperature, and set up the can as shown in Figure 3, on the next page.

Table 2   Burning a Nut

Type of nut _____	Initial	Final	Change (Δ)
Temperature of water in can			
Combined mass of nut, nut holder, and drip catcher			
Time			

Figure 2   Arranging the Nut

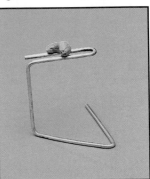

Procedure
(cont.)

5. Follow the specific directions given by your teacher to light the nut. As soon as it is lit, record the time to the nearest second. If you use a match, be sure to wet it thoroughly before placing it in the trash.

   Note: Position the bottom of the can in the flame of the burning nut, but not so close that it puts out the flame.

Figure 3   Suspending the Can

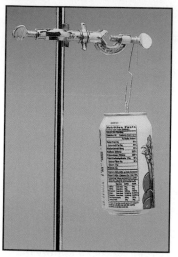

6. Allow your nut to burn until it goes out. As soon as the nut stops burning, record the time.

   Note: If you think the nut stopped burning before it was completely consumed by the flame, ask your teacher to try to re-light the nut.

7. Record the highest temperature reached by the water in the can.

8. Calculate the change in temperature of the water in the can (ΔT).

9. Measure and record the mass of the nut and nut holder again.

10. Calculate the change in mass of the nut (Δm).

11. Record your calculations on the class data table.

Part B   Measuring the Energy in Kerosene

12. Repeat Steps 2–11, using a glass fuel burner containing kerosene instead of a nut. Allow the kerosene to burn for exactly 3 minutes.

13. Record your data on the class data table.

Analysis
?

Group Analysis

1. Which was the better fuel, the nuts or the kerosene? Explain.

2. Which type of nut contained the most stored energy? Explain.

3. How would you change this experiment to more accurately compare the amount of energy stored in each type of nut?

4. What information, other than energy content, would help you determine which type of nut provides the most nutrition?

Individual Analysis

5. Would you recommend the use of nuts for food, for fuel, or both? Provide evidence for each choice, and then explain why you chose one over the other.

Extension   Carry out the experiment you designed for Analysis Question 3. Prepare any appropriate data tables and report your findings.

1.4 Lessons From a Small Island

Purpose ▶ **E**xplore the rise and fall of a once-flourishing society.

Introduction ▼

If you look at a map of the South Pacific Ocean, you'll find a small, isolated island lying more than 3,500 km (2,100 miles) off the coast of Chile and at least 1,500 km (900 miles) from the nearest inhabited island. Its land area totals 166 km², which is less than half the size of Lana'i, the smallest Hawaiian island. Its modern inhabitants call the island Rapa Nui, which is also the name of the language they speak.

The first written accounts of the island and its people came from Dutch explorers, who landed there on Easter Sunday in 1722. They named their new discovery Easter Island. When the explorers returned to Europe, they reported their amazement at finding an island dotted with hundreds of enormous stone monuments. The presence of these monuments was considered startling evidence that a much larger, much more complex society must have existed on the island in the past. The mystery of the stone monuments has prompted geologists, archeologists, and historians to develop an account of what must have happened on Rapa Nui before the arrival of the Dutch explorers.

Rapa Nui's small size and isolated location severely limit the resources available to its inhabitants. In this reading, you will explore the island and the changes that have taken place there over the past six or seven hundred years. Can the vanished society of Rapa Nui be compared to today's human global population? Can that society's experiences shed some light on dangers and opportunities that may be facing individuals and societies today? What actions are necessary to ensure that Earth's resources used today will also be available for generations to come?

Rapa Nui

The first humans to settle on Rapa Nui were Polynesian explorers who probably arrived in the third or fourth century AD. They found a fertile, uninhabited island that was dense with palm forests but supported less than 50 other types of plants. There were a few species of insects, but no animals larger than the island's two species of lizards. Rapa Nui was created by volcanic activity; the only supply of fresh water was contained in lakes that had formed in the craters of extinct volcanoes. The surrounding ocean waters contained very few species of fish. The Polynesians had brought with them the food crops they were accustomed to planting, but they found that many did not grow well on the island. Fortunately, chickens and sweet potatoes thrived, and the settlers had some success growing taro and yams, so they were able to provide themselves with an adequate variety of foods. From these crude beginnings, the people of Rapa Nui developed into a technologically and socially advanced society.

The island population eventually reached a maximum of about 10,000 people. Using a combination of their own ingenuity and the island's limited resources, they were able to build houses, grow and prepare food, make clothing, keep warm during cold and rainy seasons, and care for their children. The society developed a rich culture, including unique religious customs and social practices. Their use of technology was impressive: they built enormous temple complexes, called "ahu," as well as the more than 600 massive stone statues, called "moai," that have made the island famous. Scattered across the island, many of these statues reach a height of over 6 meters (about 20 feet) and weigh over 9,000 kg (about 10 tons). Today, the ahu and moai of Rapa Nui have become a popular tourist attraction.

Modern research shows that all of the moai were carved from similar rock. In fact, they probably all came from the same island quarry. The people of Rapa Nui had to find a way to transport these statues as far as 20 km (12 miles). Island legends tell tales of the moai "walking" to their sites. These legends have led researchers to believe that the islanders rolled the upright statues on palm logs

from the quarry to their final destinations. They not only developed the tools and technologies required for carving stone and building complex structures, but also manufactured ropes, pulleys, lubricating oil, and other devices for transporting their enormous stone creations.

The people of Rapa Nui also used palm logs for housing and for fuel. This widespread use of the palm forests resulted in gradual deforestation of the island. As the number of trees decreased, the newly exposed soil began to erode. Erosion reduced the soil's fertility, making it less suitable for the growth of food crops as well as trees. The islanders' resources became increasingly scarce, and competition over them resulted in violence—including, according to some sources, cannibalism.

By the time of the Dutch explorers' arrival in 1722, there were very few trees left on the island. The population had fallen dramatically, and the approximately 600 people who were left waged almost constant war on one another. In the end, dwindling resources were the primary factor responsible for the demise of the ancient society of Rapa Nui. Although the islanders developed a sophisticated cultural and technological society, they were ultimately unable to sustain a balance between their needs and the resources their environment could provide.

These stone statues, called moai, were carved centuries ago by the earliest inhabitants of Rapa Nui. It is believed that the massive statues were rolled across the island on logs.

Analysis

Group Analysis

1. How do you think humans were able to survive on the limited resources of an isolated island for hundreds of years?

2. What caused the island's environment to lose its ability to sustain the first society on Rapa Nui? Along with your explanation, include a diagram that shows the major relationships between the living and non-living components of the island.

3. What steps could the islanders have taken to help sustain life on Rapa Nui?

Individual Analysis

4. Could the island of Rapa Nui be seen as a model for Earth? Describe ways in which the island's ancient society seems similar to or different from today's society.

5. Describe any present or future actions people might take to help sustain life on Earth.

6. Describe any present or future human actions that could push Earth beyond its ability to sustain life.

Survival Needs: Food

2.1 Observing Producers and Consumers

Purpose ▶ **E**xamine some of the producers and consumers that live in aquatic environments.

Introduction

Energy is essential to all living things. During **photosynthesis**, green plants convert radiant energy from the sun into chemical energy. This chemical energy provides for plant growth and, directly or indirectly, serves as the primary food source for virtually all other life forms. Plants and other organisms that are capable of photosynthesis are called **producers**, because they produce food—any edible source of chemical energy—for other organisms. Organisms that obtain their energy by eating producers are known as primary **consumers**. Organisms that eat primary consumers are known as secondary consumers. Tertiary consumers are organisms that eat secondary consumers. An **ecosystem** is made up of populations of many organisms that rely upon each other, plus the air, water, and other shared features of the environment, for their sustained existence. Producers can be found in almost every environment on Earth. They are essential to the structure of virtually all ecosystems.

The open ocean, which supports populations of many fish, is one example of an ecosystem. Imagine a fishing boat in the middle of the ocean, far from any land. What do the fish eat? Where are the producers and what types of organisms are they?

Introduction
(cont.)
▼

Much of the life in the open ocean consists of tiny organisms, called plankton, that inhabit the top 20 meters of water. They are restricted to this narrow layer because sunlight can penetrate the water to this depth fairly easily. Some plankton are capable of photosynthesis and are called phytoplankton (Figure 1). Other plankton, called zooplankton (Figure 2), are consumers that feed on phytoplankton. Although plankton are very small, even microscopic, they are extremely numerous. They are essential to the health of the entire marine environment. Plankton are also essential to freshwater ecosystems, including lakes and streams.

Figure 1   Common Freshwater Phytoplankton

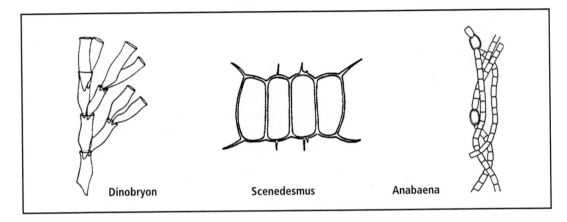

| Dinobryon | Scenedesmus | Anabaena |

Figure 2   Common Freshwater Zooplankton

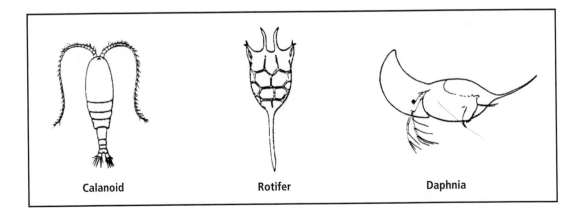

| Calanoid | Rotifer | Daphnia |

▶

Materials

■■ **For each group of four students**

1	prepared slide of phytoplankton
1	prepared slide of zooplankton
1	dropper
1	15-mL dropper bottle of methyl cellulose
1	graduated cup with water containing live plankton

■■ **For each team of two students**

1	microscope
1	microscope slide with a well
1	coverslip

Procedure

Part A   Observing Prepared Slides

1. You and your partner will be observing a prepared slide of phytoplankton and a prepared slide of zooplankton. Choose one slide and place it on your microscope stage.

   Note: Most of the plankton you will see are small, multicellular creatures similar to the ones shown in Figures 1 and 2. You may also see some one-celled organisms.

2. Center the slide so that the specimens are directly over the light aperture and adjust the microscope settings as necessary.

3. Begin by observing the specimen at the lowest level of magnification and drawing what you see.

4. Observe the specimen at each of the higher levels of magnification and draw what you see at each magnification level.

5. Repeat Steps 2–4 with your group's other prepared slide.

6. Observe and sketch the other six microscope slides as directed by your teacher.

Part B   Searching for Plankton

7. Use the dropper to place a drop of water containing live plankton into the well of a microscope slide.

8. Add one drop of methyl cellulose to the water on the slide.

9. Carefully touch one edge of the coverslip to the water, at an angle. Slowly allow the coverslip to drop into place. This helps prevent air bubbles from becoming trapped under the coverslip.

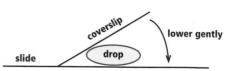

Procedure
(cont.)

10. Center the slide so that the well is directly over the light aperture and adjust the microscope settings as necessary.

11. Begin by observing the sample on the lowest objective. You may need to search the slide to find specimens, or they may move across your field of view.

12. Draw at least two different organisms that you observe. Be sure to record the level of magnification of each drawing. Use the drawings in Figures 1 and 2 to help you identify the organisms in your drawings as precisely as you can.

Analysis
?

Group Analysis

1. Identify at least two similarities and two differences between phytoplankton and zooplankton.

2. What is the role of phytoplankton in an aquatic ecosystem?

3. Did you observe evidence of the roles of phytoplankton or zooplankton in your sample of live plankton? Explain.

Individual Analysis

4. Name at least five macroscopic producers and five macroscopic primary consumers common to your local area. Identify how each one can impact human life.

5. How might plankton impact humans? Explain.

6. Do you think the ocean can provide a sustainable source of food for the world's human population? Explain why or why not.

2.2 The Web of Life

Purpose ▶ **U**se a food web to explore the transfer of chemical energy within an ecosystem.

Introduction

Every organism, dead or alive, is a potential source of food for other organisms. A **food chain** can be used to show how energy is transferred, via food, from producers to primary consumers, from primary to secondary consumers, and so on. In nature, many consumers feed on both producers and other consumers. Because of this, a food chain may not adequately describe all the pathways of energy transfer. A **food web** like the one shown in Figure 3 is a diagram that more completely illustrates the transfer of chemical energy within an ecosystem. **Decomposer** organisms, typically bacteria and fungi, feed on dead and decaying matter and are an important part of any ecosystem.

Based upon its position in the food web, each member of an ecosystem can be assigned to a **trophic level**. Producers (mostly plants and plankton) are in the first, or lowest, trophic level; primary consumers (usually herbivores) are in the second trophic level; secondary and tertiary consumers (carnivores and omnivores) are in the third and fourth trophic levels. Whenever one organism eats another, chemical energy is transferred to a higher trophic level. However, as with many energy transfers, a large portion of the energy stored in the food cannot be used by the consumer. As a result, only a small percentage

Humans and other organisms are part of food webs. These sheep are classified as primary consumers because plants are their main source of food.

Figure 3   A Simple Food Web

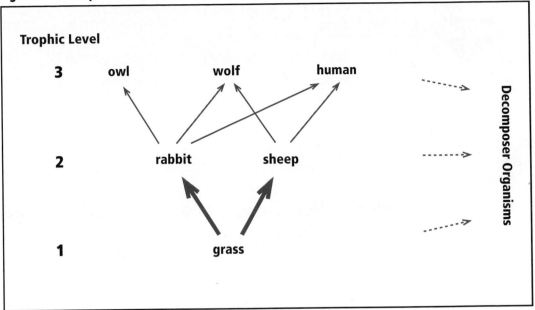

of the energy available at lower trophic levels is actually available to organisms at higher trophic levels; the majority is considered "lost" to the environment. The Law of Conservation of Energy tells us that the "lost" energy does not disappear, because energy cannot be created or destroyed. But it can take forms that are not usable by consumers. Some of the energy is transformed into heat, and some remains in the portions of the food not eaten or not digested by the consumer. The amount of energy stored in a food is measured in **Calories**.

Procedure

1. Draw a food web showing the pathways by which chemical energy is transferred among the organisms listed below.

trout	gulls
humans	catfish
minnows	bacteria
insects	phytoplankton
herbivorous zooplankton	carnivorous zooplankton

Analysis

Group Analysis

1. Rank the organisms in your food web from "most abundant in the ecosystem" to "least abundant in the ecosystem." Explain how you determined your rankings.

2. When one organism eats another, not all of the chemical energy stored in the food gets transferred to the consumer. What happens to the energy "lost" during each energy transfer?

Individual Analysis

3. In one square meter of a typical open ocean ecosystem, phytoplankton can generate 1,600,000 Calories of chemical "food" energy per year. Assume that there is an 80% "loss" of usable chemical energy during transfers from one trophic level to the next. Make a data table and a graph showing energy available at each trophic level for the first five trophic levels.
   Note: The energy "lost" between trophic levels is often greater than 80%!

4. How many more humans could be fed if everyone ate from the second trophic level rather than the fifth? Is this a reasonable possibility? Explain.

Population Estimation

Purpose ▶ **E**valuate a method used by wildlife biologists to estimate population size.

Introduction ▼

Understanding the interactions among the various organisms of an ecosystem requires a reasonable estimate of the **population** size of each species. A population is a group of organisms of the same species living in the same area at the same time. How could you determine the size of the population of animals living in the wild?

Wildlife biologists use a variety of sampling and data analysis methods to estimate population size. In this activity you will model a sampling method called "capture-tag-recapture" to investigate a population of sea otters.

Otters often float on their backs while feeding, as is shown here. In this activity you will use plastic chips to represent a population of otters.

Materials   ▣   **For each group of four students**

     1   bag containing plastic disks

     masking tape

Procedure

Part A   Capturing the Otters

1. The bag represents a bay in Alaska. Each plastic disk in the bag represents one sea otter. Capture a random group of otters by closing your eyes and removing one handful of plastic disks (10–20) from the bag.

2. Count the number of disks you removed from the bag. These disks represent your captured otters.

Part B   Tagging and Releasing the Otters

3. Tag each of your captured otters by placing a small piece of masking tape on each of the disks.

4. Release the tagged otters into the bay by placing them back into the bag.

5. Shake the bag for 30 seconds to simulate the passage of enough time to allow the tagged otters to mix back in with the rest of the otter population.

Part C   Recapturing a Sample of Otters

6. Capture another random group of otters by removing a handful of disks (10–20) from the bag.

7. Count and record the total number of otters you captured.

8. Count and record the number of tagged otters you captured.

9. Release the captured otters into the bay by placing them back into the bag.

Part D   Recapturing a Second Sample of Otters

10. Repeat Steps 6–9.

11. Remove the tape from all of the tagged otters.

12. Make a table that clearly displays the data you collected in this activity.

Analysis

Group Analysis

1. Based on the number of otters you recaptured in Part C, calculate an estimate for the total population using the following proportion:

$$\frac{\text{total number of tagged otters}}{\text{estimated total population of otters}} = \frac{\text{number of tagged otters captured in Part C}}{\text{total number of otters captured in Part C}}$$

2. Based on the number of otters you recaptured in Part D, use the same proportion to calculate an estimate for the total population of otters.

3. Which captured sample do you think gives you the better estimate for the number of otters in the bay? Explain.

4. Based on all your experimental results, predict the actual otter population in the bay. Explain.

5. Count the total number of disks in your bag. How close is the prediction you made for Analysis Question 4 to the total number of disks?

6. Use the following equation to calculate the **experimental error** for each of your samples. Convert each error value to a percent.

$$\text{error} = \frac{\text{actual population} - \text{estimated population}}{\text{actual population}}$$

Individual Analysis

7. Why are biologists and ecologists interested in determining the sizes of populations of organisms other than humans?

8. Do you think the capture-tag-recapture method would be useful for estimating the population of every type of organism? Explain why or why not.

9. You did not investigate how well the capture-tag-recapture method of population estimation would work in more realistic settings, where organisms are born and die, or migrate into and out of the ecosystem. Form a hypothesis that states whether or not the capture-tag-recapture method would still provide a reasonable estimate under those conditions. Explain why or why not.

10. Design an experiment that would test your hypothesis from Analysis Question 9.

Extension

Carry out the experiment you designed for Analysis Question 10. Prepare any appropriate data tables and report your findings.

2.4 Where Have All the Otters Gone?

Purpose ▶ **E**xplore how a change in the population of a single organism can affect other organisms in the same ecosystem and in neighboring ecosystems.

Introduction

During the early 1990s, people in Alaska noticed that there seemed to be fewer sea otters frolicking in the coastal waters. A study revealed that the sea otter population was actually declining at approximately 25% per year! Scientists were baffled by this steep decline, especially since they saw no signs of widespread disease in the otters and no decrease in the otters' food supply. In fact, the population of sea urchins, the otters' main food source in the nearshore coastal ecosystem, was larger than ever. In a seemingly unrelated study, another group of scientists documented a large decrease in the population of ocean perch living in a nearby ecosystem—the offshore open ocean ecosystem.

Procedure

1. Look carefully at the food webs for the two ecosystems shown in Figure 4 and the annual diet for each organism given in Table 1. Describe any links between the coastal ecosystem and the open ocean ecosystem.

   Note: The bottom member(s) of any food web are the producers. You observed plankton in Activity 2.1. If you have visited the ocean shore, you may have seen kelp, which is a large brown or green algae that is often referred to as seaweed.

2. Draw what you think the food web for the open ocean ecosystem would look like if the perch population fell to 10,000,000, a drop of 50%. Make sure to include estimates of the population of each of the other members of the ecosystem.

3. Orcas (killer whales) are not picky eaters. Rather than go hungry, they will actively hunt for alternative sources of food. Draw what you think the food web for the coastal ecosystem would look like if the perch population were only 10,000,000 and the orca population remained at 10. Make sure to include estimates of the populations of each of the other members of the ecosystem.

Figure 4   Open Ocean and Nearshore Coastal Food Webs

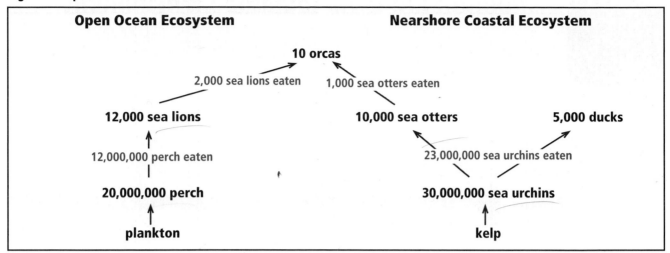

Table 1   Annual Diet for Each Organism

1 orca	eats	200 sea lions and 100 otters (a sea lion has twice the nutritional value of an otter)
1 sea lion	eats	1,000 perch
1 sea otter	eats	2,000 sea urchins
1 duck	eats	600 sea urchins

Note: The numbers provided in this simulation are used for illustrative purposes only and were not derived directly from actual studies. Numbers in black represent the population of that organism at the beginning of the year, and numbers in blue represent the number of organisms eaten during that year.

Analysis

Group Analysis

1. Describe the major differences between the open ocean food web you drew for Procedure Step 2 and the original one provided.

2. Describe the major differences between the coastal food web you drew for Procedure Step 3 and the original one provided.

3. What would happen to both ecosystems if the orca population fell to five? Draw a new food web for each ecosystem and briefly explain why the changes occurred.

Individual Analysis

4. How are humans in a similar situation to the orca? How does the human situation differ from that of the orca?

2.5 Maintaining a Sustainable Environment

Purpose ▶ **D**esign and set up a self-sustaining environment for an organism.

Prediction ◀?▶ Do you agree or disagree with the following statement? Explain.

An entire school of fish can live for years in a small pond. Therefore, if just one of those fish is put in a fishbowl, it should be able to live for years, too.

Introduction ▼ For an organism to survive, it must have a means of obtaining all materials necessary for life and disposing of unwanted waste products. Relationships among the living and non-living features of an organism's natural ecosystem have evolved so that the unwanted by-products of one form of life can be used directly by another life form or converted into materials needed by other organisms. In a pond, for example, green algae produce oxygen during photosynthesis and release it into the water. This oxygen can then be used by fish living in the pond. It is this balance of nature that has allowed Earth's finite supply of resources to recycle and continue to sustain life for billions of years.

Materials  **For the class**

supply of organisms

supply of 2-liter bottles

Procedure

Part A Designing an Ecosystem

1. List all the needs that you think your organism must have.

2. List all the wastes produced by your organism that you think could interfere with its ability to survive.

3. List all the living and non-living items that you think must be in your organism's ecosystem to ensure that the organism (or its offspring) will still be alive after one year.

Analysis

?

1. Briefly explain why you chose to include each item on the list you created for Procedure Step 3.

Procedure (cont.)

Part B Building an Ecosystem

4. Gather all the items on your list from Step 3 and set up your ecosystem.

Part C Observing an Ecosystem

5. Examine your ecosystem every week. Record your observations in your journal. Focus on those aspects that are related to the health of the organism and its environment.

Analysis (cont.)

?

2. If you think you must alter the environment to keep your organism alive, write down what you think the problem is and how it could be fixed. Then discuss the situation with your teacher.

Survival Needs: Temperature   3

3.1   Bubble-Blowing Fungi

Purpose ▶ **D**etermine the effect of temperature on yeast activity.

Introduction

Life on Earth is found in all types of environments, from the freezing cold of polar regions to the searing heat of equatorial deserts. However, environments with more moderate temperatures support many more organisms and a larger variety of species. The temperature of an organism's surroundings affects its ability to perform functions essential to survival. Organisms that have adapted to extreme climates represent a small percentage of the species found on Earth, and their populations tend to be fairly small. In this activity, you will begin to explore how and why temperature affects essential life processes by investigating live **yeast**.

Carbon dioxide bubbles produced by yeast give bread its airy texture.

Yeasts are single-celled fungi. They are commonly found on plants, on skin, in intestines, and in soil and water. Each yeast cell obtains energy to live and reproduce by digesting sugar through the process of **fermentation**. A by-product of fermentation is carbon dioxide gas. Humans have benefited, and suffered, from the growth of yeast for thousands of years. Carbon dioxide produced by some types of yeast makes bread rise and beer bubbly, but other varieties of yeast cause disease.

Prediction

At what temperature do you think yeast will be most active? Explain.

Materials

 For each group of four students

3	beakers
3	metal-backed thermometers
1	10-mL graduated cylinder
1	hot plate or other heat source
1	cup containing ~100 mL of yeast suspension
	supply of ice

For each team of two students

4	test tubes
1	packet of sugar
1	stir stick with scoop
1	metric ruler
	access to a clock with a second hand

Safety Note

Use caution when working with the hot plate and hot water.

Procedure

1. Work with your partner to design an experiment, using the materials listed above, to find out how different temperatures affect yeast.

   Hints:

   a. The more active the yeast, the more carbon dioxide produced. The production of carbon dioxide results in the formation of a layer of foam on top of the yeast solution.

   b. Some **variables** you should consider are the temperatures you will use, the amount of sugar you will provide, the length of time you will allow the yeast to ferment, and the method you will use to measure the activity level of the yeast.

2. Write up your experimental plan. It should include your prediction and a full step-by-step description of the procedure you will use.

3. Review your plan with your teacher.

4. Conduct your investigation. As you perform the experiment, be sure to carefully record your observations and measurements.

Analysis

Group Analysis

1. There are two kinds of scientific data, qualitative and quantitative. **Qualitative** data consist of general observations that can best be described using words or drawings. **Quantitative** data consist of numerical values for quantities being measured. Would you say your data about yeast fermentation are qualitative or quantitative? Explain.

Individual Analysis

2. Write at least one paragraph describing your results and stating your conclusions. Provide a summary of experimental results that relate to the purpose of your investigation, and describe experimental evidence that indicates whether or not your prediction was accurate.

3. How would you improve your investigation if you had a chance to do it again? How could you obtain more exact measurements? Describe the changes you would make and explain the reasons for the changes.

3.2 Some Like It Hot

Purpose ▶ **L**earn how temperature affects life processes and explore some of the mechanisms that enable organisms to maintain appropriate internal temperatures.

Introduction ▼

Most organisms function best when their internal temperature is in the range of about 20°C to 45°C. In the previous activity, you observed that yeast cells were most active at temperatures within this range. You also may be aware that normal human body temperature is 37°C. Every organism has a set of **optimum** conditions under which it most easily thrives. Why don't organisms function well when exposed to extremely hot and cold conditions? What adaptations allow certain organisms to survive at temperatures that would normally result in cellular death?

Seals can live in very cold water because they have a thick insulating layer of fat that helps them maintain their optimal body temperature.

Temperature and Life

Most organisms cannot survive in very hot environments because high temperatures change the shapes of proteins. **Proteins** are molecules required for the survival, growth, and repair of all cells. When a protein molecule loses its normal shape, it can no longer function properly. Albumin, the protein in egg white, provides a familiar example. Albumin is normally soluble in water, and is part of the clear liquid surrounding the yolk in a fresh egg. When an egg is heated, the albumin molecules change shape, making them insoluble. This causes the formation of a solid egg white.

Cold temperatures do not normally destroy proteins, but they do slow the rate of chemical reactions taking place in the cell. The rate of these chemical reactions determines the rate of energy use, or **metabolism**. Animals that hibernate are able to survive long periods of cold weather because their slowed metabolism reduces their need for food. Most organisms cannot survive freezing temperatures, though. At temperatures of 0°C or lower, ice crystals may form in the water inside cells, which can cause irreparable damage. Typical changes that take place in the human body when exposed to extreme temperatures are shown in Table 1.

Most organisms can tolerate only a narrow range of temperatures. As a result, they must either limit themselves to life in moderate environments or develop adaptive mechanisms that will enable them to maintain appropriate internal temperatures despite environmental extremes. The normal chemical activity that takes place within cells could be considered one such adaptation. **Cellular respiration** is one of the chemical processes that takes place within the cells of all organisms. The fermentation that takes place in yeast cells is one type of respiration. During respiration, glucose and other food molecules are broken down by chemical reactions that

Table 1   Human Responses to Body Temperature Extremes

Body Temperature (°C / °F)	Symptoms
28 / 82.4	muscle failure
30 / 86	loss of body temp. control
33 / 91.4	loss of consciousness
37 / 98.6	normal
42 / 107.6	central nervous system breakdown
44 / 111.2	death

release energy. Approximately 60% of the energy contained in glucose molecules is released in the form of heat. The remaining portion is converted into chemical energy, which may be used immediately for growth and movement or stored for future use. Thus merely being alive can generate enough heat to maintain the temperature of many organisms at or near their optimal range. Figure 1 diagrams this flow of energy in a human, beginning with digestion and cellular respiration and ending with the release of heat energy to the surrounding environment. Small animals have higher rates of metabolism—that is, cellular processes take place more quickly in small animals than in larger animals. This higher metabolic rate is needed because smaller animals lose a greater percentage of the heat they generate to the environment, as shown in Figure 2.

Depending on the organism and its environment, there may be a need to preserve warmth by decreasing heat flow from the body, or a need to avoid overheating by increasing heat flow from the body. Animals have developed a variety of adaptations for regulating internal temperature.

Figure 1   Energy Flow Diagram for Humans

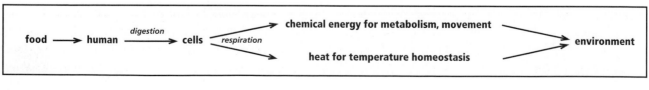

Figure 2   Metabolic Rates of Animals vs. Body Weight

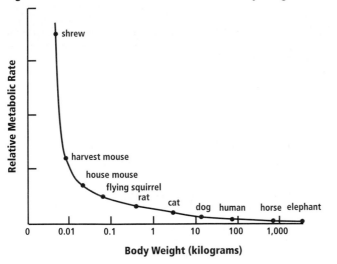

Figure 3   Rattlesnake (a poikilotherm)

You may have heard of "warm-blooded" and "cold-blooded" animals. These names are misleading, because some so-called "cold-blooded" animals can actually maintain higher internal temperatures than some that are considered "warm-blooded." The body temperature of "cold-blooded" animals varies with the surrounding environment and can therefore be quite warm. These animals, including reptiles and fish, are more accurately classified as **poikilotherms**, from the Greek words *poikilos*, which means "changeable," and *thermos*, which means "heat." The "warm-blooded" animals, primarily birds and mammals, maintain fairly constant body temperatures despite changes in environmental temperature. These animals are called **homeotherms**, from the Greek word *homos*, which means "same."

Most poikilotherms are less active when external temperatures are low and become more active as temperatures rise. Behavioral responses to changes in external temperature are the primary means by which poikilotherms maintain suitable body temperatures. For example, rattlesnakes like the one in Figure 3 bask in the sun to increase their temperature. When its body reaches an optimal temperature, the rattlesnake begins to hunt for food. If it becomes too hot, the rattlesnake will rest in the shade to cool off. At night, as temperatures fall, a poikilotherm's metabolism slows and it becomes less active. Many burrow into soil or fallen leaves for insulation when temperatures are low.

Water-dwelling poikilotherms, including fish, do not need quite such elaborate behaviors for maintaining internal temperature, because water temperatures do not fluctuate as much as land and air temperatures. When necessary, fish do swim from place to place in search of suitable temperatures. Small bodies of water, such as lakes and ponds, can freeze over during winter, but ice usually forms only at the surface. This surface layer of ice acts as an insulator that helps keep the temperature of the water beneath it high enough for the fish to survive. Some poikilotherms, including certain species of amphibians, live in very cold climates. These animals may hibernate during the coldest part of the winter. Some poikilotherms have chemicals in their cells that inhibit the formation of ice crystals, enabling them to survive even when temperatures fall below freezing.

Homeotherms maintain a relatively constant body temperature that is usually substantially higher than the temperature of the surrounding environment. As a result, homeotherms are able to remain active through a wide range of external temperatures. Consider humans, for example. We maintain an internal temperature of approximately 37°C. Although we are most comfortable when the external temperature is around 24°C (75°F), we can live and work in temperatures well below freezing (0°C). Given that heat always flows from warmer to cooler objects, we constantly lose energy to our surroundings, except on very hot days. The energy cost of maintaining a constant temperature is high. A homeotherm may require up to ten times more energy than a similarly sized poikilotherm living in the same environment. Homeotherms must accordingly consume considerably larger amounts of food.

Humans and many other homeotherms have body systems that regulate internal temperature much like a thermostat does. Part of the brain monitors internal temperature. If the temperature rises or falls too much, the brain signals other parts of the body to initiate processes like those shown in Figure 4. These processes regulate temperature in two general ways: by increasing or decreasing heat production through changes in metabolic activity, and by increasing or decreasing heat flow from the body to the environment. For example, when body temperature rises above normal, signals from the brain increase the blood flow to vessels near the skin. This is why people sometimes turn pink or look flushed when they are warm from exercise or hot weather. As long as the air is cooler than the skin, heat flows from blood to skin to air, cooling the body. Evaporation of perspiration from the skin helps to accelerate the removal of heat from the body.

Figure 5   Pelican (a homeotherm)

If the body temperature drops below normal, feedback from the brain's thermostat increases metabolic activity to generate more heat and reduces blood flow to vessels in the skin. These responses reduce heat flow from the body to the environment. Humans shiver when cold because the heat produced by this added motion helps increase body temperature. Skin coverings, including hair, fur, and

Figure 4   Temperature Regulation in Humans

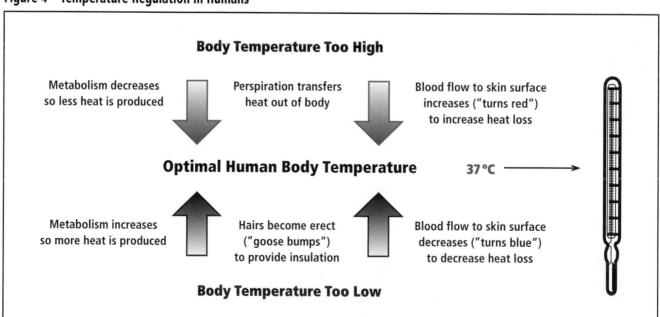

Body Temperature Too High

Metabolism decreases so less heat is produced

Perspiration transfers heat out of body

Blood flow to skin surface increases ("turns red") to increase heat loss

Optimal Human Body Temperature

37 °C ⟶

Metabolism increases so more heat is produced

Hairs become erect ("goose bumps") to provide insulation

Blood flow to skin surface decreases ("turns blue") to decrease heat loss

Body Temperature Too Low

feathers, act as insulators that also help regulate heat flow. For example, the feathers on the pelican in Figure 5 help it maintain its optimal body temperature even in cold ocean water. When body temperature falls, muscles under the skin pull the hair or feathers perpendicular to the skin to trap air, which increases the effectiveness of the insulating layer. You experience this type of temperature regulation when you get goosebumps.

Temperature is just one of the factors that must be regulated to maintain an internal environment appropriate for life processes. The maintenance of stability in an organism's internal environment regardless of external changes is called **homeostasis**. This includes regulating body chemistry and defending the body against infection, as well as regulating body temperature.

Analysis
?

Group Analysis

1. Use your knowledge of temperature regulation to propose answers for the following questions:

   a. Why do some animals hibernate in winter?

   b. Why do dogs pant on hot days?

   c. What are some other methods animals use to regulate internal temperature?

Individual Analysis

2. Some birds maintain internal temperatures of over 40°C (104°F). Why do you think this is so? What adaptive traits or behaviors might help birds maintain this temperature? (**Hint:** Look at Figures 1, 2, and 4.)

3. Suppose global climate patterns change in such a way that temperatures in some regions become consistently higher or lower than they had been in the past. Predict how these changes would affect organisms living in these regions. Would you expect all organisms to be equally affected? Explain your responses.

3.3   Heat and the Laws of Thermodynamics

Purpose ▶ **I**nvestigate the rate of thermal energy transfer between materials of different temperatures.

Introduction

Your body's ability to regulate its internal temperature is essential for your survival. Because the transfer of thermal energy (heat) occurs in a regular and predictable fashion, scientists can describe this process by the laws of thermodynamics. The **First Law of Thermodynamics** states that all heat entering a system adds that amount of energy to the system. It can be thought of as a subset of the **Law of Conservation of Energy**, which states that energy cannot be created or destroyed. The **Second Law of Thermodynamics** states that heat will always flow from areas of higher temperature to areas of lower temperature. Figure 6 provides one illustration of the Second Law of Thermodynamics. As this cat suns itself, most of the radiant heat it absorbs is transferred from its body to the air, so that the cat maintains its body temperature.

Figure 6   Cat Sunning Itself

Part A   Observing Temperature Differences

Materials    **For each team of two students**

1   plastic cup of hot water

1   plastic cup of cold water

1   plastic cup of room-temperature water

supply of paper towels

Procedure

1. a. Put one of your index fingers into the hot water and the other into the cold water for 30 seconds.

b. Move both of your index fingers into the room-temperature water. Describe what each finger feels like, focusing on the similarities and differences between the two.

2. a. Put both of your index fingers into the hot water for 30 seconds.

b. Move one of your index fingers into the room temperature water and the other into the cold water. Describe what each finger feels like, focusing on the similarities and differences between the two.

3. a. Put both of your index fingers into the cold water for 30 seconds.

b. Move one of your index fingers into the room temperature water and the other into the hot water. Describe what each finger feels like, focusing on the similarities and differences between the two.

Analysis

Group Analysis

1. Explain why you think these similarities and differences occur. Give as many details as you can.

Part B   Measuring Changes in Temperature

Prediction

Will the hot water cool down at the same rate as the cold water warms up?
Explain your reasoning.

Materials

For the class

supply of hot water (~60–65°C)

supply of cold water (~5–10°C)

supply of room-temperature water
(~20–30°C)

For each group of four students

3   clear plastic cups

1   100-mL graduated cylinder

3   metal-backed thermometers

Procedure

1. Add 50 mL of hot water to one cup, 50 mL of cold water to another cup, and 50 mL of room-temperature water to the third cup.

2. Record the temperatures of all three cups of water and note the time. You will be measuring the temperature of the water in each cup every 2 minutes for 20 minutes.

   Note: Keep close track of the time and make sure every group member has the opportunity to take some of the temperature measurements.

3. During the time between temperature readings:

   a. Prepare a clearly labeled data table in which to record your group's data.

   b. Make one clearly labeled graph that shows how the temperature of the water in each of the cups changed as time passed. When setting up your graph, put time on the x-axis and temperature on the y-axis. Separately plot the data for all three cups on the same graph, using different symbols or colors to identify the temperature measurements of the hot, cold, and room-temperature cups.

Analysis

▼
?

Group Analysis

1. Describe the shape of the curve for the hot water data. What differences are there between the portion of curve that describes the first ten minutes and the portion that describes the last ten minutes?

2. Describe the shape of the curve for the cold water data. What differences are there between the portion of the curve that describes the first ten minutes and the portion that describes the last ten minutes?

3. Use the equation below to determine the rate of temperature change for each cup for the first ten minutes and the last ten minutes.

$$\text{rate of temperature change} \quad = \quad \frac{\text{final temperature} - \text{initial temperature}}{\text{time}}$$

4. Explain how your descriptions in Analysis Questions 1 and 2 relate to your calculations from Analysis Question 3.

Individual Analysis

5. Why were you instructed to collect and graph the data for the room-temperature water?

6. If you were able to take the temperature of each cup after one hour had gone by, what do you predict the temperature of each of the three cups would be? Explain.

7. Describe the relationship between the hot- and cold-water curves and explain why this relationship exists.

8. Since the Law of Conservation of Energy states that energy cannot be created or destroyed,

    a. where does the heat from the hot water go? Explain.

    b. where does the energy needed to heat up the cold water come from? Explain.

3.4 Energy on the Move

Purpose ▶ **C**reate diagrams that describe a sequence of energy transfers.

Introduction

Energy can take many forms and is constantly being transferred from place to place. When we "use" energy, we are actually transferring it from one object to another. These energy transfers often convert energy from one form to another, as when lightning striking a tree starts a fire, or when electricity flowing through the filament wire of an incandescent bulb produces light and heat. The pathways that energy takes during any transfer or series of transfers can be illustrated with an **energy flow diagram**, such as those shown in Figure 7. (Figure 1 in Activity 3.2 illustrates a slightly different type of energy flow diagram.)

When drawing an energy flow diagram, it is very important to include all the energy sources and receivers that are components of the physical system you are describing. A **physical system** is any set of matter within a defined area.

Figure 7 Energy Flow Diagrams

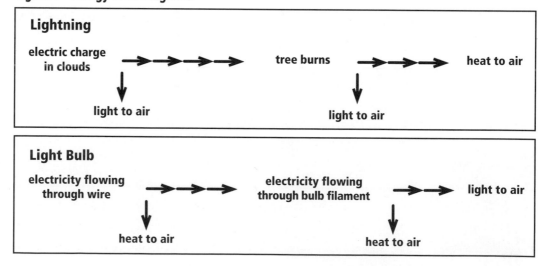

Part A Demonstrating Energy Transfer

Procedure

1. Observe the demonstration your teacher performs and describe what you see happening in your journal.

2. Draw an energy flow diagram of the type shown in Figure 7, describing the energy transfers that took place during the demonstration.

Part B Observing Energy Transfer

Materials

■ **For each student**

   1   jumbo paper clip

Procedure
(cont.)

3. Open the paper clip into a "V" shape, as shown. Touch the curve of the "V" to the skin of your upper lip. Record any observations.

4. Bend the clip back and forth rapidly 10 times, then immediately touch the curve of the "V" to the same place on your upper lip. Record any observations.

5. Touch the "V" of the clip to your lip one more time and record any observations.

6. Draw an energy flow diagram that describes all the energy transfers that took place when you carried out Steps 4 and 5.

Analysis
?

Group Analysis

1. Make a list of at least five everyday events that involve the conversion of one form of energy into another.

2. Draw an energy flow diagram for an uncovered pot of water being heated on an electric stove.

Individual Analysis

3. Would the energy flow diagram change if the pot had a lid? Explain. If you think the diagram would change, draw your new version.

Extension   Think of an everyday occurrence not mentioned in this activity that involves at least two energy transfers. Draw an energy flow diagram that represents this situation, then describe a means of decreasing the flow of energy to the environment.

Energy Transfer

4

4.1   Are You in Hot Water?

Purpose ▶ **D**iscover how to measure thermal energy and compare the amount of heat water can store relative to other materials.

Introduction

Water, in its liquid state, is essential to all life on Earth. All living cells are made primarily of water, and many organisms spend their lives surrounded by water. Controlling the transfer of heat is extremely important to an organism's survival. How do you measure the amount of heat moving into or out of a substance, such as water? How much energy does it take to keep the body of a homeothermic organism at a particular temperature?

The amount of energy transferred when the temperature of 1 gram of a material changes 1° Celsius is called the **specific heat** of that material. The abbreviation for specific heat is **Cp**. The specific heat of water is written as Cp_{water}. It is equal to $1 \frac{calorie}{gram \cdot °C}$.

This means that to raise the temperature of 1 gram of water 1°C, 1 **calorie** of energy must be transferred to the water. And vice versa: for each 1°C decrease in the temperature of 1 gram of water, 1 calorie of energy must be transferred out of the water. A **calorimeter** (Figure 1), a device often used when measuring specific heat, is designed to prevent the transfer of energy between a physical system and the environment. Because the specific heat of water is 1 calorie/gram • °C, the calorie is a convenient and commonly used unit of energy. Another commonly used unit of energy is the **joule**. One calorie is equal to 4.2 joules.

Prediction

What will be the final temperature of a mixture of 25 mL of boiling (100°C) water and 50 mL of room-temperature (20°C) water? Explain how you came up with your prediction.

Materials

 For the class

supply of room-temperature water

supply of boiling water

safety glasses

 For each team of two students

1   calorimeter

1   immersion thermometer

1   50-mL Pyrex graduated cylinder

Safety Note

Use special caution when working with boiling water.

Procedure

1. Carefully measure 50 mL of room-temperature water and pour it into your empty calorimeter. Record the temperature of the water.

2. Obtain 25 mL of boiling water from your teacher and pour it into the calorimeter. Immediately place the lid on the calorimeter. Gently slide the thermometer through the slot in the lid and record the highest temperature it reaches.

Figure 1   SEPUP Calorimeter

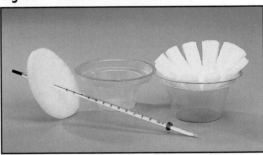

3. Empty the calorimeter.

4. Determine the change in temperature (ΔT) of the room-temperature water and the ΔT of the boiling water.

    Remember: Water boils at 100°C.

5. Repeat Steps 1–4 two more times.

6. Make a data table to display all your experimental data.

Analysis

?

Group Analysis

1. Using your results from all three trials, calculate the average final temperature of a mixture of 50 mL of room-temperature water and 25 mL of boiling water.

2. Using your results from all three trials, calculate the average ΔT of the room-temperature water and the average ΔT of the boiling water.

3. Use the average ΔTs from Analysis Questions 1 and 2 to calculate the amount of energy lost by the boiling water and the energy gained by the room-temperature water. Use the following two equations:

    a. Heat lost by boiling water =

    mass of boiling water • ΔT of boiling water • Cp of water
        (25 g)   (°C)   (1 cal/g • °C)

    b. Heat gained by room-temperature water=

    mass of room-temp. water • ΔT of room-temp. water • Cp of water
        (50 g)   (°C)   (1 cal/g • °C)

    Note: 1 mL of H_2O has a mass of 1 g.

Individual Analysis

4. Compare your calculated results to your prediction. Why do you think your prediction was not the same as your results?

5. What do you notice about the relationship between heat lost and heat gained? Explain your results in the context of the laws of thermodynamics.

6. Draw an energy flow diagram describing the energy transfers that took place during this experiment. List as many possible pathways as you can through which heat could have escaped from your experimental system.

7. If you could do this experiment again so that no heat escaped, what should be the final temperature of the water in the calorimeter?

8. The unit for the energy transfers you just calculated is the calorie. Convert your values for the energy lost and the energy gained from calories to joules.

4.2   Holding Heat

Purpose ▶ **C**alculate and compare the specific heat of a few different materials.

Introction

Materials vary in their ability to absorb and release heat. Those that have a high specific heat require more energy to heat up, and release more energy when they cool down, than do materials with a low specific heat. How does the specific heat of water compare to the specific heat of other materials? Could the specific heat of the material an animal uses for shelter affect the animal's ability to survive?

In this activity you will determine the specific heat of several different materials by measuring the temperature changes that occur when a hot sample of each material is placed in cold water. According to the Second Law of Thermodynamics, in this situation heat will be transferred from the hot material to the cold water until both reach the same temperature. According to the First Law of Thermodynamics, the amount of heat lost by the hot material will be equal to the amount of heat gained by the water.

Prediction

Predict the ranking of the five materials shown below in order from highest to lowest specific heat. Explain how you made your prediction.

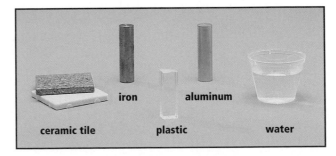

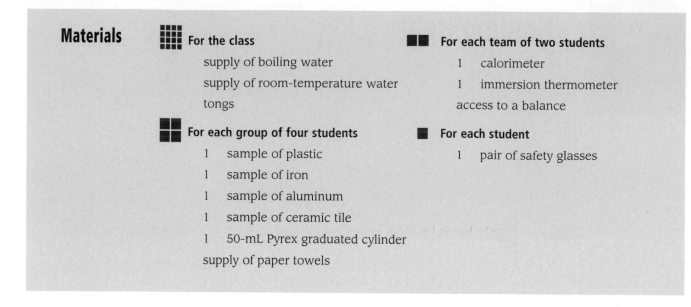

Materials

For the class

    supply of boiling water

    supply of room-temperature water

    tongs

For each group of four students

  1   sample of plastic

  1   sample of iron

  1   sample of aluminum

  1   sample of ceramic tile

  1   50-mL Pyrex graduated cylinder

    supply of paper towels

For each team of two students

  1   calorimeter

  1   immersion thermometer

    access to a balance

For each student

  1   pair of safety glasses

Procedure

1. Decide which team in your group will test which two of the samples and which team will test the other two samples.

2. Read through the rest of the Procedure Steps and prepare a data table that will allow you to record all your measurements.

3. Test each of your materials separately. To collect the data needed to determine the specific heat of each sample, follow Steps 3a–3e:

   a. Place exactly 50 mL of room-temperature water in an empty calorimeter. Measure the temperature of the water (Initial Water Temp.).

   b. Obtain a sample of one of the materials at 100°C, place it directly into your calorimeter, immediately cap the calorimeter, swirl it gently, and record the highest temperature it reaches (Final Temp.).

   c. Calculate the ΔT for the water in the calorimeter.

   d. Calculate the ΔT for the sample in the calorimeter.

   e. Remove the sample from the calorimeter. Dry off the sample and determine its mass.

4. Record the data for the other two samples from the other team in your group.

Analysis

?

Group Analysis

1. Calculate the specific heat of each sample tested (Cp_{sample}) using the following formula:

    $$\text{Heat lost by sample} \quad = \quad \text{Heat gained by water}$$

    $$mass_{sample} \times \Delta T_{sample} \times Cp_{sample} = mass_{water} \times \Delta T_{water} \times Cp_{water}$$

    $$\text{where: } Cp_{water} = 4.2 \text{ J/g} \cdot \text{°C}$$

    (**Hint:** The density of water is 1 g/mL and you used 50 mL of water)

2. Rank the materials—plastic, iron, aluminum, tile, and water—based on their specific heat. How close were your predictions?

3. Propose a theory that could explain why certain materials have a high specific heat while others do not. Your theory should relate to the composition of the materials. Use evidence from this activity to support your theory.

Individual Analysis

4. The accepted value for the Cp of iron is 0.45 joule/gram • °C. Give reasons why your experimental value is not the same as the accepted value.

5. Assuming there was error in all your calorimetry experiments, estimate what you think a more accurate value would be for the specific heat of tile and of plastic. Explain how you determined your estimates.

6. Do you think life would be easier to sustain if water had a different specific heat? Explain why or why not.

7. As a society, we use large quantities of fuel to heat vast amounts of water each day. The heated water is used for a variety of purposes, including the generation of electricity. What advantages and disadvantages are there to using solar energy to heat water?

4.3   Competing Theories

Purpose ▶ **C**ompare the evidence in support of two competing theories.

By the 1700s, most scientists agreed that all materials were made of tiny "building blocks" called **atoms**. They also agreed that these atoms bonded together to form bigger structures, called **molecules**, and that different substances contained different atoms and molecules. At that time, no one had ever seen an atom or a molecule, even under the most powerful microscope available. Even so, there was enough indirect evidence from a variety of experiments to convince scientists that atoms and molecules must exist, and that they must be extremely small.

During the 1700s and 1800s some scientists focused their attention on the study of heat and its effects on materials. What is heat? Why does the addition or removal of heat from a substance cause the volume of that substance to change? How and why does heat always move from a material at a higher temperature to a material at a lower temperature? These questions were first experimentally investigated in the mid-1700s, but scientists would not agree upon the answers until much later.

The Caloric and Kinetic Theories of Heat

Joseph Black, a well-respected scientist who lived in England in the early 1700s, was known for his research on heat. He proposed that temperature changes are due to the flow of a substance called caloric (from *calor*, the Latin word for heat). He described caloric as a fluid that fills the spaces between the atoms and molecules of all materials. According to Black's theory, an object heats up when caloric from another object flows into it; an object cools down when caloric flows from it to another object. The total amount of caloric contained in any object depends on the materials the object is made of, its size, and its temperature. The movement of caloric can easily explain the common occurrence of a substance expanding in size as it gets hotter and decreasing in size as it gets cooler. Because an object heats up when caloric is added to it, and since caloric is a fluid that takes up space, the added caloric would push the atoms and molecules of the object farther apart, causing it to expand. Caloric can also explain the laws of thermodynamics: when a hotter object touches a colder object, the caloric fluid flows from the hot object to the cold object until the temperature of both objects is the same. In this two-object system, no caloric is lost or gained.

Gustave Hirn, a Frenchman who had a flair for dramatic experiments as well as an interest in science, did a series of investigations in the mid-1800s that involved smashing lead (Figure 2). He allowed a huge iron pendulum to repeatedly swing into a small piece of lead, crushing it against a stone anvil. He found that this battering caused the temperature of the lead to rise without causing a decrease in the temperature of the pendulum or the anvil. He proposed that the increase in the temperature of the lead was caused not by adding caloric fluid, but by a transfer of energy from the moving pendulum. This theory came to be known as the kinetic theory (from *kinein*, the Greek word for movement). The kinetic theory states that the atoms and molecules of all substances are in motion and that at higher temperatures there is faster movement. This theory also explains why substances expand when they get hotter. Energy that is transferred to an object causes the atoms and molecules of the object to move faster. The faster the atoms are moving, the farther they separate after they collide. The farther apart they move, the more space they take up, thus causing the volume of the object to expand. This theory can also help explain the laws of thermodynamics: when two objects of different temperatures come in contact with each other, the faster-moving molecules of the hotter object collide with the slower-moving molecules of the cooler object. When these collisions occur, energy is transferred, causing the faster-moving molecules to slow down and the slower-moving molecules to speed up. As a result, the cooler object becomes warmer and the hotter object becomes cooler.

Let's apply each theory to the process of heating a metal pot of water over a flame until it boils. According to the caloric theory, caloric would flow from the fuel to the pot. The flame is evidence of the flow of caloric. The caloric soon fills the spaces between the metal atoms of the pot, making the pot very hot. The caloric then flows

Figure 2   Diagram of the Set-up for Hirn's Experiment

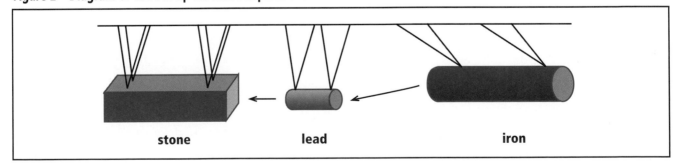

stone   lead   iron

from the hot pot to the water, making the water hot. With the burner on high, the flow of caloric fluid is very strong and this flow makes the water bubble and swirl like a fast-moving stream. As long as there is a continuous supply of fuel, there is a continuous source of caloric available to heat the water.

According to the kinetic theory, energy released from the burning fuel causes the surrounding air molecules to move very fast. These rapidly moving molecules then collide with the slower-moving atoms of metal in the pot, making the metal atoms move faster. The fast-moving atoms of the hot pot then collide with molecules of water, making the water molecules move faster, which makes the water hot. With the burner on high, the large flow of energy into the water will cause the molecules to move so fast and collide so violently that the water bubbles and swirls. As long as there is a continuous transfer of energy into the water, the water will continue to gain heat.

Analysis

?

Group Analysis

1. What is the evidence in favor of the caloric theory?

2. What is the evidence in favor of the kinetic theory?

3. Which theory do you think better explains temperature changes? What makes you prefer this theory?

4. Design an experiment that you think would provide enough evidence to convince others that one of the theories provides a better explanation than the other.

4.4 Shaking the Shot

Purpose ▶ **E**xplore the conversion of mechanical energy to thermal energy.

Introduction

You have been observing, diagramming, calculating, and considering the effects of various types of energy transfer. In this activity, you will explore the conversion of energy from one form to another, focusing on the relationship between the amount and type of energy transferred to an object and the amount and type of energy gained by that object. Specifically, you will examine the conversion of **mechanical energy**—in this case, moving muscles—to **thermal energy**. Thermal energy is not only essential to the survival of all living things, it is also a very useful form of energy. Throughout history, humans have continued to find new ways of converting many different forms of energy into thermal energy.

Figure 3 Potential and Kinetic Energy

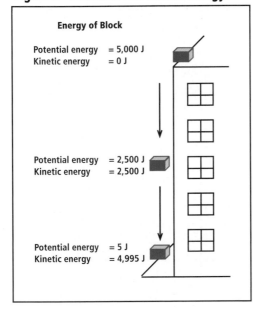

Contrary to how it might sound, "mechanical energy" does not necessarily refer to the energy needed to run a machine. Mechanical energy is energy that results from the motion or position of an object. There are two main categories of mechanical energy: **kinetic energy**, which is the energy of motion, and **potential energy**, which is the stored energy of position. For example, suppose you are standing on the roof of a tall building holding an object. Because the object is not moving, it has very little kinetic energy. However, because of its height it has a great deal of potential energy. Now suppose you drop the object over the edge of the building. As it falls, it speeds up and gains kinetic energy, at the same time losing both height and potential energy. While falling, it gains kinetic energy equivalent to the amount of potential energy it loses. Figure 3 illustrates this process.

Materials

For each group of four students

1 Shot Shaker containing metal shot
1 cap with attached thermometer
1 cup with 100 mL of cool water
supply of paper towels
access to a clock with a second hand

Figure 4 Parts of the Shot Shaker

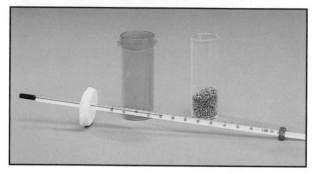

Procedure

1. Prepare a data table similar to the one below.

Table 1 Shaking the Shot

Trial	Shaking Time	Initial Temp.	Final Temp.	ΔT	Average ΔT
1	30 seconds				
2	30 seconds				
3	30 seconds				_____
4	20 seconds				
5	20 seconds				
6	20 seconds				_____
7	10 seconds				
8	10 seconds				
9	10 seconds				_____

Procedure
(cont.)

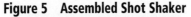

2. Take the regular cap off the Shot Shaker and replace it with the one with the thermometer attached, making sure the thermometer bulb is surrounded by metal shot. Record the initial temperature of the shot.

Figure 5   Assembled Shot Shaker

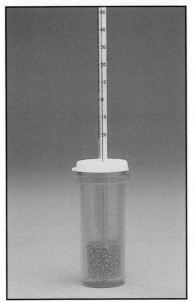

Note: If the initial temperature is more than 2°C above room temperature, remove the inner vial containing the metal shot and immerse it in a cool water bath until the temperature is within 2°C of room temperature. Dry it off and then place it back inside the larger vial.

3. Exchange caps and make sure the regular cap snaps tightly onto the Shot Shaker. Then, while holding the cap on with your thumb, shake the vial as fast as you can for 30 seconds.

4. Quickly exchange the regular cap for the one with the thermometer attached and record the highest temperature reached as the final temperature of the metal shot. Be patient: the thermometer reading rises slowly even after you stop shaking.

5. Calculate the ΔT by subtracting the initial temperature from the final temperature.

6. Repeat Steps 2–5 for Trials 2 and 3. Calculate the average ΔT for the three trials.

7. Repeat Steps 1–6, except this time shake the shot for 20 seconds.

8. Repeat Steps 1–6, except this time shake the shot for 10 seconds.

Analysis

?

Group Analysis

1. Why do you think it takes time for the temperature to rise after you have stopped shaking the shot? Explain.

2. Using your group's average data for the three different shaking times, make a graph of shaking time vs. change in temperature.

3. Describe the relationship between shaking time and temperature change as precisely as you can.

Individual Analysis

4. Does the kinetic or caloric theory better explain the results you obtained? Explain.

5. What temperature change would you expect if you shook the shot for 15 seconds? For 70 seconds? How did you decide?

6. Draw an energy flow diagram for this experiment.

7. Describe any possible sources of error in your experiment and explain how each would have affected your results.

8. Prepare a data table showing what you think the ideal results would be if this experiment could be carried out without any error.

4.5 Count Rumford's "Boring" Experiment

Purpose ▶ **D**iscover how an uncomfortable job led a man to design an experiment that gave scientists great insight into the nature of heat.

Introduction

During the 1790s, an American known as Count Rumford was employed in Germany as the overseer of work in a government military arsenal. He was amazed at how hot it became in the arsenal while cannons were being made. In those days, making a cannon meant drilling a hole with the same diameter as a cannon ball through a cylinder of solid brass. That hole formed the bore of the cannon barrel, the tube through which cannon balls would be shot. To grind out the hole, horses were used to rotate the heavy brass cylinder over a sharpened bit of harder metal, called a "borer." Count Rumford performed a series of experiments to determine how much heat the cannon-boring process generated. The following passages are excerpted from his original journals.

In Count Rumford's day, cannons were made from solid metal cylinders. The inner cannon barrel was hollowed out using a large drill bit.

An Inquiry

concerning the

Source of the Heat Which is Excited by Friction

It frequently happens that in the ordinary affairs and occupations of life, opportunities present themselves of contemplating some of the most curious operations of Nature . . . . It was by accident that I was led to make the experiments of which I am about to give an account.

Being engaged lately in superintending the boring of cannon in the workshops of the military arsenal at Munich, I was struck with the very considerable degree of Heat which a brass gun acquires in a short time in being bored, and with the still more intense Heat (much greater than that of boiling water, as I found by experiment) of the metallic chips separated from it by the borer.

The more I meditated on these phenomena, the more they appeared to me to be curious and interesting. A thorough investigation of them seemed even to bid fair to give a farther insight into the hidden nature of Heat; and to enable us to form some reasonable conjectures respecting the existence, or non-existence, of an igneous fluid [caloric]—a subject on which the opinions of philosophers have in all ages been much divided.

Now Count Rumford describes the experiment he conducted while overseeing the manufacture of cannons. To determine how much heat was generated in the cannon-boring process, he submerged a brass cylinder in water and measured the temperature of the water before boring and then at various times during the procedure. The initial water temperature was 60°F.

The result of this beautiful experiment was very striking, and the pleasure it afforded me amply repaid me for all the trouble I had had in contriving and arranging the complicated machinery used in making it.

The cylinder, revolving at the rate of about 32 times in a minute, had been in motion but a short time, when I perceived, by putting my hand into the water and touching the outside of the cylinder, that Heat was generated; and it was not long before the water which surrounded the cylinder began to be sensibly warm.

At the end of 1 hour I found, by plunging a thermometer into the water in the box (the quantity of which fluid amounted to 18.77 lb., or 2 1/4 gallons), that its temperature had been raised no less than 47 degrees; being now 107° of Fahrenheit's scale.

When 30 minutes more had elapsed, or 1 hour and 30 minutes after the machinery had been put into motion, the Heat of the water in the box was 142°.

At the end of 2 hours, reckoning from the beginning of the experiment, the temperature of the water was found to be raised to 178°.

At 2 hours 20 minutes it was 200°; and at 2 hours 30 minutes it ACTUALLY BOILED!

It would be difficult to describe the surprise and astonishment expressed in the countenances of the bystanders, on seeing so large a quantity of cold water heated, and actually made to boil, without any fire.

This reading and the figure on the next page are adapted from *Collected Works of Count Rumford,* edited by Sanborn C. Brown (Harvard University Press, 1968).

Figure 6   Diagram From Count Rumford's Journal Showing the Set-up for His Experiment

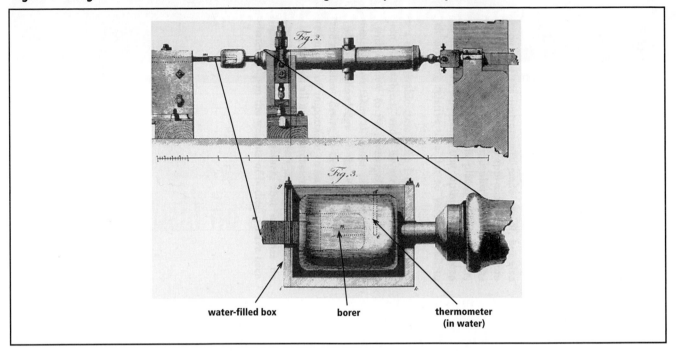

water-filled box   borer   thermometer
(in water)

Analysis
?

1. Do the results of Count Rumford's experiment provide good evidence to support either the kinetic or the caloric theory of heat? Explain your reasoning.

2. In what ways is Count Rumford's experiment similar to Activity 4.4, "Shaking the Shot," and in what ways is it different?

3. From which experiment, "Shaking the Shot" or Count Rumford's, would you expect more accurate results? Explain your reasoning.

4. Use the reading, the diagram of Count Rumford's experimental set-up in Figure 6, and your knowledge of heat to

   a. create an energy flow diagram for Count Rumford's experiment.

   b. describe how the energy flows from step to step in your diagram.

5. Before the invention of matches, one of the more common ways for humans to start fires was to "drill" a piece of wood with a sharpened stick. Explain how this process can start a fire.

Designing an Insulation System

5.1 Conducting Experiments on Insulation

Purpose **E**xamine the ability of various materials to prevent energy transfer.

Introduction

You have already observed the transfer of energy to and from the human body, water, and a variety of other materials. The movement of thermal energy through a material is called **thermal conduction**. Some materials do not conduct heat as well as others and thus slow down the rate of energy transfer. Materials that are very poor thermal conductors are very good at preventing energy transfer. These are called thermal insulators. In this activity, you will compare the thermal conduction and insulation characteristics of four materials.

This thermos has two side walls, separated by air. The air trapped between the walls is a very good thermal insulator.

Prediction Which of the four materials—water, air, sand, or styrofoam—do you think will be most effective in preventing the transfer of thermal energy? Explain your reasoning.

Materials

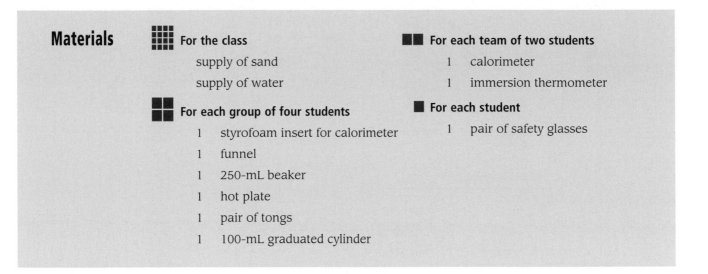

For the class

    supply of sand
    supply of water

For each group of four students

  1   styrofoam insert for calorimeter
  1   funnel
  1   250-mL beaker
  1   hot plate
  1   pair of tongs
  1   100-mL graduated cylinder

For each team of two students

  1   calorimeter
  1   immersion thermometer

For each student

  1   pair of safety glasses

Safety Note

Do not pour boiling water into the calorimeter because this can damage it. Wear safety glasses while in the laboratory.

Procedure

1. Divide the insulating materials among your group so that each team of two students has two materials to test.

2. Place approximately 200 mL of water in your beaker and heat it on the hot plate until it reaches 75°C. If it gets too hot, remove it from the hot plate.

3. While the water is heating, prepare a data table that will allow you to record the time and temperature every 30 seconds until 5 minutes have elapsed.

4. Following your teachers' instructions, place one of the materials you will be testing between the inner and outer walls of the calorimeter.

5. Carefully measure out 75 mL of 75°C water and pour it into the calorimeter. Place the lid on the calorimeter and slide the thermometer through the slot. Record the initial temperature. (It should be above 60°C. If it is not, pour the water back into your beaker and reheat it for a few minutes. Then repeat this step.)

6. Record the temperature every 30 seconds for 5 minutes.

7. Calculate the total change in temperature from start to finish.

8. Pour the water from the calorimeter back into your beaker and reheat it until it reaches 75°C.

9. Repeat Steps 3–7 for the other material you are investigating.

10. Record the data for the two materials tested by the other team in your group.

Analysis

Group Analysis

1. Propose two uses for thermal conductors.

2. Propose two uses for thermal insulators.

Individual Analysis

3. Make a single graph of temperature vs. time to display the data for each of the four materials. Use a different color or symbol to identify the data points and resulting curve for each material.

4. Which material allows the most energy to be transferred through it? Explain how you know.

5. Which material allows the least energy to be transferred through it? Explain how you know.

6. Design an experiment to compare the insulating qualities of two different fabrics used in winter jackets. Be as precise as possible when describing your experiment.

5.2 | The Can Challenge

Purpose ▶ **D**etermine which materials and design are most effective in preventing heat transfer into and out of a soda can.

Introduction

For humans, survival in many environments requires wearing clothing, finding shelter, and using other means to promote the movement of heat into or out of our bodies. Considerable research has been done on the design and construction of a wide variety of products that increase or decrease thermal energy transfer. The resulting advances in technology allow humans to live and work comfortably in conditions that are hundreds of degrees above or below 0°C. In this activity, your challenge is to design a lightweight, waterproof insulating container for a standard 355-mL soda can. Your group's goal is to create the container that prevents the most heat transfer into or out of the can.

May the best can win!

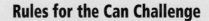

Rules for the Can Challenge

1. The insulating container can be no larger than 10 cm x 10 cm x 20 cm.

2. The dry weight of each container will be measured to the nearest 0.1 gram.

3. Each container will be tested by pouring 200 mL of cold (or hot) water into the soda can and monitoring the temperature of the water inside the soda can during a 20-minute "bath" in hot (or cold) water.

4. The temperature of the water in the soda can and in the water bath will measured, to the nearest 0.25°C, every 2 minutes during the 20-minute testing period.

5. The insulating ability of the container will be determined by calculating the total change in temperature of the water inside the soda can.

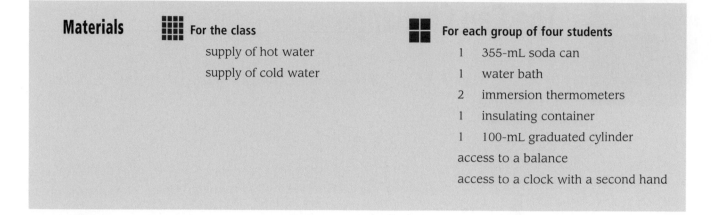

Materials

For the class

supply of hot water

supply of cold water

For each group of four students

1 355-mL soda can

1 water bath

2 immersion thermometers

1 insulating container

1 100-mL graduated cylinder

access to a balance

access to a clock with a second hand

Procedure

1. Find the mass of your insulating container.

2. Set up a data table for recording the water temperature in the soda can and the temperature in the water bath every 2 minutes for 20 minutes.

3. Measure out exactly 200 mL of the appropriate temperature water and pour it into your soda can. Fill the water bath with the opposite temperature water.

4. Put the soda can in your insulating container and then place it in the water bath.

5. Note the time and record in your data table the temperature of the water in the soda can and the temperature of the water in the water bath. Go on to Step 6, but remember to continue to take temperature readings every 2 minutes for 20 minutes.

6. Sketch the design of your insulating container, labeling the location of the various material(s) you used.

7. Go on to the Analysis Questions. Remember to continue taking temperature readings every 2 minutes for 20 minutes.

Analysis

?

Group Analysis

1. Use graph paper to plot the temperature changes in the soda can and in the water bath.

2. Explain why you chose the design and the material(s) you used.

3. What variables, other than the design of the insulating container and the materials used, could be affecting the temperature change inside your soda can when compared to the other groups testing under the same conditions?

Individual Analysis

4. Does the temperature of the water inside the soda can change at the same rate as the water in the water bath? Explain how you know. What could cause the rate of change to be similar or different?

5. How are the two curves you drew for Analysis Question 1 similar to those you drew in Activity 3.3, "Heat and the Laws of Thermodynamics"?

6. How are the two curves you drew for Analysis Question 1 different from the ones you drew in Activity 3.3?

7. Prepare a written report that includes an illustration of your design, a description of the materials used and testing procedures followed, a display of the data collected, and your analysis of the data.

Living in Today's World

6.1   Life in Other Countries

Purpose ▶ **C**ompare statistics from different countries and decide which are good indicators of day-to-day life.

Introduction

In the last few activities, you have explored the nature of heat and its importance for survival. You also designed and built technology to reduce heat transfer. Although the development of various forms of technology has made the human struggle for survival easier, there can be societal costs and drawbacks associated with its use. Because of economic and cultural differences, not all people have access to technology at the same level or in the same forms. The world's countries are often divided into two categories on this basis.

These pictures show agricultural methods in two different countries. The method on the left utilizes human labor; the one on the right utilizes a tractor fueled by petroleum products. Although they involve very different levels and forms of technology, each method has benefits and drawbacks associated with its use.

Introduction (cont.)

Countries that are economically and technologically rich are considered **more developed**; those that are economically and technologically poor are considered **less developed**.

Some of the statistics presented in this activity are reported as per capita values. **Per capita** values indicate the average amount of an item or resource available to each person living in a given country. For example, annual per capita commercial energy use is determined by calculating the total amount of energy bought and sold in a country during one year and dividing the result by the country's population. It is important to remember that commercial energy use statistics do not provide a complete measurement of all energy used in an individual family's home, especially families in less developed or rural areas. This is because "commercial energy" does not include energy obtained from burning firewood, or the energy expended by human- or animal-powered transportation. Still, this statistic provides useful data for comparing the overall energy use of different societies. In general, per capita values can be a good way to compare the lives of average residents of different countries.

Materials

■■ **For each team of two students**

1   copy of *Material World*

Procedure

1. Look through the information provided for the 16 countries in Table 1, on the next page. (Statistics for the United States are provided for comparison.) Compare the statistics for the eight more developed countries with the statistics for the eight less developed countries. Of the statistics listed, which one do you think is the most helpful in determining whether a country should be classified as more developed or less developed?

2. Choose one country classified as more developed and one classified as less developed. Compare the statistics given for each country. List the two statistics that you think are the best indicators of what day-to-day life is like for people living in each country.

3. Find the same two countries in *Material World*. Look through the photographs of the families from those countries and compare their possessions. Describe the five most obvious differences between the possessions of the family from the more developed country and those of the family from the less developed country.

4. Now go back to Table 1 and look at the commercial energy use statistics for the two countries that you chose. What correlations do you see between per capita energy use and your observations about each family's possessions?

Figure 1 Locator Map

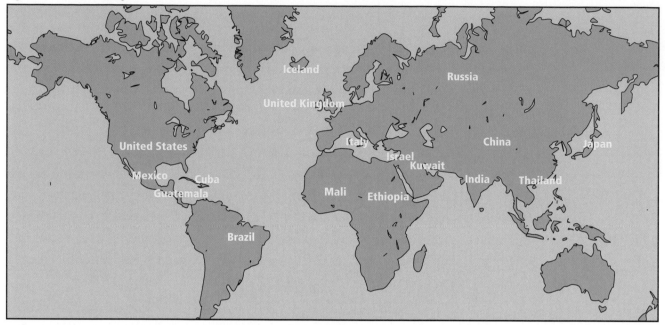

Table 1 Selected Statistics for 16 _Material World_ Countries

	Area (km²)	Population (x 10⁶)	Average Years of Schooling* (female/male)	% of Income Spent on Food	Annual Per Capita Commercial Energy Use (kg oil equivalent)
United States	3,618,000	293.6	12.4 / 12.2	9	7,996
1. Brazil	8,511,930	179.1	3.8 / 4	35	1,074
2. China	9,572,825	1,300.1	3.6 / 6	61	896
3. Cuba	110,929	11.3	7.7 / 7.5	n/a**	1,216
4. Ethiopia	1,221,991	72.4	0.7 / 1.5	50	291
5. Guatemala	108,999	12.7	3.6 / 4.4	36	626
6. Iceland	102,972	0.3	9 / 8.8	n/a	11,926
7. India	3,287,974	1,086.6	1.2 / 3.5	52	515
8. Israel	20,324	6.8	9 / 10.9	21	3,291
9. Italy	300,997	57.8	7.3 / 7.4	19	2,981
10. Japan	377,997	127.6	10.6 / 10.8	16	4,099
11. Kuwait	17,819	2.5	4.7 / 6	29	7,195
12. Mali	1,240,189	13.4	0.1 / 0.5	57	n/a
13. Mexico	1,957,986	106.2	4.6 / 4.8	35	1,532
14. Russia	17,074,868	144.1	n/a	60	(4,293)
15. Thailand	512,997	63.8	3.3 / 4.3	30	1,235
16. United Kingdom	244,999	59.7	11.6 / 11.4	12	3,982

* for adults 25 years or older

** n/a = not available

Analysis

?

Group Analysis

1. What other statistics would be useful in helping you understand what day-to-day life is like in other countries? Explain.

2. What other statistics would it be useful to have in "per capita" form? Explain.

3. What do you think is the biggest obstacle to survival for residents of each of the two countries you chose to examine? What evidence do you have to support your opinion?

4. What do you think is the biggest obstacle to sustainability for residents of each of the two countries you chose to examine? What evidence do you have to support your opinion?

5. What currently available technological innovation—or what possible future invention—would best help overcome the obstacles that you described in Analysis Questions 3 and 4? Explain.

Individual Analysis

6. In Procedure Step 1, you identified one statistic as the best piece of data to use in determining whether a country is less developed or more developed. What reasons do you have for choosing this statistic?

7. What do you think are the main advantages and disadvantages of living in a less developed country?

8. What do you think are the main advantages and disadvantages of living in a more developed country?

6.2 Energy Use and the Atmosphere

Purpose ▶ **E**xplore one of the global environmental issues associated with increased combustion of fossil fuels.

Introduction ▼

Carbon dioxide (CO_2) is a colorless, odorless gas. It is a natural part of Earth's atmosphere and is important to all life on our planet. All plants and animals release CO_2 to the atmosphere during respiration. Green plants and other producers remove some of this CO_2 for use during photosynthesis. Large amounts of atmospheric CO_2 become dissolved in Earth's bodies of water, especially the oceans. Nature has evolved to create a balanced situation in which annual CO_2 inputs to the atmosphere and CO_2 outputs from the atmosphere are roughly equal.

In recent years, standards of living and levels of technology use have risen considerably throughout the world, particularly in the United States. As a result, energy use has also risen considerably, as shown in the graph in Figure 2. Much of the energy used in the United States comes from the burning of **fossil fuels**, which include coal, natural gas, and petroleum. Burning these fuels produces not only heat and light, but also considerable amounts of CO_2.

What are the possible global consequences of this increase in CO_2 production? Have increases in CO_2 production affected the natural balance of CO_2 in the atmosphere?

The "smoke" from this factory's chimneys consists primarily of water vapor and carbon dioxide, two products of the combustion of fossil fuels.

Figure 2   Energy Use in the United States

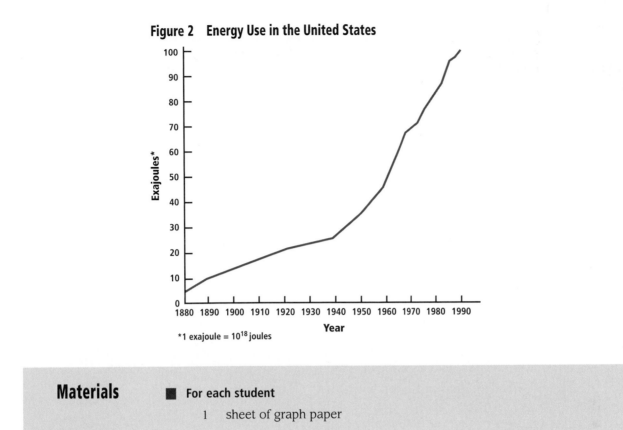

*1 exajoule = 10^{18} joules

Materials   ■ **For each student**

1   sheet of graph paper

Procedure

1. Based on the data provided in Table 2, make a line graph that shows the amount of CO_2 released into the atmosphere as a result of fuel combustion from 1860 to 1990.

2. Based on the data provided in Table 2, make a **scatterplot** that shows Earth's average surface temperature from 1860 to 1990.

3. Draw a line of best fit that allows you to predict future average surface temperatures. Record your predicted temperature for the year 2050.

Table 2   Global Carbon Dioxide Emissions and Surface Temperatures

Year	CO₂ Emissions from Fuel Combustion (in billions of tons)	Average Surface Temperature (in °C)	Year	CO₂ Emissions from Fuel Combustion (in billions of tons)	Average Surface Temperature (in °C)
1860	1	13.5	1930	4.5	13.8
1870	1	13.7	1940	5	14.0
1880	1.5	13.7	1950	6	13.8
1890	2	13.6	1960	10	14.0
1900	2.5	13.9	1970	13	14.0
1910	3	13.5	1980	18	14.1
1920	3.5	13.8	1990	22	14.4

Analysis
▼
?

Group Analysis

1. Briefly describe what you think could be a possible global consequence of an increase in Earth's average surface temperature.

2. Some people claim that having access to many goods and services can lead to a decrease in health and well-being. Could the information presented in this activity be used to support or refute this claim? Explain your reasoning.

Individual Analysis

3. Compare your two graphs. Explain whether or not they provide any evidence to support the claim that increased levels of CO_2 in the atmosphere lead to increased surface temperatures.

4. What additional information would you like to have before you would be confident in saying that increased levels of CO_2 in the atmosphere definitely do or do not lead to increased surface temperatures? Explain why this information is important.

5. Do you think society should take steps to reduce the emission of CO_2 into the atmosphere? Explain your reasoning.

6.3 **Materials, Energy, and Sustainability**

Purpose ▶ **C**ompare how inhabitants of different countries use materials and energy.

Introduction

When you built a soda-can insulator in Activity 5.2, "The Can Challenge," you had a variety of materials to choose from. You probably based your decision on each material's physical properties and availability. When faced with a choice between items that serve the same purpose, we often choose the item that is easiest or safest to use or the one with the lowest price. However, these criteria are based on short-term needs. We often fail to take into account the **hidden costs** involved in the manufacture, use, and disposal of an item. Could any of the items we regularly use be replaced with more effective or more sustainable alternatives?

These three containers—a plastic graduated cylinder, metal measuring cup, and glass beaker—can all be used to measure volume. Why do you think a different material was used to make each of these containers?

Materials

 For each team of two students

   1   copy of *Material World*

Procedure

1. Turn to Table 1 (page 68) and choose one less developed country and one more developed country (other than the United States) pictured in *Material World*. Look over the photographs and written descriptions for each country. Find two possessions—one in each country's photograph—that serve the same purpose but are made from different materials, such as wood and metal spoons.

2. Many household items are designed to alter the transfer of thermal energy (heat) from one place to another. For each of your two chosen countries, list at least 5 possessions that in some way increase or decrease the rate of thermal energy transfer. For example:

   A wall with glass windows allows more heat from the sun to enter a house than a wall without any windows.

   A wool rug decreases the flow of heat from your body to the floor.

   A metal pot increases the flow of heat from the stove to the food.

3. For each family depicted in the two countries you selected,

   a. list the daily activities that account for the majority of energy use.

   b. name the source(s) of the energy used for each activity.

   > **Example: The Skeen Family (United States)**
   >
   > | Activity | Energy Source |
   > |---|---|
   > | watching television | electricity* |
   > | driving cars | gasoline |
   > | cooking food | electricity* (and maybe gas) |
   > | refrigerating food | electricity* |
   > | doing laundry | electricity* (and maybe gas) |
   > | creating light | electricity* |
   > | listening to music | electricity* |
   > | working | gasoline, human labor, electricity* |
   >
   > * Electricity is actually not an energy source; it is an energy carrier. The major source of electricity in the U.S. is the combustion of fossil fuels.

4. Find two possessions—one from each of your two chosen countries—which serve the same purpose but use different energy sources.

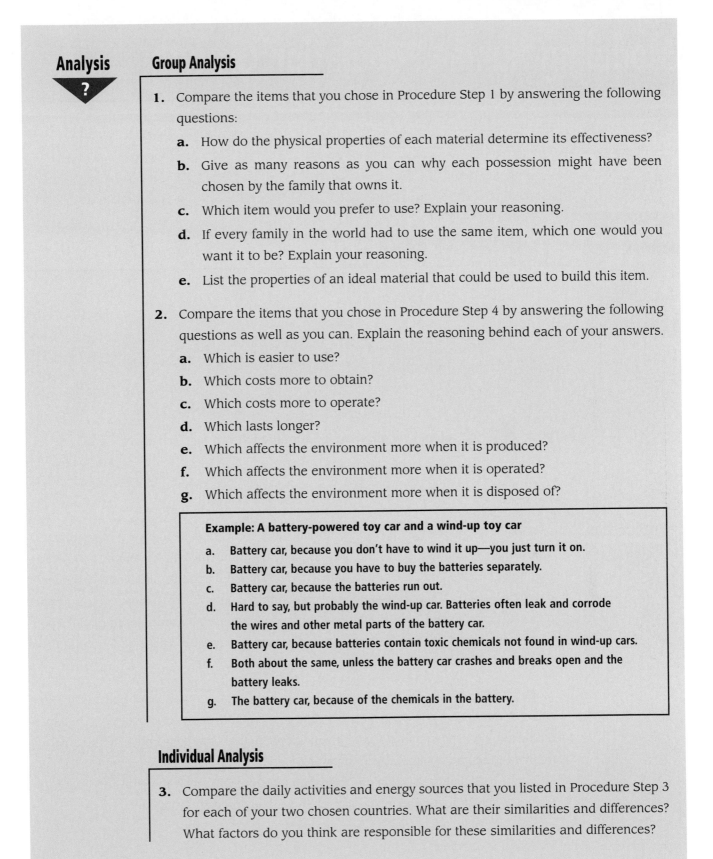

Analysis

?

Group Analysis

1. Compare the items that you chose in Procedure Step 1 by answering the following questions:

   a. How do the physical properties of each material determine its effectiveness?

   b. Give as many reasons as you can why each possession might have been chosen by the family that owns it.

   c. Which item would you prefer to use? Explain your reasoning.

   d. If every family in the world had to use the same item, which one would you want it to be? Explain your reasoning.

   e. List the properties of an ideal material that could be used to build this item.

2. Compare the items that you chose in Procedure Step 4 by answering the following questions as well as you can. Explain the reasoning behind each of your answers.

   a. Which is easier to use?

   b. Which costs more to obtain?

   c. Which costs more to operate?

   d. Which lasts longer?

   e. Which affects the environment more when it is produced?

   f. Which affects the environment more when it is operated?

   g. Which affects the environment more when it is disposed of?

   > **Example: A battery-powered toy car and a wind-up toy car**
   >
   > **a.** Battery car, because you don't have to wind it up—you just turn it on.
   > **b.** Battery car, because you have to buy the batteries separately.
   > **c.** Battery car, because the batteries run out.
   > **d.** Hard to say, but probably the wind-up car. Batteries often leak and corrode the wires and other metal parts of the battery car.
   > **e.** Battery car, because batteries contain toxic chemicals not found in wind-up cars.
   > **f.** Both about the same, unless the battery car crashes and breaks open and the battery leaks.
   > **g.** The battery car, because of the chemicals in the battery.

Individual Analysis

3. Compare the daily activities and energy sources that you listed in Procedure Step 3 for each of your two chosen countries. What are their similarities and differences? What factors do you think are responsible for these similarities and differences?

Modeling Human Population Growth 7

7.1 Oodles of Models

Purpose **E**xplore how different kinds of models can help scientists.

Introduction

A **model** is a type of research tool commonly used by scientists. There are many different kinds of models, and they can be used in many different ways. For example, an environment like the one you designed in Activity 2.5, "Maintaining a Sustainable Environment," could be used as a model to study the conditions of an animal's habitat or of an entire ecosystem. In Activity 2.3, "Population Estimation," you used plastic chips to model a population of otters. Recently, computers have begun to play an increasingly important role in modeling.

Models are especially helpful for doing research on subjects that cannot be easily duplicated or manipulated in the laboratory. One of the primary uses of models is to help predict what will happen under different conditions. Actual data collected from the environment can then be compared with data provided by the model. This comparison enables scientists to evaluate how well the model imitates present environmental conditions and suggest ways to revise the model to more accurately predict future conditions.

What Are Models?

Imagine building a model car out of plastic, glue, and paint. The finished model would look like a real car, only smaller. The scale of a model tells how big it is in relation to the real thing; if your model car were at a 1:10 scale, it would be one-tenth the size of an actual car. Put another way, a real car would be ten times bigger than your model. Some model kits come with motors that allow the finished model to move. When fully assembled, a model car with a motor can look and behave somewhat like a real car. That is all that any model is: a representation of something else.

A model car is called a **physical model**. Not all physical models are smaller than the thing they represent. For example, a scientist might use three styrofoam balls to make a physical model of a carbon dioxide molecule—two balls representing oxygen atoms and one ball representing a carbon atom—in order to illustrate the shape of the molecule. Not all physical models look like the thing they represent. The plastic disks you used in Activity 2.3, for example, were physical models of sea otters.

Scientists also use **conceptual models**. Conceptual models are often used to describe relationships among different parts of a system and define any rules governing those relationships. Examples include the energy flow diagrams you made in Activity 3.4, "Energy on the Move," and the food web diagrams you made in Activity 2.2, "The Web of Life." Illustrations of natural cycles—such as for water, rocks, or carbon—are other examples of conceptual models.

The food web diagram shown in Figure 1 is a conceptual model that describes the way in which populations of

Figure 1 Conceptual Model

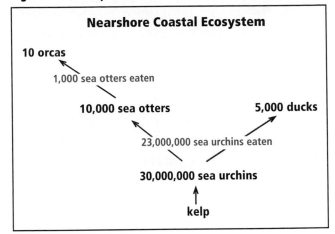

organisms in an ecosystem are interrelated. This food web includes the amount of food per year transferred along each of the various food web pathways. An even more complicated model might try to show how these different rates of food transfer depend upon one another. In that case, instead of a series of words and numbers the model might consist of a series of mathematical equations. This would be a **mathematical model**.

Figure 2 provides an example of two very simple mathematical models. When trying to model more complex mathematical relationships, scientists might enter the equations into a computer program, creating a **computer model**. A computer model takes advantage of the computer's ability to do rapid calculations. Computer models can quickly show how a situation might change as time goes by or as conditions change.

Figure 2 Mathematical Models

$$\text{final population} = [(\text{birth rate} - \text{death rate}) \cdot \text{initial population}] + \text{initial population}$$

$$\text{death rate of sea urchins} = \frac{\text{\# eaten by otters} + \text{\# starved due to lack of kelp} + \text{\# who die of old age}}{\text{initial population of sea urchins}}$$

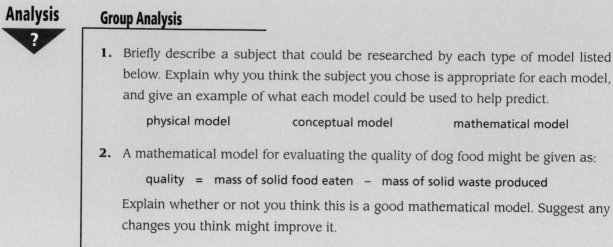

Analysis

?

Group Analysis

1. Briefly describe a subject that could be researched by each type of model listed below. Explain why you think the subject you chose is appropriate for each model, and give an example of what each model could be used to help predict.

   physical model   conceptual model   mathematical model

2. A mathematical model for evaluating the quality of dog food might be given as:

   quality = mass of solid food eaten – mass of solid waste produced

   Explain whether or not you think this is a good mathematical model. Suggest any changes you think might improve it.

3. Make a conceptual model and a mathematical model to describe a food web in an imaginary game preserve that includes grass, deer, wolves, and hunters.

7.2 Deer Me!

Purpose ▶ **U**se a computer model to represent changes in a deer population.

Introduction

In the last activity, you read about physical, conceptual, mathematical, and computer models. In this activity, you will use a computer model to represent changes in a deer population in a game preserve. The size of any population is determined in part by its **birth rate**, which is typically reported as the number of births per year per 1,000 organisms, and its **death rate**, typically reported as the number of deaths per year per 1,000 organisms.

The computer model that you will use allows you to set both the birth rate and the death rate for the deer population. Based on this information, it will produce a graph showing how the population changes over time. In this model, the birth and death rates will be reported as decimals. A birth (or death) rate of 0.55, or 55/100, means that for every 100 deer in the herd at the beginning of the year, 55 deer will be born (or die) by the end of the year.

Materials

For each group of four students

1   computer with STELLA® software and NCSA deer population models installed

Procedure

Part A   Examining the Model

1. Launch the first population model by double-clicking on the "Deer Herd" icon (Mac) or "deer.stm" (PC). You will know that you have done this successfully when a graph appears on the screen.

2. Close the graph by clicking in the box in the upper left hand corner of the graph. The following diagram should then appear on your screen.

Figure 3   Deer Herd Model

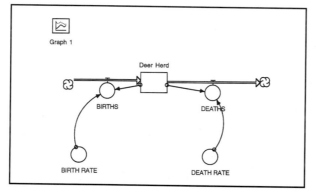

3. Each of the objects in the model represents an important component of the game preserve. Each of the lines in the model represents a connection between components. Double-click on each of the following objects to reveal the initial set-up of the model:

    Deer Herd: Record the initial population of the deer herd, then click OK.

    Birth Rate: Record the birth rate for the deer herd, then click OK.

    Death Rate: Record the death rate for the deer herd, then click OK.

    Births: Record the equation used to calculate the number of deer born each year, then click OK.

    Deaths: Record the equation used to calculate the number of deer that die each year, then click OK.

4. Double-click on the icon labeled "Graph 1," then run the model by selecting "Run" from the Run menu. You will know the program is running if you see a graph being drawn.

5. The model's projected changes in the deer population will be shown in Graph 1. Make a sketch of Graph 1 in your journal.

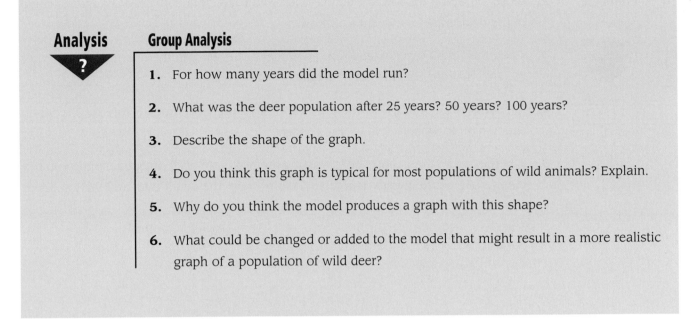

Analysis

Group Analysis

1. For how many years did the model run?

2. What was the deer population after 25 years? 50 years? 100 years?

3. Describe the shape of the graph.

4. Do you think this graph is typical for most populations of wild animals? Explain.

5. Why do you think the model produces a graph with this shape?

6. What could be changed or added to the model that might result in a more realistic graph of a population of wild deer?

Part B   Manipulating the Model

Prediction

The model you just ran had an initial deer population of 100, a birth rate of 0.50, and a death rate of 0.45. Predict what you think the deer population will be after 100 years

    a. if the birth rate is 0.50 and the death rate is 0.50.

    b. if the birth rate 0.50 and the death rate is 0.55.

Procedure

1. Prepare a data table similar to the one below.

Table 1   Deer Population Over Time

Birth Rate	Death Rate	Deer in 25 Years	Deer in 50 Years	Deer in 100 Years
0.50	0.50			
0.50	0.55			
0.50	____			
____	____			
____	____			
____	____			
____	____			

Procedure
(cont.)

2. Close the Graph window and double-click on the icon for death rate. Change the value of the death rate to 0.50, then click OK.

3. Open the Graph window and run the model again. Record the deer population after 25, 50, and 100 years in your data table.

4. Repeat Steps 1 and 2 two more times, first changing the death rate to 0.55, then changing it to whatever value you choose.

5. Run the model three more times, varying the birth rate each time. Each time you run the model, record in your data table the birth rate, the death rate, and the total deer population after 25, 50, and 100 years.

Note: A population size of 2.00 e+8 is equal to 2.00×10^8 or 200,000,000.

Analysis
?

Group Analysis

1. What other variables could be added to this computer model to make it more realistic? **Hint:** What factors control birth and death rates?

Individual Analysis

2. How accurate were your predictions?

3. What are the advantages and disadvantages of using this computer model rather than the conceptual model and mathematical model you created for Analysis Question 3 in Activity 7.1, "Oodles of Models"?

7.3 World Population Growth

Purpose ▶ **E**xamine changes in human population growth.

Introduction

Ecology is a branch of the life sciences that focuses on the interactions among organisms and between organisms and their surroundings. One field of study within ecology is **population dynamics**. The study of population dynamics focuses on how populations change over time. A single population can be studied independently, such as examining the changes over time in a herd of deer, or several populations can be studied simultaneously, such as observing how populations of deer and wolves change in relation to each other.

In this activity, you will consider how the global human population has changed over time.

The global human population has increased rapidly over the past 100 years. What are the trade-offs associated with population growth?

Procedure

1. Study the graphs shown in Figures 4 and 5.

2. List the similarities and differences between the two graphs.

3. List a set of conditions that could cause

    a. the UN high projection.

    b. the UN middle projection.

    c. the UN low projection.

Figure 4   Historical World Population Growth   **Figure 5   Future World Population Projections**

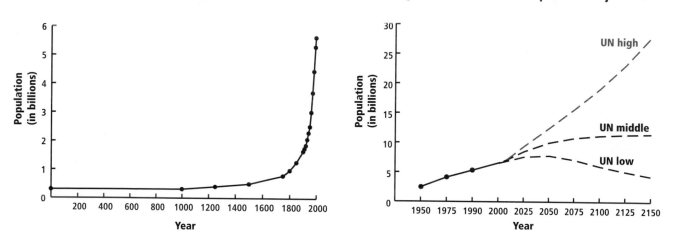

Analysis

?

Group Analysis

1. There appear to be significant slope changes on the graph shown in Figure 4 at the years 1000, 1500, 1750, and 1900.

    a. What do these slope changes indicate?

    b. List at least five possible events that could have brought about these changes.

2. What conclusions can you draw from the graph shown in Figure 4?

Individual Analysis

3. Which long-range population projection from the graph shown in Figure 5 do you think is the most likely to occur? Explain.

4. Can graphs, in general, be considered a type of model? Why or why not?

7.4 Plants and People

Purpose ▶ **C**ompare human population density with global vegetation patterns.

Introduction

The maximum population size an environment is capable of sustaining is called the **carrying capacity** of that environment. During the last 100 years, the world's human population has increased rapidly. Today, the size of the human population far exceeds carrying capacity in many regions of the world. This situation is able to continue year after year only because people do not have to rely on local resources for many of their needs; humans are able to trade for goods and services from all over the world. In terms of global sustainability, a discussion of the environment's carrying capacity for humans involves considering the resources of the entire planet, not just the resources of a specific country or region. Nonetheless, studying the regional distribution of human populations can provide important information about each region.

One way to understand the distribution of human populations is to study **population density**, the number of organisms inhabiting a given area at a given time. Human population density varies from one region to another. To imagine differences in population densities, consider two places you already know something about, but may never have visited: New York City and Antarctica. How many people do you think live in one square kilometer of New York City versus one square kilometer of Antarctica?

Prediction

Look at the map in Figure 6, on the next page. Consider what you already know about where people live and where food is grown, then make a prediction about each of the following questions:

    a.   Which part of the world do you think has the greatest number of people per square kilometer? Why?

    b.   In which parts of the world do you think the most food is grown? Why?

Figure 6   Locator Map

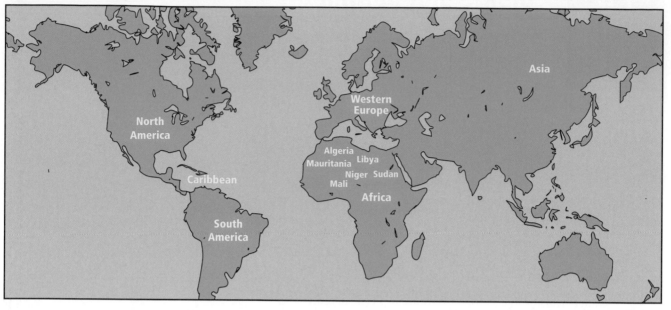

Materials      **For each group of four students**

    1   set of 40 plastic disks

    1   global vegetation map

    1   copy of *Material World* (optional)

Procedure

1. Examine the global vegetation map carefully. List the three regions of highest vegetation and the three regions of lowest vegetation.

2. Prepare a data table similar to the one on the next page.

3. To create a three-dimensional population density map that uses a stack of plastic disks to represent the number of people living on each continent, you will need to determine an appropriate scale. You have only 40 plastic disks, and you must first determine the number of people per square kilometer that each disk will represent. **Hint:** You don't have to use all 40 disks.

4. Use the scale you have determined to complete the data table.

5. Build a three-dimensional population density map by stacking the appropriate number of disks on the appropriate region of the vegetation map.

6. Record your observations regarding similarities and differences between the distribution of vegetation and the distribution of humans.

Procedure
(cont.)

Table 2   Modeling Population Density

Region	2004 Population Density (people per km^2)	Number of Disks Used
Africa	29	
Asia	122	
Caribbean	166	
North America*	16	
Western Europe**	167	

*United States and Canada

** Austria, Belgium, France, Germany, Liechtenstein, Luxembourg, Monaco, Netherlands, and Switzerland

Analysis

?

Group Analysis

1. a. How many people per square kilometer did each of your disks represent?

    b. What was the total number of disks that you needed to build your map?

2. What relationship, if any, do you see between human population density and the amount of vegetation in a region? Explain.

3. The average population density on the African continent is 29 people/km^2. The average population density of many northern African countries (Mauritania, Mali, Niger, Chad, Sudan, Algeria, and Libya) is less than 15 people/km^2.

    a. What geographical feature might explain the lower-than-average population density for the northern part of Africa?

    b. What characteristics of a region do you think most affect its carrying capacity? Explain.

Individual Analysis

4. How close were your predictions to your observations? Explain.

5. The population density of Australia is 2.6 people/km^2. How would you change the scale of your map so that it could accurately represent this region?

6. Over 30,000 people live in Monaco, a small western European country with an area of only 2 km^2.

    a. Calculate the population density of Monaco.

    b. How do you suppose such a small area of land can support so many people? Explain.

Population Dynamics

8.1 Population Curves

Purpose  **G**raph population data to analyze trends in population changes.

Prediction

Choose a wild animal that is widely distributed in your community. Reproduce the axes shown in Figure 1, then create a graph by drawing a curve to show how you think the size of the local population of this animal will change from year to year for the next 15 years. Write a short paragraph explaining why your graph looks the way it does.

Figure 1   Population Size Over Time

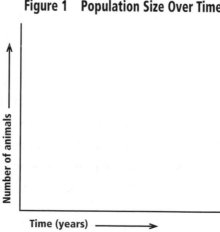

Introduction

Given plentiful resources and a safe environment, the population of any organism will grow. You may be familiar with population growth if you've had rabbits, hamsters, or guppies as pets. These species reproduce quickly. If you kept all the babies, you would soon have a large number of animals to care for.

Suppose a population increases by the same amount at regular intervals, such as each month, year, or decade. This type of growth pattern is called **linear growth**. An example would be a population that increases by two individuals every month. Population data collected monthly would follow this pattern: 2, 4, 6, 8, 10, 12, 14, . . . When plotted on a graph, this data set produces a straight line which can be described by the mathematical equation $y = 2x$, where **y** represents the population size, **x** represents time (in this case, the number of months that have passed), and **2** is the number of individuals added during each time interval.

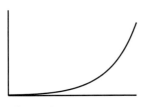

**Figure 2
Linear Curve**

In fact, a graph of any population undergoing linear growth will be a straight line like the one shown in Figure 2. Data for any linear population growth can be described by the equation $y = mx+b$, where **y** represents the population size, **x** represents time, **m** represents the number of individuals added during each time interval, and **b** represents the initial population.

When a population exhibits **exponential growth**, the number of individuals added to the population gets larger during each time interval. A data set for a population that is growing exponentially might follow a pattern such as this one: 1, 2, 4, 8, 16, 32, 64, 128, . . . When plotted on a graph, the curve for exponential population growth resembles the letter "J" and is often called a **J-curve** (Figure 3). The equation for the sample data set just given, in which the population doubles during each time interval, is $y = 2^x$. In this equation, **y** represents the population size, **x** represents the time elapsed, and the **2,** called a "base," causes the population to double during each time interval. The equation describing a population that triples during each time interval would use base 3 and would be written $y = 3^x$.

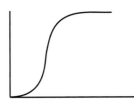

**Figure 3
Exponential Curve
(J-Curve)**

No population can continue to grow forever. Eventually it will reach, or even temporarily exceed, the ecosystem's carrying capacity. Given the right conditions, the size of a population will reach the ecosystem's carrying capacity, then remain relatively stable at that level. Suppose a small population of a given species is introduced into an ecosystem for the first time. Typically—if there are ample resources and few predators—there will be an initial period of slow population growth followed by a period of rapid exponential growth, which will then level off when the population reaches the carrying capacity. Data for this pattern of population growth—exponential, then stable at or near carrying capacity—produces a graph with a curve shaped somewhat like the letter "S." This is known as an **S-curve** (Figure 4).

**Figure 4
S-Curve**

Materials

■■ **For each team of two students**

1   computer or graphing calculator (optional)

■ **For each student**

1   sheet of graph paper

Scenario

Imagine spending your spring break kayaking out on the ocean!

Unfortunately for you, it turns out to be an El Niño year and the weather suddenly becomes much stormier than you expected. You are blown off course and land on a remote island. The only other animal living on the island is an unusual type of turquoise lizard. To while away the hours, you begin counting the number of lizards on the island. As years go by, you realize that someday—when you're rescued—you'll be able to share the results of your study of lizards and perhaps achieve fame and fortune!

After many years, you have gathered the data shown in Table 1. To make some general conclusions, you decide to graph the data (in the sand, of course!).

Table 1   Population of Lizards Over Time

Year	Population Size	Year	Population Size
1	21	14	421
2	58	15	410
3	101	16	409
4	138	17	414
5	177	18	431
6	205	19	483
7	209	20	532
8	216	21	539
9	228	22	540
10	264	23	535
11	399	24	550
12	560	25	544
13	456	26	530

Procedure

1. With your partner, decide what type of graph (bar, pie, line, scatterplot, etc.) is most appropriate for this data.

2. Title your graph, label its axes, and plot the data.

Analysis

?

Group Analysis

1. Explain why you chose to make this type of graph. Do you think you made the best choice?

2. The lizard population goes through five major phases of population growth. Identify the time period for each phase and describe the type of growth taking place during each phase.

Individual Analysis

3. Look back at the graph you made for the Prediction at the beginning of this activity.

   a. Describe the type(s) of growth your curve represents.

   b. What conditions could have brought about this pattern of growth?

4. What environmental factor(s) on the island do you think would be most important for determining its carrying capacity for turquoise lizards?

5. In 1798, Thomas Malthus, an English economist, published his "Essay on the Principle of Population." In it, he proposed that the human population grows exponentially, while the food supply grows linearly.

   a. Sketch a graph of Malthus' theory. Draw one curve that represents his prediction for the growth of the human population. Draw a second curve that represents his prediction for the growth of food supplies.

   b. Based on this graph, why do you think Malthus predicted in his essay that within 50 years there would not be enough food to feed the human population?

   c. Why do you think Malthus' predicted food shortage did not occur?

Extension

Use data supplied by your teacher to graph human population growth in your city or county. Identify the time period for each phase of growth and describe the type of growth taking place during each phase.

8.2   A Population of Fruit Flies

Purpose ▶ **G**row a population of organisms in the laboratory and plot a population curve based on your laboratory data.

Introduction ▼

In Activity 8.1, "Population Curves," you explored the patterns of some common population growth curves. In this activity, you will collect actual data from a growing population of *Drosophila*, the fruit fly, and create a population curve. To track population changes for at least three generations of flies, you will collect data once each week over a period of eight weeks.

Figure 5   Life Cycle of the Fruit Fly

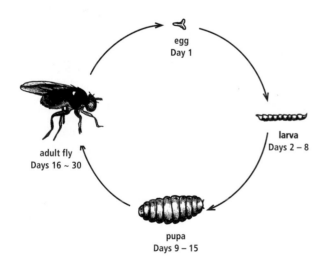

There are four stages in the life cycle of the fruit fly: egg, larva, pupa, and adult. These stages are shown in Figure 5. The time required to complete one life cycle is mainly dependent on the temperature of the surroundings. At a typical classroom temperature of 21°C, it takes about two weeks for a fruit fly egg to mature into an adult *Drosophila*: one day as an egg, seven days in the larval stage, and six days in the pupa stage. The life span of an adult fruit fly is only several weeks.

Prediction ◀?▶

What type of population curve do you predict that you will see? Sketch your predicted population curve, making sure to label the graph axes.

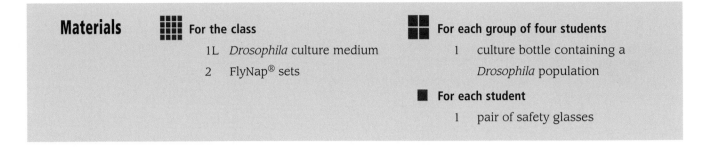

Materials

▦ **For the class**

    1L *Drosophila* culture medium

    2   FlyNap® sets

◪ **For each group of four students**

    1   culture bottle containing a
          Drosophila population

■ **For each student**

    1   pair of safety glasses

Procedure

Part A Getting Started

1. Label your culture bottle.

2. Create a data table to record and track your population each week for 8 weeks.

Part B Establishing Your Baseline

3. Following these guidelines and your teacher's instructions, use the FlyNap to anesthetize your flies. To administer the FlyNap without killing your flies or letting them escape, push the wand carefully along the side of the bottle plug until it is clearly visible in the bottle. As soon as one fruit fly appears anesthetized, immediately remove the wand. Wait until all the flies are anesthetized before proceeding.

> **Caution:** Overexposure to FlyNap will kill small organisms. Fruit fly exposure to FlyNap should be relatively short and flies should not be anesthetized more than once a day! Do not wet the wand so much that it drips the anesthetic into your fruit fly bottle.

4. Carefully remove (if necessary) and count the number of fruit flies. Record the population in your data table.

5. Carefully return the flies to the bottle (if necessary) and make sure the foam plug is securely in place.

Part C Tracking the Population (To be done once a week for 8 weeks)

6. Repeat Steps 3–5.

7. Add additional culture medium to the bottle as necessary.

Analysis

Group Analysis

1. What is the purpose of putting the blue culture medium in the bottles?

2. Create a graph of time versus population size for your fruit fly population.

Individual Analysis

3. Do your data support your prediction? Explain.

4. Does your graph resemble any of the population curves discussed in Activity 8.1, "Population Curves"? If so, which one?

5. Interpret your graph. What happened to your fruit fly population over the eight-week study period? Why do you think your population behaved the way it did?

6. What environmental limiting factor do you think is most important in setting the carrying capacity for fruit flies in this experiment? Explain your reasoning.

8.3 Sharing an Environment

Purpose ▶ **E**xamine how one species' population size can affect the carrying capacity for a different species living in the same environment.

Introduction ▼

There are many limiting factors that help determine an ecosystem's carrying capacity for a particular species. There are physical factors, such as water, air, and available space, as well as biological factors, such as food supplies and the presence of other species. Different species not only compete with each other for available physical resources, but also often have relationships in which one species depends on the other. In such interdependent relationships, the size of one population is frequently closely tied to the size of the other. One of the more obvious examples of interdependence is the **predator-prey relationship**, in which one species serves as the food source for another species.

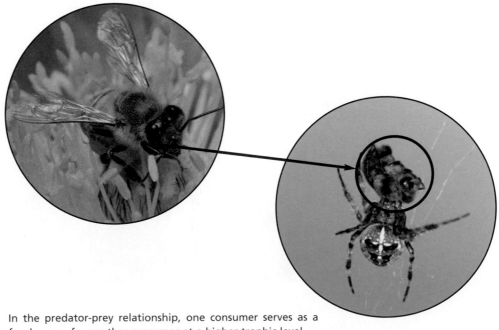

In the predator-prey relationship, one consumer serves as a food source for another consumer at a higher trophic level.

Population Dynamics   **95**

Materials   ■ **For each student**
1   sheet of graph paper

Procedure

1. Examine the graph shown in Figure 6. Describe the relationship between the two curves.

2. Imagine that for the next 50 years, the food supply for the rabbits is 50% lower than it was for the 50 years shown in the graph in Figure 6. Make a data table predicting the size of the rabbit population and the bobcat population every five years for the next 50 years.

3. Using the data from the table you created in Step 2, make a double line graph showing what you predict will happen to both populations during the 50-year period.

Figure 6   Population Curves for Bobcats and Rabbits

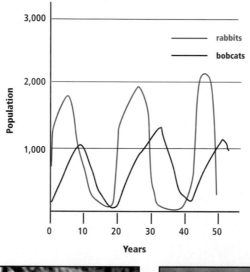

Populations of animals coexisting in an ecosystem are often interdependent. The bobcat population depends upon rabbits for food; this predator-prey relationship keeps both populations from becoming too large.

Analysis
?

Group Analysis

1. Estimate this ecosystem's carrying capacity for rabbits and for bobcats. Explain how you made your estimates.

2. Describe the shapes of the curves for bobcats and rabbits and the relationship between the two curves. Explain, in as much detail as you can, why the curves for bobcats and rabbits create this pattern.

3. Look back at the food webs in Figure 4 in Activity 2.4, "Where Have All the Otters Gone?" Suppose the orcas began to eat twice as many otters. Draw a graph showing how large you think the otter and sea urchin populations would have been during the 20 years before and the 20 years after the orcas first began eating more otters.

Individual Analysis

4. Describe a relationship between two species, other than the predator-prey relationship, that you think would produce a set of population curves similar to or different from the one shown in Figure 6. Explain why you think this relationship would create this pattern.

5. How does the size of populations of other species relate to human sustainability? Think of at least two species that affect human survival and describe the relationship of each to the human population.

6. Some bacteria reproduce by cloning themselves once every 20 minutes.

   a. If you started with one bacterium, how many would you have in 24 hours?

   b. What mathematical equation would describe the hourly population growth of the bacteria?

   c. Suggest at least two limiting factors that keep bacteria from taking over Earth.

8.4 Deer Me, Deer Me!

Purpose ▶ **U**se a computer model to explore factors that affect population growth and decline, including the availability of resources and the presence of predators.

Introduction

In Activity 7, "Modeling Human Population Growth," you learned about models used for estimating population changes over time. Now you have the opportunity to use a computer model to investigate some of the factors that cause a population to increase or decrease. You will use a model similar to the one you used in Activity 7.2, "Deer Me." This model also represents a population of deer on a game preserve, but this time you will be able to investigate the effect of changes in the size of the game preserve, an increase or decrease in a population of wolves, and the presence of deer hunters.

Population density, the number of organisms inhabiting a given area at a given time, has an important influence on the size of a population over time. Any defined area has a limited number of resources. The higher the population density, the greater the competition among individuals for those resources. Greater competition typically results in more deaths and fewer births.

Materials   ■ For each group of four students
          1   computer with STELLA® software and NCSA deer population models installed

Procedure   **Part A   Carrying Capacity**

1. Launch the first population model by double-clicking on the "Carrying Capacity" icon (Mac) or "capacity.stm" (PC). You should see a graph on the screen. Close the graph by clicking in the box in the upper left hand corner. You will see that in addition to the objects from the model you examined in Activity 7.2, this model has three new ones:

    Area: The size of the area in which the deer live

    Animal per area: The population density

    "Dapa":   The time delay before birth rates adjust to a new population density

2. Double-click on each of the three new objects and, for each object, record the initial value or the equation used to calculate the value.

3. Double-click on the icon for birth rate and make a sketch in your journal of the graph that appears. When you're done, click OK.

4. Double-click on the icon for death rate and make a sketch of the graph that appears. When you're done, click OK.

5. Double-click on the graph icon, then run the model.

6. Make a sketch of Curve 1, which describes the changes in the deer population. Then make a separate sketch of Curve 2 (deer births) and Curve 3 (deer deaths).

    Note: Different curves can have different scales on the y-axis. Use the key found at the upper left corner to determine the appropriate scale for each curve.

7. Prepare a data table similar to the one on the next page. Use the first row of your data table to record the information generated when you ran the model in Step 5.

8. Change the values for area and delay time ("dapa") in the model, one variable at a time, until you can describe what effect each variable has on the deer population. Record each change you make and the resulting change in the deer population.

    Note: If you decide you'd like to get back to the original model but have trouble doing so, quit the program and then reopen the population model.

Procedure

(cont.)

Table 2   Deer Population Over Time

Model Parameters		Time when deer population stabilizes	Size of stable deer population
Area	Delay Time		

Analysis

?

Group Analysis

1. If the size of a growing population stabilizes, this defines the carrying capacity of the ecosystem for that species. Look back at the sketches of the graphs you made for Procedure Step 6. Describe the relationship that exists between births and deaths when carrying capacity is reached. Why does this relationship exist?

2. Describe the effect that changing the area has on the deer population.

3. Describe the effect that changing the delay time ("dapa") has on the deer population.

4. What other variables could be added to the model so that it would more realistically describe limiting factors for the deer population?

Part B Predator-Prey

You will now explore a model similar to the one you used in Part A, only slightly more complex. In this model, wolves can be introduced into the game preserve. Like the deer, the wolf population is subject to limiting factors in the ecosystem. Because wolves eat deer, the two populations are connected. An increase in the number of wolves will cause a decrease in the deer population; however, if there are too few deer some wolves will die of starvation, and the wolf population will decrease.

The relationship between the deer and the wolves in your model is a predator-prey relationship. Other common relationships among species include competition, in which two species attempt to obtain the same resource, and symbiosis, in which the daily lives of two species are closely connected. The relationship between a parasite and its host is an example of symbiosis. A parasite is an organism that uses another living organism as a food source; parasitism is a type of symbiosis in which one organism benefits and the other is harmed. In another type of symbiosis, mutualism, both species benefit. A sea anemone attached to the shell of a living crab is an example of mutualism. The crab benefits from the camouflage; the anemone benefits because it becomes mobile, increasing its potential access to food.

Prediction

What effect do you think the addition of wolves will have on the population of deer in the game preserve?

Procedure

1. Open the model titled "Predator-Prey" (Mac) or "pred-prey.stm" (PC). Prepare a data table with headings similar to those shown below.

2. Record the initial deer and wolf populations for this model in the first row of your data table.

Deer and Wolf Data Over Time

Model Parameters		Time when population stabilizes		Carrying Capacity	
Initial Deer Population	Initial Wolf Population	Deer	Wolves	Deer	Wolves

Procedure
(cont.)

3. Open the graph window, then run the model. Make a sketch of the resulting graph.

4. Use the information on the graph to complete the first row of your data table.

5. Vary the initial populations of deer and wolves, one variable at a time, in at least six ways. Use your data table to record what happens to each population each time.

Analysis
?

Group Analysis

1. Describe the effect of changing the initial population of deer. Explain why this effect occurs.

2. Describe the effect of changing the initial population of wolves. Explain why this effect occurs.

Individual Analysis

3. How is the graph that you sketched in Procedure Step 3 similar to the graph of rabbits and bobcats shown in Figure 6 in Activity 8.3, "Sharing an Environment"? How is it different?

4. What additional variables could be added to make this model more realistic?

Part C Hunters

Using the same software that you used in Parts A and B, you will now model the impact of deer hunters on the deer and wolf populations.

Prediction

What effect do you think the addition of deer hunters will have on the population of deer and wolves in the game preserve?

Procedure

1. Open the model titled "Hunters" (Mac) or "hunters.stm" (PC). You can double-click on any element to find out what it is.

2. Run the model and sketch the changes shown on the graph.

3. Prepare a data table with headings similar to those shown below. Use the first row of your data table to record the information from the first graph.

4. Change the model, one variable at a time, in at least six ways. Record what happens to each population each time.

Deer, Wolf, and Human Data Over Time

Model Parameters			Time When Population Stabilizes		Carrying Capacity	
Initial Deer Population	Initial Wolf Population	Deer Killed by Hunters	Deer	Wolves	Deer	Wolves

Analysis

Individual Analysis

1. Describe the types of population curves produced by the model. Describe the conditions that caused each curve, and explain why each set of conditions created each curve.

2. What additional variables could be added to make this model more realistic?

Changing Populations

9

9.1   Population Projections

Purpose ▶ **I**nvestigate how changes in birth and death rates affect population size.

Introduction

A population **growth rate** describes how the size of a population changes over time. Any population that increases in size is said to have a positive growth rate. A decreasing population has a negative growth rate. A population that remains the same size has a zero growth rate. What are the variables that determine a population's growth rate? The answer to this question might seem complicated at first, but population growth can be determined by the interactions of three major variables: birth rate, death rate, and the number of organisms migrating from one region to another. In this activity, you will use the equations and variables shown in Figure 1 to investigate the growth patterns in populations.

Figure 1   Calculating Population Size

$$Pop_f = (Pop_i + \text{\# of births}) - \text{\# of deaths}$$

where

Pop_f	=	final population
Pop_i	=	initial population = last year's Pop_f
# of births	=	birth rate • Pop_i
# of deaths	=	death rate • Pop_i

Materials

■■ **For each team of two students**

    supply of graph paper

   1   computer or graphing calculator (optional)

Procedure

1. Make a data table similar to the one below.

Table 1   Constant Birth Rate Exceeds Constant Death Rate

Year	Initial Population (Pop$_i$)	Birth Rate (per 1000)	# of Births	Death Rate (per 1000)	# of Deaths	Final Population (Pop$_f$)
1	1,000,000	250	250,000	25	25,000	1,225,000
2		250		25		
3		250		25		
4		250		25		
5		250		25		
6		250		25		
7		250		25		
8		250		25		
9		250		25		
10		250		25		

2. Perform the necessary calculations to fill in the data for years 2–10 in Table 1. Notice that in this case the birth rate and death rate remain constant, and the birth rate is higher than the death rate.

Note: If you use a spreadsheet program or a graphing calculator to complete these calculations, you may wish to extend the data table for significantly more years (more than 50). The additional data will give you a more complete view of the population trend.

3. Using the data from Table 1, make a graph that plots time vs. final population. Title it "Graph 1: Constant Birth Rate Exceeds Constant Death Rate."

▶

Procedure
(cont.)

4. Make two data tables, one similar to Table 2, the other to Table 3.

5. Fill in the data for years 2–10 in your versions of Tables 2 and 3. Notice that these tables provide the number of births and deaths that occur each year, so you do not need to calculate these values.

6. Plot two more graphs (time vs. final population), one using the data from Table 2 and the other using the data from Table 3.

Table 2 Birth Rate Equals Death Rate

Year	Initial Population (Pop$_i$)	Birth Rate (per 1000)	# of Births	Death Rate (per 1000)	# of Deaths	Final Population (Pop$_f$)
1	1,000,000	50	50,000	50	50,000	1,225,000
2		50	50,000	50	50,000	
3		50	50,000	50	50,000	
4		50	50,000	50	50,000	
5		50	50,000	50	50,000	
6		50	50,000	50	50,000	
7		50	50,000	50	50,000	
8		50	50,000	50	50,000	
9		50	50,000	50	50,000	
10		50	50,000	50	50,000	

Table 3 Decreasing Birth Rate, Increasing Death Rate

Year	Initial Population (Pop$_i$)	Birth Rate (per 1000)	# of Births	Death Rate (per 1000)	# of Deaths	Final Population (Pop$_f$)
1	1,000,000	50	50,000	30	30,000	1,225,000
2		48	48,960	32	32,640	
3		46	47,671	34	35,235	
4		44	46,145	36	37,755	
5		42	44,400	38	40,172	
6		40	42,455	40	42,455	
7		38	40,332	42	44,578	
8		36	38,057	44	46,514	
9		34	35,655	46	48,239	
10		32	33,155	48	49,732	

Analysis ?

Group Analysis

1. Use the following equation to calculate the population growth rate for each of your three data sets:

$$\frac{\text{final population in year 10} - \text{initial population in year 1}}{\text{initial population in year 1}}$$

2. Sketch a population curve that you think describes a population in which the birth and death rates are constant and the death rate exceeds the birth rate.

3. Using the curve you sketched in Analysis Question 2, estimate the population growth rate.

Individual Analysis

4. Figures 2 and 3, below, provide graphs of birth- and death-rate data for France and Nigeria from the mid-1950s to the late 1990s and projected up to 2020. Based on the data from each graph, sketch the approximate shape of the population curve, from 1960 to 2020, that these birth and death rates should produce.

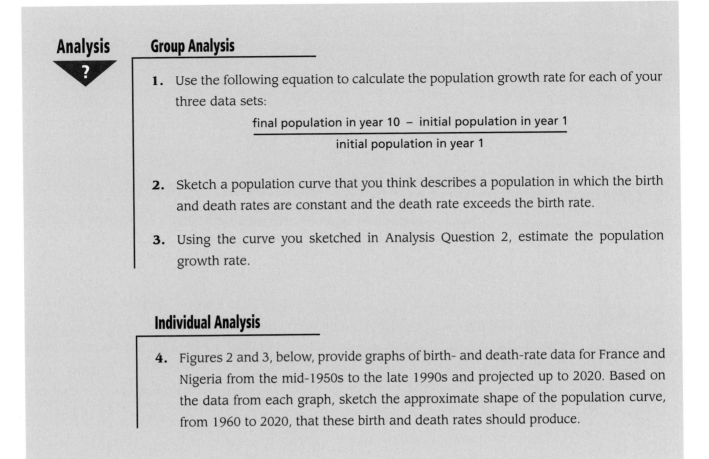

Figure 2   France: Birth and Death Rates

Figure 3   Nigeria: Birth and Death Rates

9.2 Comparing Countries

Purpose ▶ **U**se international statistics to discover global population trends and investigate the causes of these trends.

Introduction ▼

Population growth rates vary from one region to another around the world. International organizations like the United Nations gather statistics on each country's growth rate, not only to keep track of the worlds' changing population, but also to make projections for the future. **Demography** is the study of human populations, including the factors that influence population size and growth patterns.

Population doubling time is the number of years, given present trends, that it will take for a country to double its current population size. This statistic is commonly used to compare rates of population growth from country to country and to forecast future population size. Another statistic that is strongly linked to population doubling time and growth rate is the **fertility rate**, the average number of children born to each woman. For many reasons, some countries have higher fertility rates than others. You might assume that a country with a high fertility rate would also have a high population growth rate, but this is not always true. Many countries with high fertility rates actually have lower population growth rates than countries with lower fertility rates. For example, a country with high fertility rates might also have high emigration. The number of people who emigrate from, or leave, a region and the number who immigrate to, or enter, a region can significantly affect total population size. Do statistics other than migration data have a significant impact on population size?

Table 4   Population (in millions) by Region: 1650–2050

Year	World	North America	Latin America	Africa	Europe	Asia	Oceania
1650	545	1	12	100	103	327	2
1950	2,555	166	166	228	572	1,411	12
2004	6,396	326	549	885	728	3,875	33
2025*	7,934	386	685	1,323	722	4,778	41
2050*	9,276	457	778	1,941	668	5,385	47

*projected data

Prediction

Which of the following statistics do you think is most closely associated, either positively or negatively, with fertility rate? Explain your reasoning.

- daily caloric intake (average number of calories eaten per person per day)

- per capita income (average number of US$ earned per person per year)

- infant mortality (number of children per 1000 births who die before reaching age 1)

- population per physician (total number of people ÷ total number of doctors)

Materials

■■ **For each team of two students**
  1   copy of *Material World*

■ **For each student**
  2   sheets of graph paper

Procedure

1. Decide which two of the four statistics listed in the Prediction you will investigate further and which two your partner will investigate.

2. Find the statistics provided in the table on pages 248–249 of *Material World*. Create a graph, or graphs, to illustrate the relationship between each of the statistics you have chosen and the fertility rate for the sixteen *Material World* countries listed in Table 1 in Activity 6.1, "Life in Other Countries" (page 68). Be sure to consider what type of graph (bar graph, pie graph, scatterplot, or line graph) will best illustrate this relationship.

Analysis

Group Analysis

1. Which type of graph(s) did you choose to make? Explain your choice.

2. How does graphing this information help you make additional conclusions from the data?

3. What relationship do you see between each of the four statistics and fertility rate? Be as precise as possible.

Individual Analysis

4. Based on your investigation, which of the four statistics correlates most closely with fertility rate? Use evidence to explain your choice.

5. Why do you think this particular statistic has the closest correlation?

6. Name two statistics, other than the four you studied in this activity, that you think might have a significant correlation with a country's population growth rate. Explain why you chose these two statistics.

9.3 Evaluating a Theory

Purpose ▶ Explore historical trends in birth rates, death rates, and population growth for two countries.

Introduction

Scientists develop theories by analyzing data from many sources. In previous activities you have studied birth and death rates as a means of describing population changes over time. In addition to the factors you have already studied, there are many other influences that affect birth and death rates, including social, economic, and environmental factors. The importance of each of these factors changes within a society over time. In this activity you will consider a theory that attempts to explain the relationship between these factors and changes in population growth rates.

Figure 4 Demographic Transition

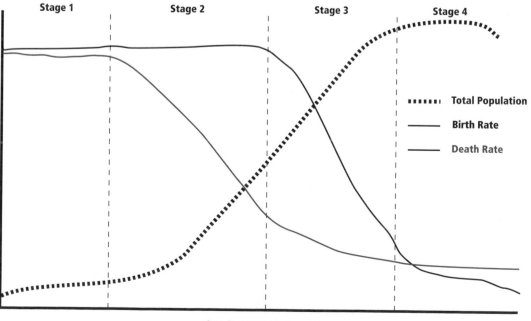

Time (not to scale) ⟶

The Theory of Demographic Transition

Compare the birth rates, death rates, and total population in the graphs you plotted for Activity 9.1, "Population Projections." Consider how the changes that take place in a human society over time can cause different types of population curves. Such observations have led to a theory of human population change called **demographic transition**. This theory proposes that the growth of human populations goes through the four stages illustrated in Figure 4. In Stage 1, birth rates and death rates are high and the population remains low. Stage 2 begins when death rates start to decline, but birth rates remain relatively high. During this stage the population grows. Stage 3 begins as the birth rate decreases, which slows the growth of the population. Finally, during Stage 4 the death rate exceeds the birth rate and the population begins to decline.

The theory of demographic transition is based primarily on data from more developed countries, such as those in western Europe. The theory suggests that over time, as a country becomes more developed, its population will go through each of the four stages indicated on Figure 4.

But people who are critical of the theory of demographic transition are not convinced that today's less developed countries will follow this growth pattern. The theory assumes that the decrease in birth rates seen in Stage 3 is a natural response to Stage 2's lower death rates. Since there is a greater likelihood of children surviving

to adulthood, one could conclude that families will no longer choose to have as many children. However, the theory's critics point out that there are many cultural and social factors involved in the decision to have children. They argue that some families will continue to have just as many children and that, as a result, Stage 3 may not occur in the developing countries of today. If— for whatever reason—a country does not reach Stage 3, neither will it reach Stage 4.

Analysis

Individual Analysis

1. Consider the graphs in Figures 2 and 3 in Activity 9.1, "Population Projections." Based on these graphs, do you expect that population growth rates in either France or Nigeria will follow the theory of demographic transition? Why or why not?

Population Size, Standard of Living, and Ecological Impact

Purpose ▶ Explore the relationship between population size and standard of living, and compare the per capita environmental impact of people living in different countries.

Introduction

▼

Due to environmental pressures, in nature most organisms can survive only in a very limited geographical area. All individual organisms in a local population tend to have similar access to resources and similar chances of survival; each also has about the same impact on the ecosystem. The situation is different for much of the world's human population. Through the use of science and technology, we have managed to insulate ourselves from many environmental pressures, such as availability of food and shelter, that would otherwise limit the size and distribution of the human population. However, not all individual humans have similar access to resources. In today's world, one family's access to energy, materials, and technology may be very different from that of its neighbors or of families in other countries.

The residents of this house have a high standard of living. They are also likely to have a large ecological impact.

One measure used to compare resource use from one society to another is the statistic called **standard of living**. A society's standard of living is determined by analyzing quantitative indicators, such as average income or the number of cars per household. Residents in societies with a higher standard of living have greater access to energy and materials on a per capita basis than those who live in societies with a lower standard of living.

Introduction
(cont.)

Another measure used to compare one society with another is **quality of life**. Quality of life takes into account the overall health and well-being of members of a society. Quality of life is difficult to measure because it depends, in part, on matters of opinion, such as job satisfaction, stress levels, and points of view on the definition of a "good" life.

Ecological impact is a relative measure of how many resources an individual or country uses, and how much waste that individual or country creates, during the course of a year. Determining ecological impact is very difficult because there are many variables to consider. One method of comparing the relative ecological impact of different countries is to compare the total commercial energy use of each country. Because energy is required to produce, operate, transport, and dispose of the food and materials used by people and societies, energy use can provide a rough correlation to ecological impact.

Materials

■■ **For each team of two students**

   1   copy of *Material World*

■ **For each student**

   1   sheet of graph paper

Procedure

1. Using the data provided in Table 1 in Activity 6.1, "Life in Other Countries" (page 68), place the 16 countries in order from highest to lowest population.

2. Using the data provided in Table 1 in Activity 6.1 and information from *Material World*, rank the 16 countries in terms of what you would consider highest to lowest standard of living.

3. Using the data provided in Table 1 in Activity 6.1, rank the 16 countries in order from highest to lowest annual per capita energy use.

4. Based on the data provided in Table 1 in Activity 6.1, use the following equation to estimate each country's annual commercial energy use:

   total annual commercial energy use = annual per capita energy use x population

Procedure
(cont.)

5. Create a single data table that clearly presents your results from Steps 1–4.

6. Use your rankings for each country to make a double bar graph showing each country's standard of living rank and its energy use rank. A sample graph is provided in Figure 6.

7. Make a second double bar graph showing each country's annual per capita energy use and its total annual commercial energy use.
   Hint: Use two different scales.

Figure 6   Standard of Living and Energy Use Ranking

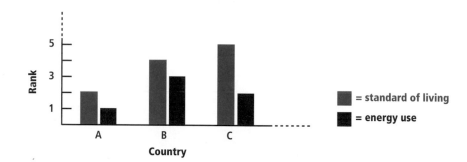

The mobile produce markets shown in these pictures use very different energy sources. The truck on the left is powered by petroleum; the boats on the right are powered by food eaten by the produce vendors.

Analysis

Group Analysis

1. What information did you use to assess each country's standard of living? Why did you choose to use these data?

2. What does your graph indicate about the relationship between annual per capita energy use and standard of living? Why do you think this relationship exists? Is it possible for this relationship not to exist? Explain.

3. What relationship, if any, exists between the size of a country's population, its standard of living, and its ecological impact? Provide evidence for your response and explain why relationships among these three variables do or do not exist.

4. Should society work toward providing everyone in the world with the standard of living of the average person in the United States? Justify your answer by briefly describing the trade-offs involved.

Individual Analysis

5. Many people do not find that having a higher standard of living results in a better quality of life. Write an essay that addresses each of the following:

    a. Do you think people with a low standard of living can have a good quality of life? Why or why not?

    b. Describe the three most important areas that you would evaluate to determine the level of your quality of life.

    c. What do you think are the minimum requirements for a "good" quality of life that everyone in the world should be able to achieve?

    d. How does the country one lives in affect one's ability to attain a good quality of life?

6. Describe the quality of life that you hope to achieve when you are 40 years old.

7. Write an essay describing the potential effects of continued population growth on Earth's resources and environmental health.

Providing for the Population 10

10.1   What Is Sustainable Development?

Purpose ▶ **E**valuate the purpose, practices, and desirability of sustainable development.

Introduction

The processes that transform a less developed country into a more developed country often create unintended by-products. In the past, economic development has typically involved an increased use of money, commercial exchange, manufactured goods, and the services of others, replacing barter, subsistence farming, and self-sufficient living. These economic changes are typically accompanied by considerable changes in social organization. As large-scale economic organizations, such as firms and corporations, become increasingly important, social relationships—within families, local communities, and traditional social networks—are often dramatically altered. Development has also put increasing demands on the environment and on government and public policy. For future development to be sustainable, all of these changes must be considered.

The production of material goods and energy can cause significant damage to the environment. The last 25 years have seen growing concern about how to meet the essential needs of human life on Earth—a goal that requires consideration of both our desire for continued industrial and economic development and our need to reduce the amount of harm done to the environment. The term **sustainable development** is typically used to refer to the achievement of industrial and economic growth in ways that do not cause environmental degradation.

Sustainable Development and Sustainability

In 1990, the World Commission on Environment and Development, a group sponsored by the United Nations, defined sustainable development as "development that meets the needs of the present without compromising the ability of future generations to meet their own needs." Two assumptions underlie this definition: that development must continue, and that future development efforts should differ from those of the past. Although development has historically led to increased resource use and significant ecological impact, it can be carried out in ways that preserve natural resources. Sustainable development ensures that future generations of humans—and all other life on Earth—will continue to have access to needed resources. Unfortunately, there is no single, agreed-upon way to maximize sustainability.

The sustainability of a society depends on day-to-day decisions made by its members. The decisions they reach will depend to some extent on how well-informed they are and on how many options are available to them. For example, an individual will be more likely to participate in a recycling program if he or she understands that recycling not only conserves dwindling resources and landfill space, but can also reduce energy consumption. If most or all of the individuals in a community understand the impact of their actions, they are more likely to design and implement community-wide efforts that promote sustainability, like recycling programs. To create a sustainable culture, both public policy and individual decision-making must be based on accurate information. Public policy in particular must be mindful of a wide range of issues—some immediate and obvious and others distant (in space or time) and uncertain—involving the assessment of risks and the availability and distribution of resources. Ultimately, however, it is the daily actions of individuals that determine whether or not a society's practices will be sustainable.

Currently, many of the practices that allow modern societies to function, such as burning fossil fuels to heat water and generate electricity, limit the resources that will be available to future generations. Recent development has resulted in depletion of resources, extinction of entire species of organisms, and reduction in the environment's natural ability to recycle and purify resources such as air, water, and soil that

are necessary for life on Earth. Unfortunately, many sustainable alternatives, such as solar power, are considered to be more expensive, making them less competitive in today's world economy. When companies and individuals calculate the cost of producing products and services, they usually do not include in their calculations an estimate of the cost of environmental damage and resource depletion. As a result, in many cases the apparent cost of production—and the price paid by consumers—does not accurately represent a product's true long-term cost. Hidden costs are borne by the public at large, locally or globally, and will continue to be borne in the future.

For this reason, many sustainable alternatives are rejected on strictly economic grounds, because they seem to be more expensive than current practices. Other sustainable alternatives may be less convenient and are thus rejected because they represent a reduction in the standard of living, despite being less expensive in both the short and long term. Most people would prefer to maintain their current standard of living in a sustainable way, provided that the out-of-pocket cost is not too high. Cost, environmental damage, and standard of living must each be weighed, and the trade-offs between them considered, in any decisions involving sustainable development.

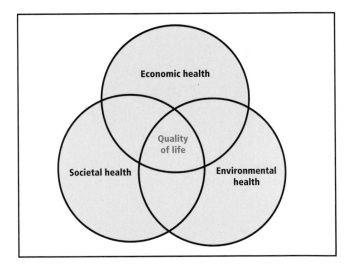

When considering which alternatives are sustainable, trade-offs involving quality of life issues must be evaluated. Quality of life is affected by the economic, social, and environmental health of a community.

Analysis

?

Individual Analysis

1. Why do you think sustainable development has only recently been proposed as a goal for society?

2. Describe a sustainable method and a non-sustainable method currently available to meet each of the following survival needs. Explain why each method is sustainable or non-sustainable.

   a. food

   b. shelter

   c. heat

3. Transportation, although not a survival need, is critical to the functioning of almost all societies.

   a. What governmental or industrial changes would make transportation more sustainable in your community?

   b. What can you and other members of your community do to make transportation more sustainable in your community?

4. Should sustainable development be a goal in our society? Explain your point of view.

5. Describe an example of non-sustainable development in the world today.

   a. Propose changes that would make this development more sustainable.

   b. What are the trade-offs you would have to consider when deciding whether or not to make these changes?

10.2 Taking a Closer Look From Far Away

Purpose ▶ **U**se satellite images to gather data related to land use.

Introduction

Sustainable development requires the wise use of limited resources. In this activity, you will explore the use of a limited resource with many possible uses—land. Land use has become an increasingly important political, social, economic, and ecological concern in many parts of the world. You are going to assess changes in land use in and around Beijing, China, by comparing remote-sensing satellite images taken in 1976 and in 1991.

Remote sensing is a term used to describe methods of obtaining information about objects that are too distant for direct observation. The history of remote sensing includes the use of telescopes, hot-air balloons, sonar, radar, aircraft, satellites, spacecraft, and even carrier pigeons and dolphins equipped with cameras. Today, Earth-orbiting satellites are used to gather important data about the surface of our planet. Earth-orbiting satellites detect light and other types of electromagnetic radiation that are reflected from Earth's surface. Because different substances reflect radiation differently, the reflections can be analyzed and interpreted. A common method of interpreting satellite data is to have computers generate **false-color images**—images that look like color photographs, but are not. On a false-color image, objects do not appear in their natural colors, but

This satellite is designed to collect energy with solar panels on each of its arms. Some of this energy is used to power the satellite, and some is used to transmit useful information to Earth.

▶

Introduction
(cont.)

are instead color coded to indicate the characteristics of the radiation reflected from them. False-color images emphasize features of interest, such as plant cover, water, and structures built by humans. The colors on the images of Beijing that you will use in this activity are related to the composition of materials covering the Beijing region. Bluish-gray tones show areas of urban development (buildings, pavement, etc.), while red and pink tones represent natural vegetation and agricultural crops. Soils with no vegetation or sparse vegetation range in color from white (for sand) to greens or browns, depending on moisture and organic-matter content. Dark blue to black represents water.

The ancient city of Beijing consisted of a square, walled Inner City and a rectangular, walled Outer City that extended to the south. The original walls no longer stand, but in their place the Second Ring Road now outlines the Inner City, and canals outline the Outer City. Within the Inner City are six lakes, seen on the satellite images as dark, almost black, regions. In the Outer City, three parks are visible as red areas. These areas appear darker than the agricultural vegetation outside of the city.

Materials

■ **For each group of four students**

1	blue filter gel
1	red filter gel
1	yellow filter gel

■■ **For each team of two students**

1	set of satellite images of Beijing
1	transparency grid
1	transparency marker
1	copy of *Material World*

Procedure

1. View the two satellite images of Beijing through each of the colored filter gels. Pay close attention to how the different gels affect your ability to locate and define different features. Prepare a data table and record your observations with each gel.

2. Describe three differences between the image taken in 1976 and the image taken in 1991.

3. Use the appropriate gel, the transparency grid, and a marker to outline, on the grid, the area of urban development in 1976 and in 1991.

4. Estimate the area of urban development in 1976 and 1991 by counting the number of grid squares within each of your outlines. For squares that are only partially included, estimate the fraction of each square to the nearest tenth.

Analysis

Group Analysis

1. Which gel did you find most useful in your analysis of the images? Which did you find least helpful? Why do you think the gels made a difference?

2. Propose a cause for each of the differences you described in Procedure Step 2.

3. Based on the satellite evidence, what changes in agricultural production do you think took place in the region of Beijing between 1976 and 1991? Describe your evidence and any potential impacts of these changes.

4. Do you think the development in Beijing between 1976 and 1991 was sustainable? Explain.

Individual Analysis

5. Did the size of the area of urban development in and around Beijing change between 1976 and 1991? If so, by what percent?

6. In your opinion, what decisions were made in the past regarding the use of land in the Beijing area? What might be the positive and negative consequences of these decisions? Explain.

7. What additional information would help you provide a better answer to Analysis Question 4? Explain how this information would help.

10.3 Development Decisions

Purpose ▶ **A**ssess a variety of options for the future use of undeveloped land.

Introduction

Recently, to stimulate business investment and promote China's emerging market economy, the Chinese government has allowed land previously controlled by the government to be used by industry. Your challenge is to determine the best use for approximately 200 acres of undeveloped land located just outside the Outer City of Beijing. The property's proximity to Beijing makes it desirable for many commercial ventures, but it is also valuable in its undeveloped state as a potential nature preserve. The property cannot be developed without approval by the city's land-use committee. Eight groups have come forward with proposals to make use of the land.

Proposals Submitted to the Land-Use Committee

1. Residential area

2. Car factory

3. Park / Nature preserve

4. Nuclear power plant

5. Agriculture

6. Water reservoir

7. Wastewater treatment plant

8. Amusement park

Procedure

1. Your group is responsible for preparing and making a presentation to promote one of the eight development proposals. In addition to giving the presentation in front of the land-use committee, group members should be prepared to answer any questions concerning their proposal that might arise at the meeting. The presentation must be based on evidence and explained clearly so that the committee can be fair in its consideration of each proposal's potential benefit to the city and its residents.

2. Review the instructions for group work given by your teacher.

3. On the following pages you will find a summary of each group's proposal for how to use the land. These summaries have been written to help you direct your research and construct a reasonable argument. Using the information on these pages and any additional information you can find, outline the evidence supporting the case for your group's proposed use of the land.

   Note: The summaries provided do not necessarily reflect the position of your teacher or the developers of this course.

   You may wish to gather general information about China and more specific information about Beijing on the Internet. Go to the *Science and Sustainability* page on the SEPUP website to find updated sources.

4. Divide the main points of your outline among the members of your group. Each member is responsible for presenting this information at the land-use committee meeting.

5. Prepare any charts, graphs, or other visual aids that will help get your points across. Rehearse what you plan to say so that your presentation stays within the time limits set by your teacher and the land-use committee.

6. Participate in the meeting as instructed by your teacher and the land-use committee.

1. Residential Area

This group would like to construct a series of small apartment buildings and single-family homes on the land. Due to the large increase in China's population over the past few decades, adequate housing is always in demand. Using this land to house people who are currently unable to find comfortable homes would clearly be of benefit to the region. Small buildings pose less of a threat to the environment than large ones, because they create less pollution and fewer traffic problems. Also, comfortable dwellings will attract people who can afford to rent or purchase these buildings, which would create a more affluent suburb and raise the standard of living in the outlying areas around Beijing.

2. Car Factory

This group would like to build a car factory on the land. One of the reasons that the people in the area around Beijing are so poor is that agriculture is the only available occupation, and there are very few land owners. Constructing a factory on this land would create much-needed jobs, making it possible for people living in the region to have a higher standard of living. A car factory is a particularly good idea because demand for cars is increasing in this area due to of population growth, economic development, and expansion of the urban area.

3. Park/Nature Preserve

This group would like to make the land into a nature preserve and public park. Since this is the only undeveloped land near Beijing, people should be permitted to use and enjoy it in its natural state. The land is frequented by migrating birds; it is the only undeveloped area along a migration route that lies within 20 miles of the city. This area is also home to over 10,000 species of insects, frogs, turtles, and small mammals. Developing the land would mean jeopardizing their survival, because these animals cannot move elsewhere—the land is surrounded by other development. To make the land into a park, picnic benches and trash containers could be placed throughout the area, a playground could be constructed in one corner, and walkways could be built through the wooded areas to protect the wildlife. This plan allows for preservation of the land as a wildlife refuge and as a place for people to visit and enjoy nature.

4. Nuclear Power Plant

This group would like to build a nuclear power plant on the land. Beijing is constantly growing, and with this growth comes the need for additional electric energy. A nuclear power plant can generate electricity cheaply. At present China has two nuclear power plants, one near Shang-hai and one near Hong Kong. Beijing would benefit from the construction of a new power plant not only because the area needs additional electric power, but also because the plant's construction and continued operation of the power plant would create many short- and long-term jobs. Nuclear power has an excellent safety record: there has been only one major accident that caused any harm to the public. During normal operation, nuclear power plants release no water or air pollutants. They produce only solid waste, which can be buried in such a way that it will cause no harm to people or the environment.

5. Agriculture

This group would like the land to be used for agriculture. Due to the area's drastic increase in population and construction over the last two decades, agricultural land has become more and more difficult to find and farmers have had to produce more crops on less land. This land should be used to produce food for the area's residents and to supplement the nation's food supply. We know that the land is fertile, because there is currently ample vegetation growing wild. A large barn would be built to house animals and to store the necessary farming machinery. Crop yield would be maximized through the use of fertilizer and pesticides.

6. Water Reservoir

This group would like to build a reservoir on the land. The growing population and high agricultural demand in and around Beijing require a large volume of water every day. Currently the demand for water is met with natural streams and with a reservoir system made up of a series of artificial lakes and other water storage facilities. This system is limited, however. During periods of drought, the city has only enough water stored to meet two days' worth of demand. An additional reservoir holding up to 10,000 gallons of water would help ensure adequate supplies of life-giving water in times of trouble.

7. Wastewater Treatment Plant

This group would like to build a wastewater treatment plant on the land. Population growth in and around Beijing has resulted in an increase in solid and liquid wastes. Wastewater typically consists of fouled water removed from residences and commercial establishments, along with runoff that enters storm drains during wet weather. Untreated wastewater usually contains dissolved chemicals, disease-causing organisms, and solid particles that pose hazards to human and environmental health. Currently there is only one wastewater treatment plant near Beijing, and it treats only 40% of the wastewater generated in the entire metropolitan area. Adding a second treatment plant would allow almost 90% of the wastewater to be treated, resulting in much less contamination of the drinking-water supply. The existing treatment plant was constructed almost 10 years ago and does not use the most up-to-date technologies. The proposed new plant would take advantage of newer technologies, improving the quality of water that is released after treatment.

8. Amusement Park

This group would like to build an amusement park on the land. There is currently no amusement park in this part of Beijing. Such a park would create jobs and revenue for a part of the city that desperately needs both. The park's theme could be the land on which it is built, with each ride featuring a wild animal native to the region. Caged specimens of the larger animals could be kept in a safe area for viewing and controlled breeding. Because the animals would be bred in captivity, none of the area's native species would become endangered. Additional buildings would be constructed to house administrative offices, ticket offices, animal cages, fast-food establishments for patrons, and training space for staff.

Analysis

Group Analysis

1. Imagine that you are a resident of Beijing. What are the three most appealing alternatives? Explain the trade-offs of each.

2. Imagine that you are a resident of a neighboring city. What are the three most appealing alternatives? Explain the trade-offs of each.

3. From the perspective of a visitor from the United States, what are the three most appealing alternatives? Explain the trade-offs of each.

Individual Analysis

4. Taking into account all the perspectives you considered above, which is the single most appealing alternative? Explain the trade-offs involved in your decision.

Feeding the World

For one-half of the people in the world, the cycle of life revolves around rice. Death is symbolized in Taiwan by chopsticks stuck into a mound of rice. A good job in Singapore is an "iron rice bowl," and unemployment is a "broken rice bowl." The Chinese greeting "Ni chir gwo fan mei," means "Have you eaten your rice today?"

W. W. Williams, "From Asia's Good Earth: Rice, Society, and Sun"
Hemispheres, December 1996

Today, rice feeds more people than any other crop. Although some grains, such as wheat, are harvested in greater amounts than rice, a significant portion of those grains is used to feed animals. In the 1950s, the supply of rice was no longer keeping pace with the increasing human population. Attempts to increase the rice harvest by increasing fertilizer use failed, because the stalks of fertilized rice plants grew so tall they collapsed under their own weight. Many rice-dependent nations appeared to be facing a future of starvation.

During the 1960s, in what is now known as the Green Revolution, new strains of wheat, corn, and rice were developed and introduced into the world market. These strains were known for their high yields (more food per plant), quick maturation times (more harvests per year), and short stems (less damage from wind and weather). For a while it seemed that the Green Revolution would be the technological solution to the world's growing need for food. Many regions enjoyed record harvests and for the first time produced enough grain to support their populations. World rice yields in the early 1990s were nearly double those for the early 1960s.

However, as farmers began to rely on the new strains of high-yield grains to increase food production, problems emerged. The new strains required nutrient-rich soils and an abundant water supply. In some regions, costly dam building and river diversion projects were required, and these projects significantly altered natural patterns of water flow. Countries with nutrient-poor soils needed to add high levels of expensive fertilizer to grow the new crops. Furthermore, in areas of poor soil, young grain plants were often killed by weeds or insects that thrived with the addition of fertilizer and water. This made it necessary to apply expensive herbicides or pesticides. After harvesting, the introduced grains held more moisture than the naturally-occurring local grains, so they spoiled faster when stored using traditional methods. All of these issues made it difficult for small-scale family farms to produce a good crop, and many were put out of business. Today, large corporate farms, which have access to fertilizers, herbicides, pesticides, and water, have become the mainstay of modern agriculture.

The world's rice production has been growing at a slower rate since the initial boom period, which lasted until the 1980s. At the same time, the human population has continued to expand. The World Bank projects that the number of people who rely on rice as their daily food staple will increase from 2.7 billion today to 3.9 billion by the year 2025.

Part 2

Feeding the World

In Part 1 of *Science and Sustainability*, "Living on Earth," you were introduced to the basic material and energy needs of humans and other organisms. Part 2, "Feeding the World," focuses on one of these basic needs—food. Sustaining the existence of Earth's human inhabitants requires an adequate supply of nutrients and energy for each individual. Food production and food consumption are extremely important issues in today's world and are major concerns for the future. In this part of *Science and Sustainability*, you will investigate a number of questions related to food production:

- What can be done to increase the amount of food produced each year?

- How can we, as individuals, help ensure that all humans have access to adequate food resources?

- How have food production efforts affected the environment?

- How do our cells use the energy and nutrients contained in food?

- What are the key elements and molecules found in living things?

- How do plants convert energy from the sun into chemical energy?

- What characteristics make some organisms better food sources than others?

- How do genes determine the characteristics of an organism?

- What can humans do to influence the characteristics of an organism?

- Should humans attempt to influence the characteristics of an organism?

Food Production

11

11.1 Growing Plants

Purpose ▶ **I**nvestigate how population density and nutrient availability affect plant growth.

Introduction

Producers are organisms that produce their own food, usually by photosynthesis. They occupy the first trophic level of virtually all food webs on Earth. Although producers include algae and other organisms that live in oceans, lakes, and streams, most of the producers eaten by humans or fed to livestock are green plants grown on land. As with any organism, plant populations are influenced by many factors in an ecosystem, such as the availability of living space, the supply of nutrients, and the presence of predators. Because of factors like these, the number of organisms of each species an ecosystem can sustain is limited. This limit, the ecosystem's carrying capacity, results from complex interactions among all parts of an ecosystem. Left undisturbed, energy and nutrient transfer within an ecosystem tends toward a steady state, resulting in populations of the various species remaining relatively constant over time. As changes in an ecosystem occur, its carrying capacity for a species may also change. Farmers deliberately modify the ecosystems of agricultural fields to increase carrying capacity for crops and decrease carrying capacity for competing organisms.

Materials

▦ **For the class**

20	plant pots
	supply of radish seeds
	supply of potting soil
	light source (natural or artificial)
	supply of water
	supply of liquid fertilizer of varying concentrations

▪▪ **For each team of two students**

1	plant pot
2	large cotton balls (optional)
1	graduated cylinder
1	ruler

Procedure

Part A   Planning

1. Design an experiment to determine the fertilizer concentration and/or pattern of seed distribution that will maximize radish growth. Your experimental design should use a total of 20 plant pots.

2. Make a sketch indicating your seed distribution.

3. Create a table showing your fertilization schedule.

Part B   Planting

1. Begin your experiment by planting your allotment of seeds.

2. Once the seeds germinate, start your schedule of fertilization.

Analysis

?

Individual Analysis

1. What fertilization concentration do you predict will produce the most plant material? Which will produce the least? Explain.

2. What seed distribution pattern do you predict will produce the most plant material? Which will produce the least? Explain.

Part C   Harvesting

Materials   ■■ **For each team of two students**   ■ **For each student**
1   ruler   supply of graph paper
access to a balance

Procedure

1. Collect ongoing plant growth measurements and observations.

2. At the end of the experiment, following your teacher's instructions, carefully uproot your plants and determine the mass of each one.

3. Report all of your data to your teacher so that a class data table can be prepared.

4. Record all of the data collected by the entire class.

Analysis

?

Group Analysis

1. Explain how adding fertilizer to a field affects the carrying capacity for

   a. plants grown in the field.

   b. predators of the plants grown in the field.

   c. weeds growing in the field.

2. What impact could changing a field's carrying capacity for crop plants have on other organisms that make up the field's food web?

Individual Analysis

3. Analyze and graph your data as directed by your teacher.

4. Write a lab report for the investigation. Make sure you address the following questions:

   a. What is the effect of fertilizer concentration on the yield of radishes?

   b. What is the effect of seed density on the yield of radishes?

   c. What combination of fertilizer and seed density optimizes the yield of radishes?

11.2 Will There Be Food Enough?

Purpose ▶ **I**dentify the issues involved in feeding the world's population.

Introduction

Every day people in all countries of the world go hungry. About one billion people—1/6 of the Earth's population—suffer from hunger and malnutrition. Almost 10 million people die from these causes each year. Three-fourths of the deaths are children under the age of five. The world's farmers produce enough food, but people are starving because they don't have access to the food they need. Transportation, poverty, politics, economics, technology, and personal habits all contribute to world hunger.

Today there are over 6 billion people in the world. By 2030, it is expected that there will be about 8 billion people. As the population continues to grow, many population and agricultural scientists believe that more and more people will go hungry. In 1996, world leaders met at the World Food Summit in Rome, Italy to discuss strategies to cut in half the number of hungry people by the year 2015. However, when they reconvened in 2002, they recognized that progress toward this goal was falling short.

Advances in technology have increased global food production. Will future increases in food production be able to keep pace with increases in the world's population?

11.2 Will There Be Food Enough?

Part A How Much Is Needed?

Materials
- ■ **For each student**
  - 1 calculator

Procedure

1. Use the information in Table 1 to calculate the total number of Calories needed each year for one person.

2. Use the information in Table 1 to calculate the total number of Calories needed to supply the world's population in 2004.

3. Calculate the total number of Calories needed to supply a population of 8.5 billion people.

Table 1 Human Calorie Requirements

Human Energy Requirement* (Cal/day/person)	2004 World Population	Energy Available From World's 2004 Food Harvest (Cal)
2,500	6.4 billion (6.4×10^9)	7,000,000 billion (7×10^{15})

*Energy expended by an average human at a moderate activity level

Analysis

Group Analysis

1. Were enough Calories harvested in 2004 to supply the world's population?

2. By what percent would the world's food harvest have to increase (from the 2004 level) to provide enough Calories for a population of 8.5 billion?

Part B   Food for Our Future

Access to Food

Currently enough food is produced to feed everyone on Earth. Yet people living in many areas of the world suffer from widespread undernourishment and hunger, because they don't have access to this food. In many cases, too little food is grown in these areas and too little can be brought in or stored without spoiling. Loss and spoilage can greatly reduce the amount of food available for people to eat. For example, food is considered lost when it is eaten by rodents and insects, or it can be spoiled by becoming rotten or moldy.

Wealthy countries reduce the loss and spoilage of their food by using appropriate pest control in addition to technologies for food storage and preservation, such as sealed containers and refrigerators. However, people in poorer nations or in very remote areas often do not have access to these products, because they cannot afford them or they do not have an adequate supply of electricity (or other fuels) to operate them. Sometimes, even in areas where plenty of food is available, people may be starving for the simple reason that they are too poor to afford the food. Many experts argue that reducing poverty is one of the biggest steps toward reducing undernourishment. This will require changes in governments' social, political, and economic policies.

Changes in Food Production

In the 19th century, Thomas Malthus predicted that food production would not be able to keep up with the world's population, and in the 1970s Paul Ehrlich predicted massive food shortages due to overpopulation. The widespread famines predicted by Malthus and Ehrlich were avoided by technological advances. Although the population continued to grow, advances in fertilizers, pesticides, farming techniques, and high-yield crops allowed farmers to increase their harvests to keep pace with the needs of the growing population.

In order for world agriculture to feed the rising number of people, grain production will have to continue to increase, most likely through increased fertilizer use. A 2003 report by the United Nations Food and Agriculture Organization states that fertilizer accounts for 43% of the nutrients used by global crop production and that to keep up with the projected 60% increase in food demand by 2030, this may have to double.

In the past, increased use of chemical fertilizers and pesticides has caused environmental problems. Some scientists think that genetically modified organisms (GMOs) are the key to ending chronic food shortages. For example, researchers believe that genetically modified seeds can be used to fight parasites that cause major crop losses to farmers in Africa. However, others fear that GMOs will bring new environmental problems. Some organizations, such as Greenpeace, want to stop field experiments involving GMOs.

But many of today's experts do not think that future technological advances can increase food production fast enough to keep everyone adequately fed. Beginning about 1950, the world saw huge increases in crop production. Modern agricultural techniques and the development of drought-and disease-resistant plants, dramatically improved the yield—the amount of food produced on a unit of land. Farmers grew more food on less land, with the result that the amount of land used to grow crops has been steadily decreasing since 1981. Unfortunately, much of the remaining land cannot be used for growing future crops because it is now being used for housing, roads, industry, and recreation.

Also, there is less water available for irrigation as the supplies of groundwater are depleted and water previously used for agricultural purposes is diverted to urban areas for residential and industrial uses. Finally, the land available for growing food crops has decreased due to erosion and the increase in the amount of farmland planted with non-food crops such as cotton and coffee.

Changes in Habits

The amount of available food per person is directly related to the number of people living on Earth. Fewer people means more food per person. Some areas with high rates of population growth, such as sub-Saharan Africa, have very low food supplies. Reducing the population growth in these areas could help reduce food shortages and starvation. Another means of increasing the food supply without increasing food production is to change eating habits. Today many people, particularly those in more developed countries, eat much more than their bodies need, and they often throw out plenty of perfectly good food. By eating less and wasting less, these people could make more food available to those who don't have enough.

Even those people who don't eat too much or waste too much can help by changing what they eat. In 2002, each person in the U.S. consumed an average of almost 100 kg (222 lbs) of meat while the average person in the world ate only about 40 kg (88 lbs). It is estimated that about 40% of the world's grain harvest is fed to livestock, and even a 10% reduction in global meat consumption could save enough grain to feed almost 1 billion people. The grain currently used to feed animals for meat can instead be used to feed hungry people.

Analysis

Group Analysis

1. Briefly describe the main issues involved in providing food for future generations.

2. Choose one of the countries depicted in *Material World*. How might issues of food distribution and preservation affect a family living in this country?

Individual Analysis

3. Imagine that you are responsible for making sure that all people in the world are adequately fed.

   a. Describe the major trade-offs you would need to consider.

   b. What would you propose that individuals and/or governments do? Provide evidence to support your ideas.

11.3 Can Satellites Help Feed the World?

Purpose ▶ **A**nalyze a series of false-color satellite images to identify changes in an ecosystem and propose possible reasons for those changes.

Part A   Where's the Water?

Introduction ▼

The Aral Sea is located in the former USSR, between Kazakhstan and Uzbekistan, as shown in Figure 1. Satellite images of the Aral Sea can be used to study the effects of recent agricultural practices in the region. The images you will use in this activity show the region in 1973 and 1987. During this 14-year period, much of the fresh water from two rivers that flow into the Aral Sea, the Amu Darya and the Syr Darya, was diverted for the irrigation of cotton fields. In these images black and purple areas indicate water; other colors indicate different biological and chemical components of the land.

Figure 1   Locator Map

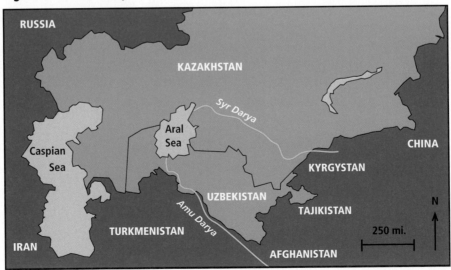

Materials

For each group of four students

1   color copy of satellite images of the Aral Sea region
1   transparent grid
2   transparency markers
1   copy of *Material World*

Procedure

1. Examine the two images and list at least two differences between them.

2. Use the grid and markers to estimate the surface area of the Aral Sea in 1973 and in 1987.

3. Read the following passage, then answer the Analysis Questions.

Today, approximately 55 million people live in the Aral Sea region. Over-use of water from local rivers has put the future of their livelihood at risk.

Reduced water flow into the Aral Sea, coupled with evaporation, has had three primary effects:

1. There is less water in the Aral Sea, and the water that remains has become extremely salty. The salinity increased from about 12% in the early 1970s to 23% in 1987, and has continued to rise since then.

2. The moderating effect of the Aral Sea on the local climate has diminished, resulting in hotter summers, colder winters, and a shorter growing season.

3. Over 20,000 square kilometers of salty sea sediments that were once submerged are now exposed to the air. Dust storms blow this salt-rich dirt onto surrounding crop lands. The added salt reduces crop yields and requires farmers to increase irrigation to maintain crop production levels. Increased irrigation further depletes the waters of the Aral Sea.

The higher salinity of the Aral Sea has reduced fish populations. As a result, one local industry—commercial fishing—has already ended. International attention is now focused on the problems faced by the people living in this area, and work is being done to find solutions.

Analysis

Group Analysis

1. What was the percent change in the surface area of the Aral Sea from 1973 to 1987?

2. Do your observations support the statements made in the reading in Procedure Step 3 about how the Aral Sea has changed? Explain.

3. What substance do you think is most responsible for the bright red color in the images? Explain your evidence.

4. Why does the salinity of the water affect the health of fish? Explain.

Individual Analysis

5. Give an example, other than salinity, that shows how changes in a non-living component of an ecosystem could affect the organisms living in that ecosystem. Explain.

6. Do you think a continuing increase in the use of river water for crop irrigation is sustainable in the Aral Sea region? What role can science play in addressing this issue?

Part B   What's in the Water?

Introduction

Streams, lakes, ice sheets, and oceans cover over 70% of Earth's surface. Many of these water bodies contain valuable food resources. Throughout history, humans worldwide have harvested fish, crustaceans, and even seaweed for food. Today, over-harvesting and environmental degradation are causing wild fish populations to decline. As a result, aquaculture, or fish farming, is becoming an increasingly important source of food for the world's human population.

The species present in a body of water, and the population size of each species, is heavily influenced by the temperature and chemical characteristics of the water, especially nutrient and pollutant levels. Satellite-based technology can be very useful in monitoring biological activity in Earth's bodies of water. In this activity, you will examine and compare satellite images of Earth made during different seasons. These images indicate where the major areas of phytoplankton growth are located. As you learned in Part 1, phytoplankton are primary consumers. They are capable of photosynthesis due to the presence of **chlorophyll** in their cells. Areas rich in chlorophyll-containing phytoplankton are regarded as very productive, because they serve as the basis of ocean and freshwater food webs. The colors on the satellite images indicate the concentration of chlorophyll in the water. The color scale shows the correlation between the colors on the images and the chlorophyll concentration.

These satellite images can be used not only to locate regions of high and low productivity, but also to help monitor the impact of human activity on the health of Earth's surface waters. Pollutants created by human activities can have two very different effects on phytoplankton. Many pollutants, including industrial wastes and pesticides, may contain toxic substances that can cause serious declines in phytoplankton populations. On the other hand, nutrient-rich pollutants, such as sewage or fertilizers in runoff from agricultural fields, can cause phytoplankton to reproduce over-abundantly. When phytoplankton populations become too large, they have a harmful effect on other organisms. For example, a large population of phytoplankton in a lake or pond may consume so much oxygen that fish and other water-dwelling organisms cannot survive there.

The equipment on this boat is used to harvest food from the sea.

Materials

For each group of four students

1   set of satellite images of Earth
1   chlorophyll color scale

Procedure

1. Study the images and use the color scale to identify which ocean region(s) have the highest productivity and which region(s) have the lowest productivity.

2. Compare the images for the two seasons. Describe the major similarities and differences between them.

Analysis

?

Group Analysis

1. What do you think are the major factors that cause an ocean region to be productive or non-productive? Explain.

2. What do you think caused the similarities and differences you described in Procedure Step 2? Explain.

3. How could these maps be used to help provide the world with more food?

4. How could these maps be used to help monitor the impact of human activity on the health of Earth's surface waters?

Individual Analysis

5. In which areas would you expect to find the most complex food webs? Explain your reasoning.

6. What additional information would help you better respond to the Group Analysis Questions?

Extension

Explore additional issues facing the people who live near the Aral Sea, including problems caused by an increase in erosion as well as the limited availability of fresh water.

Eating Patterns Around the World

Purpose ▶ **I**nvestigate the types of food supplies available to people living in different parts of the world and determine how food choices affect energy transfer in a food chain.

Introduction

As you learned in Activity 2.2, "The Web of Life," when a consumer eats another organism, much of the energy contained in that organism is not transferred to the consumer. It is considered "lost" from the food web. In any ecosystem, the amount of energy that is lost increases with each step in the food chain. For this reason, the food choices made by consumers, including humans, can influence the amount of energy that is lost.

All humans have the same dietary requirements, yet human societies differ in the amounts and types of food they consume. Why do these differences exist? Are different diets equally healthy? Are they equally sustainable?

Materials

■ **For each student**

   calculator (optional)

   supply of graph paper (optional)

Procedure

1. Use Table 2 to list the top five countries for availability of the following:

   a. total Calories   **b.** animal protein   **c.** vegetable protein   **d.** total fat

2. The recommended daily intake per person of calories, protein, and fat is as follows:

   2,000 Calories   75 grams of protein   65 grams of fat

   Use Table 2 to make a list of the countries in which residents do not have access to enough food to meet the recommended daily intake of Calories. Make similar lists for protein and fat.

3. Calculate the ratio of protein to fat for each of the 16 countries listed in Table 2.

Procedure continued on page 146→

Table 2 Summary of Available Food by Country in 1995

	Food Type	Calories per capita (Kcal/day)	Protein per capita (grams/day)	Fat per capita (grams/day)
Recommended Daily Amount	Total	2,000	75	65
Brazil	Total	2,834	71.1	80.7
	Vegetable	2,290	36.2	43.2
	Animal	543	34.9	37.4
Cuba	Total	2,291	53.6	55.6
	Vegetable	1,934	31.1	34.2
	Animal	357	22.5	21.4
Ethiopia	Total	2,144	51.6	43.2
	Vegetable	2,008	41.7	34.4
	Animal	135	9.9	8.8
Guatemala	Total	2,300	57.8	40.9
	Vegetable	2,127	46.2	29.3
	Animal	172	11.6	11.6
Iceland	Total	3,159	127	120.2
	Vegetable	1,896	41.9	32.2
	Animal	1,263	85.1	88.0
India	Total	2,388	59.0	42.7
	Vegetable	2,218	49.6	31.4
	Animal	170	9.3	11.3
Israel	Total	3,271	106.9	116.4
	Vegetable	2,641	53.8	76.7
	Animal	630	53.0	39.7
Italy	Total	3,458	108.2	141.5
	Vegetable	2,560	50.5	73.7
	Animal	898	57.7	67.8
Japan	Total	2,887	95.7	80.3
	Vegetable	2,292	42.4	44.0
	Animal	596	53.3	36.3
Kuwait	Total	3,160	96.9	107.2
	Vegetable	2,362	44.9	54.7
	Animal	798	52.0	52.5
Mali	Total	2,149	59.6	44.5
	Vegetable	1,960	45.3	33.1
	Animal	189	14.3	11.5
Mexico	Total	3,136	83.9	85.0
	Vegetable	2,627	51.1	48.7
	Animal	509	32.7	36.3
Russia	Total	2,926	89.4	78.9
	Vegetable	2,174	45.7	22.9
	Animal	752	43.7	56.0
Thailand	Total	2,296	49.8	44.4
	Vegetable	2,066	31.3	29.5
	Animal	231	18.5	14.9
United Kingdom	Total	3,149	92.7	136.8
	Vegetable	2,115	39.7	55.1
	Animal	1,034	53.0	81.7
United States	Total	3,603	110.5	141.8
	Vegetable	2,613	40.1	74.4
	Animal	989	70.4	67.5

Procedure
(cont.)

4. Assuming that beef cattle eat a diet of pure corn, use Tables 3 and 4 to calculate

    a. the amount of corn eaten by beef cattle in 1996.

    b. the number of Calories consumed by beef cattle in 1996.

    c. the number of Calories supplied to humans by beef in 1996.

    d. the number of additional people whose annual Calorie needs could have been met if all the corn eaten by beef cattle was instead eaten by humans.

Hint: You calculated the number of Calories needed per person per year in Activity 11.2, Part A.

Each of these corn stalks will produce several ears of corn.

Table 3   Some Major Food Resources "Harvested" in 1996

Food Item	Quantity (billions kg)	Energy Content (Cal/kg)
sugar cane	1,187.442	480
wheat	582.043	3,000
corn	577.724	1,100
rice	567.312	3,000
milk	539.767	600
fish	101.433	1,100
pork	83.188	2,000
beef	53.672	2,000

Table 4   Grain Consumed by Feedlot Animals

Food provided by animal	Animal	Kg of grain eaten by animal to produce each kg of food
beef	cattle	7
pork/bacon	pigs	4
chicken	chickens	2
fish	fish	2
milk/cheese	cows	3
eggs	chickens	2.6

Analysis

?

Group Analysis

Use appropriate data, graphs, and/or calculations to support your answer to each of the questions below.

1. People in more developed countries usually have access to more food than people in less developed countries. Does this imply that more developed countries are more sustainable than less developed countries? Explain why or why not.

2. Why do you think 500 million people are undernourished each day?

3. Compare the ratio of protein to fat available to people in more developed countries with the ratio available to people in less developed countries. Do less developed or more developed countries come closer to the recommended ratio?

4. Do you think that all humans should consume the diet of an average U.S. resident? Why or why not?

Individual Analysis

5. Eating sustainably involves many factors and decisions. Of the countries listed in the table below, which do you think has the most sustainable eating habits? Explain your reasoning, using evidence to support your answer.

Average Per Capita Food Consumption (in kg) in 1990

Country	Grain Equivalent*	Beef	Pork	Poultry	Mutton	Milk	Eggs
United States	800	42	28	44	1	271	16
Italy	400	16	20	19	1	182	12
China	300	1	21	3	1	4	7
India	200	0	0.4	0.4	0.2	31	13

*The amount of grain that is consumed to produce the entire year's food supply

Necessary Nutrients

12

12.1   Soil Nutrients and Fertilizer

Purpose ▶ **C**ompare the nitrogen content of two different soils and discover how well each soil absorbs nitrogen from fertilizer.

Introduction

Like all living organisms, plants need nutrients to survive and grow. The soil in which a plant grows is the source of most of these nutrients. Soil fertility refers to the ability of soil to store nutrients in a form that can be absorbed by plant roots. Some soils are naturally more fertile than others. **Fertilizers** are substances rich in plant nutrients that are often used to increase the fertility of soils, especially soils used to grow agricultural products.

Fertilizers commonly contain nitrogen (N), phosphorus (P), and potassium (K). These three elements are essential to plant growth. For centuries, farmers have used naturally occurring decaying organic matter, such as manure and compost, as fertilizers. Compost and animal manure contain abundant supplies of N, P, and K as well as other essential nutrients. However, the increased crop yields obtained by modern farming methods are due in large part to the use of manufactured fertilizers, which are mixtures of chemicals produced in factories. These fertilizers are expensive, because fertilizer production is a costly, energy-intensive process. Many farmers, especially those with small farms in less developed countries, cannot afford manufactured fertilizer.

Introduction
(cont.)

In this activity you will test fertilized and unfertilized soils for the presence of nitrates (NO_3^-), the form of nitrogen most easily used by plants. You will also test the nitrate level found in the **runoff** water that is not absorbed by the soil. Runoff often ends up in streams, lakes, and other water bodies, including reservoirs that provide drinking water. Unfortunately, there is no easy way to "see" nitrates. The procedure you will use involves converting the nitrates to nitrites (NO_2^-) and then allowing the nitrites to react with an organic dye to produce a visible pink color.

Materials

For the class

supply of local soil
supply of potting soil
supply of nitrate testing powder
supply of fertilizer solution
supply of weighing paper (optional)

For each group of four students

1   90-mL bottle of nitrate extraction solution
2   pieces of masking tape
2   vials with caps
2   SEPUP trays
2   stir sticks
1   SEPUP filter funnel
1   filter-paper circle
1   pipet
1   nitrogen color chart
1   10-mL graduated cylinder
access to a clock with a second hand
access to a balance

Procedure

Part A   Testing Soil

1. Use the masking tape to label one of your vials "A" and the other "B."

2. Place 1 gram of potting soil into Vial A and 1 gram of local soil into Vial B.

3. Add 5 mL of nitrogen extraction solution to each vial, then cap the vials and shake each one for 1 minute.

4. Set the vials aside until most of the solids settle out and the liquid is clear enough to see through.

Procedure

(cont.)

Figure 1   SEPUP Tray

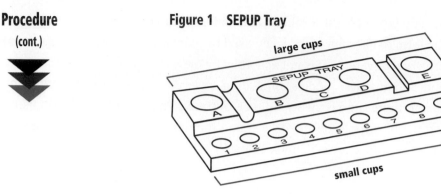

5. Use the pipet to transfer 3 mL of the liquid from Vial A to Cup A on the upper row of your SEPUP tray.

6. Rinse the pipet and then use it to transfer 3 mL of the liquid from Vial B to Cup B.

7. Add 10 stir-stick scoops (0.5 gram) of nitrate testing powder to Cup A and stir to mix.

8. Rinse and dry the stir stick, then add 10 stir-stick scoops (0.5 gram) of nitrate testing powder to Cup B and stir to mix.

9. While these mixtures stand for 5 minutes, clean up all your other dirty labware so that it is ready for use in Part B. After 5 minutes have elapsed, use the nitrogen color chart to estimate the nitrogen content of each soil sample.

Part B   Testing Runoff

1. Use a clean SEPUP tray, the funnel, and the filter paper to prepare a filtering setup that will drip into Cup D on the upper row of your SEPUP tray.

2. Fill the filter-paper cone with local soil, then slowly pour 5 mL of fertilizer solution over the soil. Allow the fertilizer solution to percolate through the soil and drip into Cup D.

3. Transfer 1 mL of the runoff liquid that collects in Cup D to Vial A, then transfer 1 gram of the fertilized soil in the filter cone to Vial B.

4. Add 5 mL of nitrogen extraction solution to each vial, then cap the vials and shake each one for 1 minute.

Procedure
(cont.)

5. Set both vials aside until most of the solids settle out and the liquid in Vial B is clear enough to see through.

6. Use the pipet to transfer 3 mL of the liquid from Vial A to Cup A.

7. Rinse the pipet and then use it to transfer 3 mL of the liquid from Vial B to Cup B.

8. Add 10 stir-stick scoops (0.5 gram) of nitrate testing powder to Cup A and stir to mix.

9. Rinse the stir stick, then add 10 stir-stick scoops (0.5 gram) of nitrate testing powder to Cup B and stir to mix.

10. While these mixtures stand for 5 minutes, clean up all your dirty labware. After 5 minutes have elapsed, use the nitrogen color chart to estimate the nitrogen content of the soil sample and the runoff liquid.

Analysis
?

Individual Analysis

1. Rank the samples tested in order of highest to lowest nitrate content. Explain how you determined your ranking.

2. Of the samples tested, which do you think has the greatest carrying capacity for plants?

3. What evidence do you have that adding fertilizer to soil increases the nutrients available to plants?

4. Does adding fertilizer to soil have any drawbacks? Describe the evidence that supports your answer. Are these drawbacks relevant to all types of fertilizer?

5. Do you think that adding fertilizer to soil in order to increase agricultural output is a sustainable practice? Explain.

12.2 Dirty Differences

Purpose ▶ **D**iscover how soils form and explore some of the important components of soil.

Introduction

Soil is a complex mixture of living and non-living matter that contains air, water, minerals, particles of rock or sand, organic material in the process of decay, and a host of soil-dwelling organisms, including bacteria, fungi, worms, insects, and even burrowing mammals such as gophers and moles. Soil could be considered the foundation of any terrestrial ecosystem, because the ecosystem's producers rely on it for essential nutrients. Many of the organisms living in and on the soil recycle nutrients by breaking down the tissue of dead organisms. In this way the materials that the dead organisms were made of can be used again by other living organisms. Major factors that determine a soil's characteristics include the type of rock from which the soil was formed, the species of organisms living in and on the soil, the age of the soil, and the climate of the region.

Early stages of soil development are characterized by moss and lichen growth on rocks.

The Evolution of Soil

Soil Structure

Soils begin to form when weather or other forces break rock into smaller and smaller particles. Mosses and lichens growing on the surface of the rock also help begin the process of soil formation. As mosses and lichens grow, they slowly break down the rock beneath them. As they die, their decomposing bodies leave organic material behind. Young soils are thin and rocky, and their ability to support plant growth is generally limited to small, hardy plants like grasses, wildflowers, and small shrubs.

Mature soils are built up over very long time periods. As plants grow, then die, they continually break down rocks and leave behind organic material that gradually decomposes. Over many years, a thick layer of soil can build up. Mature soils can support the growth of many plants, including large trees.

In general, mature soils consist of layers. The top layer, called **topsoil**, is made almost entirely of fresh and partially decomposed organic material along with decomposer organisms such as bacteria, insects, and fungi. Topsoil is very rich in plant nutrients and contains a large percentage of dark-colored organic matter, called **humus**. It is produced by the decay of plants, animals, and other organisms. Topsoil also includes microorganisms, worms, insects, the roots of living plants, minerals, and rock particles. It may even contain pebbles or larger rocks. As water drips down through the topsoil, it dissolves some of the nutrients and other chemicals found in the topsoil and carries them away in a process called leaching. Topsoil gradually gives way to a lower layer, the **subsoil**, that is composed primarily of the rock from which the soil originally evolved. It also contains some organic material, roots of larger plants, and nutrients leached from the soil layers above it.

The overall characteristics of a soil include the composition and size of its inorganic particles, the percentage of decaying organic material it contains, and the amount of water it is able to hold. Soils with a higher percentage of organic matter tend to be more fertile; organic material is the source of many essential nutrients and also helps soil absorb and retain water.

The pH of the soil directly affects the availability of mineral nutrients to plants. Soils that are too alkaline (pH>8) or too acidic (pH<6) can convert nutrients to chemical forms that do not dissolve in water or cannot be absorbed by plant roots. A pH that is too high or too low can also inhibit the action of soil microorganisms that play an essential role in the production of certain nutrients.

Soil Nutrients

As you learned in Activity 12.1, "Soil Nutrients and Fertilizer," nitrogen, phosphorus, and potassium are essential for plant growth. All three nutrients are naturally present in fertile soil. Nitrogen is a part of every living cell; it is a component of amino acids, which are the building blocks of proteins, and it is directly involved in photosynthesis. Nitrogen stimulates above-ground plant growth and is responsible for the rich green color of healthy plants. Fruit size is also related to the amount of available nitrogen. Although about 80% of Earth's atmosphere is nitrogen gas (N_2), nitrogen cannot be absorbed by plants unless it is present in the soil in the form of nitrate (NO_3^-) compounds. Since nitrates dissolve easily in water, rain and irrigation can cause nitrates to move into and out of soils.

Phosphorus is essential for the growth of new seedlings, early formation of roots, and the development of seeds and fruits. It plays a major role in processes that require a transfer of energy, such as the formation of fats within plant cells. Phosphorus cannot be absorbed by plants unless it is present in the soil in the form of phosphate (PO_4^{3-}) compounds.

Potassium is a catalyst necessary for plant metabolism. It strengthens a plant's resistance to disease and also appears to play a role in the synthesis of starch molecules.

As nutrients cycle through ecosystems, they accumulate naturally in the soil, mainly through the decay of organic material. A wide variety of chemical, physical, and biological processes are involved in the recycling of soil nutrients. For example, the decomposition of dead plants and animals or manure returns nitrogen to the soil in the form of nitrate (NO_3^-) and ammonium (NH_4^+) compounds.

Nitrogen-fixing soil bacteria convert nitrogen gas (N_2) from the air into ammonia (NH_3), and other soil bacteria convert the ammonia to nitrates. Because they are water-soluble, nitrates are accessible to plant roots. Cycling processes similar to these occur with all of the nutrients required for life. Stable ecosystems maintain a balance between the nutrients added to a soil and the nutrients removed from the soil. The availability of nutrients is a limiting factor that affects the carrying capacity of an ecosystem.

Nutrient Balance

The biological, chemical, and physical processes at work in natural ecosystems maintain a balance between the nutrients removed from soil by plant growth and the nutrients added to the soil by decomposition and other biological processes. Agriculture can upset this balance. Nutrients are removed from an ecosystem when natural vegetation is replaced with crops that are harvested and taken away. If too much plant cover is removed, bare soil may be exposed to erosion by wind and rainfall; water from rain or irrigation can leach nutrients from the soil and carry them out of the area. To offset these nutrient losses and to increase productivity, farmers often add fertilizers to the soil of their fields.

Fertilizers are effective in enhancing plant production, but excessive or improper use can have negative effects. For example, fertilizers that leach down through soil and into the groundwater may contaminate water supplies with nitrates. Excessive nitrate concentration in drinking water can contribute to a blood disorder in infants that may lead to respiratory failure. In a process known as **eutrophication**, high levels of phosphate or nitrate in surface waters can promote excessive growth of plants and algae. This lush growth deprives other water-dwelling organisms of sunlight and essential nutrients. In extreme cases, eutrophication can turn a healthy pond or lake filled with a variety of plants and animals into a swampy, plant-choked bog.

Analysis

Group Analysis

1. How do agricultural practices affect ecosystems?

2. As you learned in Activity 7.1, "Oodles of Models, " models can help us understand relationships between organisms and various aspects of the ecosystems in which they live.

   a. Draw a model that shows how a single nutrient is cycled among living and non-living components of an ecosystem.

   b. Draw a model that shows how a single nutrient is transferred from one organism to another in a food web.

   c. Explain how the two drawings you made are related.

3. Plant life has existed on Earth for hundreds of millions of years. Fertilizer has been manufactured for only about 100 years. How did plants survive for so long without manufactured fertilizer? Why do we need to use so much of it now?

12.3 <u>Soil Components and Properties</u>

Purpose ▶ **E**xplore variability in the composition of different soils and discover how a soil's characteristics affect its ability to deliver water and nutrients to plants.

Introduction ▼

Soils vary in many ways. In this activity you will relate soil properties to soil composition as you examine three different soils and measure and describe their properties. **Porosity** is a measure of the volume of space that lies in the openings between soil particles. In most soils, these spaces are filled with air or water. **Permeability** is the rate at which fluids can move through the soil. **Water retention** is the amount of water the soil can absorb. Each of these properties affects the soil's ability to support plant life and is directly related to the composition of the soil.

Sandy desert soils often have high permeability and porosity.

Materials

For the class

supply of pea gravel
supply of local soil
supply of potting soil

For each group of four students

2   magnifiers
1   profile tube with hole in the bottom
1   snap cap to fit profile tube
1   30-mL graduated cup
1   100-mL graduated cylinder
access to a clock with a second hand

Procedure

1. Prepare a data table to display the overall composition, porosity, permeability, and water retention of the three soil samples: pea gravel, potting soil, and local soil.

2. Describe the overall appearance and apparent composition of each soil sample in as much detail as you can. Record your descriptions in your journal.

3. Use the graduated cup to place 30 mL of gravel in a profile tube and then put 100 mL of water in the graduated cylinder.

4. Have one group member use a thumb to block the hole in the bottom of the profile tube while another slowly pours the water into the tube until the water level reaches the top of the gravel. Record the volume of water added, then pour in the remaining water.

5. Position the graduated cylinder directly beneath the hole in the bottom of the profile tube, then measure the time it takes for the water to flow out of the profile tube into the graduated cylinder.

6. Record the volume of water that flowed out of the profile tube.

7. Repeat Steps 3–6, using local soil instead of gravel.

8. Have one group member use a thumb to block the hole in the bottom of the profile tube and pour the water back into the tube.

9. Cap the tube, invert it (cap end down, hole end up) and then carefully shake the tube so that the soil and water mix thoroughly.

10. Place the tube, cap side down, on the table and let it stand. Observe and describe what happens to the mixture until the water is clear enough to see through.

11. Describe the appearance of the contents of the tube in as much detail as possible. Sketches and/or measurements may be helpful.

12. Repeat Steps 7–11, using potting soil instead of local soil.

Analysis

?

Group Analysis

1. What are the major differences in the overall composition of the three soil samples?

2. Which of the three samples had the highest porosity? Which one had the lowest?

3. If a soil has a high porosity, does that necessarily mean it will also have a high permeability? Why or why not?

Individual Analysis

4. What soil characteristics have the greatest effect on water retention? Explain your reasoning.

5. What components and what properties would you expect all fertile soils to have? Explain.

6. Do you think all soils can be made fertile by adding fertilizer? Explain your reasoning.

12.4 Mineral Mania

Purpose ▶ **I**dentify foods that contribute to the minimum daily requirements for specific nutrients. Use this information to plan a healthy menu of snacks for a class party.

Introduction ▼

You have explored how plants obtain essential nutrients from the soil. All organisms, including humans, must consume minimum amounts of many types of nutrients to survive. Most ecosystems and organisms have evolved so that sufficient quantities of necessary nutrients can be obtained from the ecosystem. Technology has not only made many different foods available, but has also allowed human populations to survive in ecosystems that would not otherwise support so many people. Humans today have many options when deciding how to obtain nutrients.

Proteins, carbohydrates, and fats are essential components of the diets of humans and other animals. Vitamins and minerals are also essential. In this activity, you will focus primarily on **minerals**, which when referring to foods are nutrients made of a single element. (The word "mineral" has different meanings in different contexts.) You will identify which minerals are supplied by different foods and select an assortment of foods that not only make good party snacks, but also provide half of the recommended daily allowance (RDA) of several important minerals. (Assume that your guests will obtain the other half of their RDA during other meals on the day of your party.)

Fruits and vegetables supply many needed nutrients.

Materials

■■ **For each team of two students**

at least 8 nutritional labels from various party foods

Table 1   A Quick Mineral Guide

Mineral	Role in Body	Source	Adult RDA**
Calcium	Maintains healthy bones and teeth; especially important during growth. Also involved in blood clotting and muscle contraction. Osteoporosis is a disease associated with too little calcium in the diet.	milk and dairy products, whole grains, dark green vegetables, kelp & other sea vegetables, some fish	1,000 mg
Chromium	Assists in regulating blood sugar. Important for diabetics and those with Turner's syndrome, because it is necessary for the conversion of carbohydrates to energy.	brewer's yeast, broccoli, cottage cheese, chicken, corn oil, grapes, milk, rice, wheat germ	male: 35 μg* female: 25 μg*
Copper	Required for the production of hemoglobin, ATP, and many hormones. Helps sustain healthy nerves and blood vessels.	almonds, beans, cocoa, mushrooms, potatoes, seafoods	900 μg
Iodine	Fundamental to the structure of thyroid hormones, which regulate metabolism and stimulate immune functions.	iodized salt, kelp, milk, shellfish, other seafoods	150 μg
Iron	Essential to the structural integrity of hemoglobin. Enhances overall resistance to disease and assists in production of ATP.	asparagus, beans, fruits, grains, meats, molasses, nuts, potatoes	male: 8 mg female: 18 mg
Magnesium	Needed for proper nerve and muscle action. Assists in the production of new cells, bone, protein, and fatty acids.	grains, grapefruit, almonds, oysters, shrimp, nuts, beans	male: 400 mg female: 310 mg
Manganese	Assists in the production of large organic molecules in the body. Necessary for nervous system function.	beet tops, beans, cocoa, milk, egg yolks, shellfish, tea, pineapple	male: 2.3 mg* female: 1.8 mg*
Phosphorus	Integral to the structure of DNA, therefore necessary for cell reproduction and repair. Also plays a role in nervous system functions.	milk products, grains, meat, fish, poultry, nuts	700 mg
Potassium	Essential for fluid balance, blood pressure, and muscle tone in the body, so plays a role in kidney, blood, and brain function.	bananas, yogurt, grains, apricots, sweet potatoes, beans, citrus fruits, green vegetables	4.7 g*
Selenium	Aids in some metabolic processes and in normal body growth and fertility by stimulating thyroid hormones. A well studied antioxidant that reduces cancer risk.	Brazil nuts, celery, cucumbers, broccoli, onions, tomatoes	55 μg
Zinc	Fundamental to immune function, repair mechanisms, and protein synthesis.	meat, eggs, yogurt, brewer's yeast, beans, nuts	male: 11 mg female: 8 mg

*Note that for these minerals there is not enough information to establish an RDA. Figures given are the AI (adequate intake).   **μg = microgram = 1 x 10^{-6} gram

Safety Note   A science lab is not a recommended area for food consumption. If you use the lab area for this activity, cover all surfaces with tablecloths.

Procedure

1. Use Table 1, "A Quick Mineral Guide," to select three minerals you want to make sure your party food will supply to your guests.

2. Prepare a data table with headings similar to those shown below. Title it "Data Table 1: Minerals in Foods" and enter the name and RDA of each of your three chosen minerals.

Data Table 1: Minerals in Foods

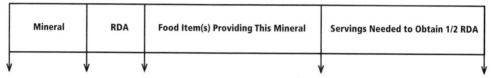

Mineral	RDA	Food Item(s) Providing This Mineral	Servings Needed to Obtain 1/2 RDA

3. Obtain eight food labels and record in Data Table 1 which foods provide each of your 3 target minerals.

   Note: One food item may provide several of your target minerals.

4. Determine how many servings of each food item are needed to provide half of the RDA for that mineral. Record this information in Data Table 1.

5. Prepare a data table with headings similar to those shown below. Title it "Data Table 2: Party Menu." Compare your foods to those selected by the other team in your group. Using all the minerals and foods investigated by both teams, prepare a group menu for a party which will provide half the RDA for three minerals. Include the number of servings (you may use fractions) of each food item that each of your friends at the party must consume in order to obtain half the RDA for each of your three target minerals. Record your answers in Data Table 2.

Data Table 2: Party Menu

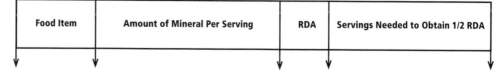

Food Item	Amount of Mineral Per Serving	RDA	Servings Needed to Obtain 1/2 RDA

6. At the direction of your teacher, share your ideas with the entire class. Help the class plan the party, and volunteer to bring one of the snacks. Make sure that you volunteer for something you can actually bring. (If you have specific food allergies, make sure that your teacher knows about them and that the party menu includes foods that you will be able to consume and that will provide the minerals for half of your RDA.)

Analysis

?

Group Analysis

1. How did you select your three target minerals?

2. Which mineral or minerals do you think are the most important to the diet of humans? Explain.

3. Do you think the nutrient requirements of people and plants are similar? Explain.

4. What effect, if any, will a continuing increase in human population have on the nutrients available to other organisms? Explain.

5. Should world leaders attempt to provide the RDA of every nutrient to every person on Earth? Discuss the trade-offs involved. Include in your answer what you think is the solution that best balances the benefits and costs of providing adequate nutrients to the world's population.

Extension   Include other nutrients—such as vitamins, fats, and protein—or calorie content in your party menu planning.

Cell Structure and Function

13.1 In and Out Nutrients

Purpose ▶ **O**bserve the structure of plant and animal cells.

Introduction

In Part 1 of this course you produced a list of essential resources needed for human survival. Your list undoubtedly included food. All living organisms need a continuous supply of the energy and essential nutrients contained in food. Where do these nutrients end up? How are they used? The answers to these questions lie within the cell—the building block of every living organism. This activity begins with a brief look at some cells and their contents.

A cell can be thought of as a very small bag of chemicals inside of which all life processes take place. Each cell is enclosed in a **cell membrane** that forms a boundary between the chemicals inside the cell and the environment outside it. Inside the membrane of every cell is a complex mixture of chemicals and water, called **cytoplasm**, plus genetic material in structures called **chromosomes**. This genetic material directs all cell functions, including how cells process and use food for growth, maintenance, and repair of cellular structures. In many organisms, including humans, the chromosomes are contained in a compartment within the cell, called the **nucleus**, that is separated from the cytoplasm by a membrane. Cells often contain other membrane-bound compartments, or **organelles**. For example, cells of green plants contain organelles called **chloroplasts**, which contain all the materials needed for photosynthesis.

Materials  **For each group of four students**

1	30-mL dropper bottle of water
1	30-mL bottle of Lugol's solution
1	piece of onion
1	prepared microscope slide
	paper towels

For each team of two students

2	microscope slides
2	coverslips
1	wooden toothpick
1	microscope

Procedure

Each team of two students will prepare two slides. One member of your team should prepare a slide of his or her own cheek cells. The other team member should prepare a slide from the onion. Every student should observe both of these slides, plus the prepared slides of blood, liver, plant leaf, and plant stem.

Cheek Cells

Safety Note

Take care when using the toothpick to obtain a sample of cheek cells. Be careful not to poke yourself or anyone else. Avoid areas of your mouth where you may have cold sores or other cuts. Handle only your own toothpick and be sure to throw it away immediately after use. Carefully clean your own slide with soapy water after you and your partner have both observed the cheek cells.

1. Gently scrape along the inside of your cheek with the side of a clean, new wooden toothpick.

2. Place a drop of water on a clean slide, then transfer the cheek scrapings to the drop of water and stir a little with the tip of the toothpick. Discard the toothpick immediately.

3. Carefully place a coverslip over the cells. Begin by holding the coverslip at an angle over the water droplet, and then gradually lower the coverslip, as shown in Figure 1.

Figure 1   Placing a Coverslip on a Microscope Slide

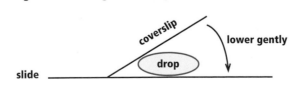

Procedure
(cont.)

4. Observe the cells with the microscope. Begin at low power. Make a drawing of what you see.

5. Center some cells in the field of view and switch to higher power. Again, sketch what you see.

6. Return to low power and stain the cells by placing one drop of Lugol's solution at one side of the coverslip. Touch the liquid at the other side of the coverslip with the edge of a small piece of paper towel or tissue. This will draw the stain under the coverslip.

7. Observe the cells again at low and high power and add any new details to your sketches.

8. Can you find the cell membrane? Does this type of cell have a nucleus? Do you see other small structures within these cells? Label the parts of the cell shown in your drawing.

Onion Cells

1. Use forceps or your fingernail to peel off a piece of the very thin inner layer of the onion.

2. Place a drop of water on a clean slide, then place your piece of onion in the drop of water.

3. Carefully place a coverslip over the cells. Begin by holding the coverslip at an angle over the water droplet, and then gradually lower the coverslip, as shown Figure 1.

4. Observe the cells with the microscope. Draw what you see at low power and also at high power.

5. Return to low power and stain the cells by placing one drop of Lugol's solution at one side of the coverslip. Touch the liquid at the other side of the coverslip with the edge of a small piece of paper towel or tissue. This will draw the stain under the coverslip.

6. Observe the cells again at low and high power and add any new details to your sketches. Can you find the cell membrane? Does this type of cell have a nucleus? Are there any other small structures within the cell? Label the parts of the cell shown in your drawing.

Procedure

(cont.)

Liver Cells

1. Observe the prepared slide of liver cells at low and high power. Draw one of the cells, labeling the membrane and nucleus. Draw any other structures you see.

Blood Cells

1. Observe the prepared slide of blood. The most common kind of cell is the red blood cell. Draw and label (as best you can) a typical red blood cell.

2. Scan around the slide until you find another kind of blood cell. Draw and label (as best you can) a typical example of this type of blood cell.

Plant Leaf and Stem Cells

1. Observe each of the prepared plant cell slides. Draw and label a typical cell from each slide.

Analysis

?

Group Analysis

1. Compare the three animal cells—cheek, liver, and red blood cell. What do they have in common? How are they different?

2. Compare the three plant cells—onion, leaf, and stem cell. What do they have in common? How are they different?

3. Compare your observations of the animal cells and the plant cells. What do these two types of cells have in common? How are they different?

4. You stained the cheek and onion cells with Lugol's solution. How did the onion cells look before and after staining? Explain the purpose of the stain.

13.2 Inside the Membrane

Purpose ▶ **I**nvestigate basic structures found in plant, animal, and bacteria cells. Develop models of different kinds of cells.

Introduction

One of the most important concepts in biology is cell theory, which states that

- all living matter is composed of one or more cells, which form the basic unit of organization of all organisms.

- all chemical reactions that take place in a living organism, including reactions that provide energy and synthesize biological molecules, occur within cells.

- all cells arise from the division of parent cells, and all cells contain hereditary information that is passed from parent to daughter cells.

Figure 2 Timeline: History of Cell Theory

1665 Robert Hooke makes an improved microscope and uses it to examine cork and other plants. He observes that all of these plant materials are made of small compartments separated by walls. He names these compartments "cells," meaning "little rooms."

1824 Rene Dutrochet is the first scientist to hypothesize that cells form the basic unit of all life.

1833 Robert Brown names the nucleus and proposes that it has an important role in the control of cellular processes.

1838 Matthias Schleiden concludes that all plants are made of cells.

1839 Theodore Schwann concludes that all animals are made of cells. Schwann, Schleiden, and other scientists propose that all living organisms are made of cells.

1858 Pathologist Rudolph Virchow proposes that all cells arise from the division of parent cells: "Where a cell exists, there must have been a preexisting cell."

Introduction
(cont.)

According to this theory, cells are a basic building block of life. As the timeline in Figure 2 illustrates, it took almost 200 years and the work of many scientists to develop and accept basic cell theory. During the past century, scientists have continued to add details to cell theory that have increased our understanding of how living organisms function. In addition, theories have also been proposed about how the first cell may have come to exist and how cell structure has evolved.

What Is in a Cell?

Each cell has at least three parts: a cell membrane that separates it from the external environment; genetic material that directs cell functions; and cytoplasm that contains proteins, sugars, fats, salts, and other important molecules. There is a great deal of variation in the structure and function of cells. You have already observed some of the important differences between plant and animal cells. Even within the body of one multicellular organism, such as a human, there are many different kinds of cells with different structures and functions.

The **cell membrane** is composed primarily of two kinds of molecules: proteins and phospholipids. The phospholipids line up in two layers, so that one end of each phospholipid molecule faces either the outside or inside of the cell, and the other end faces the middle of the membrane, as shown in Figure 3. Proteins are arranged within this double layer in different ways. Some proteins extend all the way through the membrane, whereas others are found only on the outer or inner surface of the membrane. Cell membranes play an important role in the movement of materials into and out of cells, in the recognition of foreign materials and organisms, and in

communication among cells. Materials that pass through the cell membrane include water molecules, sugars, proteins, oxygen, and waste materials.

The cell membranes of plants and some single-celled organisms are surrounded by a cell wall. The **cell wall** is more rigid than the cell membrane. It provides structural support and helps prevent the movement of excess water into or out of the cell.

The **cytoplasm** contains protein, carbohydrate, and lipid (fat) molecules, as well as many other dissolved molecules. These molecules may take part in chemical reactions that release energy to fuel cellular processes, or they may serve as building blocks for larger biological molecules. The cytoplasm always contains water, but its consistency is not always watery: the presence of large, complex protein molecules often gives the cytoplasm a consistency that is thicker than water.

All cells contain genetic material that directs cellular processes. This genetic material is called deoxyribonucleic acid, or DNA. DNA is an extremely long molecule that is usually packaged into structures called **chromosomes**.

Figure 3   Model of the Cell Membrane

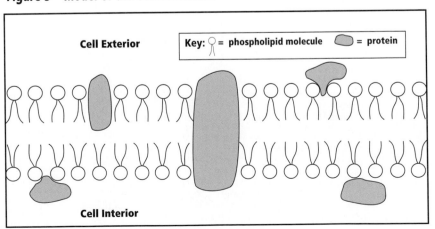

Cells reproduce by dividing. During cell division, the chromosomes are duplicated and passed on to the daughter cells. Chromosomes are not usually visible with the light microscope, except during cell division.

In the cells of most organisms, the chromosomes are found in a separate, membrane-bound structure called the **nucleus**. Cells containing a nucleus are called eukaryotic cells; organisms that are made up of eukaryotic cells are called **eukaryotes**. In bacterial cells, the chromosomes are not separated from the cytoplasm. Cells that do not contain nuclei are prokaryotic cells; organisms made up of prokaryotic cells are called **prokaryotes**.

Eukaryotic cells often contain cell structures in addition to the nucleus that are separated from the cytoplasm by membranes. These structures are called **organelles**, which means "little organs." Each organelle has a specialized function within the cell. For example, you may have seen chloroplasts in some of the green plant cells you observed under the microscope during Activity 13.1. Chloroplasts contain the materials needed for photosynthesis. Many organelles are so small that they are difficult or impossible to see with a light microscope, and can be revealed only with an electron microscope. Later in this course you will learn more about cell organelles.

Analysis

Group Analysis

1. Using information from the reading and from your observations of plant and animal cells as your guide, draw a diagram of each of the following:

   a. a prokaryotic cell

   b. a plant cell

   c. an animal cell

   Each diagram should fill up more than half of a normal-sized notebook page and should include all the structures discussed that are found in each type of cell. Be sure to label each structure.

2. In Activity 13.1 you observed many different kinds of cells. Which of these cells are prokaryotic and which are eukaryotic? Describe the evidence that supports your answer.

Extension

During this course, you will learn about additional cell organelles. Each time a new organelle is introduced, add it to your diagrams.

13.3 Moving Through Membranes

Purpose ▶ **I**nvestigate the movement of substances across membranes.

Introduction

If a cell is to function properly, it must contain the correct balance of water, sugars, salts, proteins, fats, and other substances. Maintaining a chemical environment suitable for proper cell function is called chemical homeostasis. The cell membrane plays an important role in chemical homeostasis because it regulates the movement of substances into and out of the cell. The cell membrane is **semi-permeable**, which means that certain substances can pass through it easily, but others cannot.

Imagine two separate solutions of water separated by a membrane through which water molecules can readily pass. One solution contains a very high concentration of dissolved sugar, the other contains a very low concentration of sugar. In the absence of other forces, water will move from the solution with a lower concentration of sugar (hypotonic) and into the solution with a higher concentration (hypertonic). This movement will continue until both solutions have the same sugar concentration (isotonic). **Osmosis** describes this natural tendency for water to pass through a membrane in a way that equalizes different concentrations of a dissolved substance, as is shown in Figure 4. If a cell is to survive, the movement of water and other substances into and out of it must be regulated. Too much or too little water, or too much or too little of a dissolved substance, can cause a cell to die.

Figure 4   Schematic Diagram of Osmosis

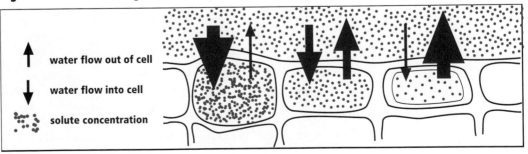

↑ water flow out of cell

↓ water flow into cell

⋰ solute concentration

Materials

For each group of four students

1	SEPUP tray
3	pieces of plastic dialysis tubing (to model cell membranes)
2	9-oz clear plastic cups
1	180-mL bottle of liquid glucose solution
1	180-mL bottle of liquid starch solution
2	droppers
1	50-mL graduated cylinder
1	30-mL bottle of Lugol's solution
4	glucose test strips
1	fine-point marker

supply of 20% salt solution

access to running water

Procedure

Within each group of four students, one team will perform Part A and the other team will perform Part B.

Part A investigates what happens when the concentration of a dissolved substance within a cell is different from the concentration of that substance in the environment surrounding the cell. Part B investigates the movement of glucose and starch across the cell membrane. Each student team should be sure to understand the procedure and make observations of the work done by the other team.

Part A Membranes and Water Balance

1. Obtain two pieces of dialysis tubing and wet them by dipping them into a beaker of water. Squeeze one end of each membrane between your thumb and forefinger to make an opening. Use a clean dropper or a faucet to rinse water through the membrane tubes.

Figure 5 Filled and Tied Membrane Tubes

2. Tie a very tight knot in one end of each membrane tube. To one tube, add enough 20% salt solution to fill approximately 3–4 cm of the tube. Tie a tight knot in the top end of the tube, leaving a little space in the tube above the level of the liquid, as shown in Figure 5. Rinse the tied membrane tube in running water, then place it in a cup half-filled with tap water. Label the cup "hypotonic."

Procedure
(cont.)

3. Add 3–4 cm of tap water to the second membrane tube. Tie a knot in the top end of the tube, leaving a little space in the tube above the level of the liquid. Rinse the tied membrane tube in running water, then place it into a cup half-filled with 20% salt solution. Label the cup "hypertonic."

4. Allow the membranes to remain in the cups for 15–30 minutes. Every 5 minutes, observe and record any changes in the membrane tubes or their contents. Also observe Part B of this investigation.

Part B   Membranes and the Movement of Small and Large Molecules

1. Add two or three drops of glucose solution to Cup 1 on the lower row of your SEPUP tray. Add two or three drops of starch solution to Cup 2.

2. Dip the colored end of one glucose test strip into Cup 1 and the colored end of a second glucose test strip into Cup 2. Observe what happens to each test strip and record your observations.

3. Add one drop of Lugol's solution to Cup 1 and Cup 2 and observe and record any changes that take place.

4. Fill the 50-mL graduated cylinder with 35 mL of water. Test the water with an unused glucose test strip to be sure that no glucose is present.

5. Obtain a piece of dialysis tubing and wet it by dipping it into a beaker of water. Squeeze one end of the membrane between your thumb and forefinger to make an opening. Use a clean dropper or a faucet to rinse water through the membrane tube.

6. Tie a knot very tightly near one end of the membrane tube.

7. Pour approximately 3 mL of liquid starch and approximately 3 mL of glucose solution into your membrane tube.

8. Pinch the top of the membrane tube closed and carefully rinse the tube thoroughly under running water. This is very important, because you do not want to have any glucose or starch on the outside of the tube.

9. Tie the top of the membrane tube very tightly just above the level of the liquid, as shown in Figure 5. Rinse the tube again to be sure there is no glucose or starch on the outside.

Procedure
(cont.)

10. Slowly and carefully put the filled membrane tube into the 50-mL graduated cylinder so that it is surrounded by water, as shown in Figure 6.

11. After 5 minutes, test the liquid surrounding your membrane tube with a glucose test strip. Record your observations.

12. Add 3 drops of Lugol's solution to the liquid surrounding your membrane tube.

13. Observe for 5–10 minutes. Record your observations. Also observe Part A of this investigation.

Figure 6 **Putting the Membrane Tube into a Graduated Cylinder**

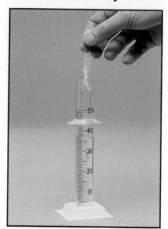

Analysis
?

Group Analysis

Part A

1. What changes do you observe in the simulated cell when the external environment surrounding the cell is less salty than the cell's contents (hypotonic)?

2. What changes do you observe in the simulated cell when the external environment is more salty than the cell's contents (hypertonic)?

3. How can you explain the changes that you observe?

Part B

4. Describe the ability of each of the following molecules to move through the plastic membrane: glucose, Lugol's solution, starch. Give evidence for your answers.

5. Why do you think some substances can pass through the membrane but other substances cannot?

Individual Analysis

6. Based on your observations in Parts A and B of this investigation, describe how the cell membrane helps to control the environment inside the cell.

7. Imagine a single-celled organism living in a pond. What would happen to the organism if runoff from crop irrigation made the pond significantly saltier?

13.4 Nature's Crossing Guards

Purpose ▶ **E**xplain the role of the cell membrane in controlling the chemical environment within the cell.

Introduction
▼

Every living organism has cells, and every cell has a membrane that separates it from the external environment. The cell membrane controls the entry and exit of molecules into and out of the cell. It can maintain an environment inside the cell that is quite different from the environment outside it.

A saltwater environment can kill many organisms. How do fish survive in salty ocean water?

Membranes and the Cellular Environment

Living organisms must be able to maintain the proper balance of water and dissolved substances, called solutes, within their cells. Otherwise, the cells would shrink or swell with changes in the external environment, as shown in Figure 7. **Diffusion** describes the natural tendency of substances to move from areas where they are more highly concentrated to areas where they are less concentrated. Some small particles can easily diffuse across the cell membranes of living creatures. Diffusion is critical to a cell's survival. The diffusion of water into and out of cells, called osmosis, was described in Activity 13.3. Another example of diffusion involves oxygen. Oxygen is constantly moving from red blood cells, in which the concentration of oxygen is relatively high, into adjacent tissue cells that have a lower concentration of oxygen. As a tissue cell consumes oxygen during metabolism, the concentration of oxygen inside it drops, so more oxygen diffuses into it from the oxygen-rich blood.

Not all substances used or produced during life processes can freely diffuse through a cell membrane, especially those that are electrically charged or very large in size. These molecules must be transported by proteins that are part of the cell membrane. (See Figure 3 on page 167.) In some cases, these proteins transport substances into or out of a cell against the normal direction of diffusion. This process, which moves molecules from regions of lower concentration to regions of higher concentration, is called **active transport**, and it requires energy.

The cells of some organisms have specialized mechanisms that help maintain chemical homeostasis. For example, organisms that live in fresh water must be able to maintain solute concentrations in their cells that are higher than the water that surrounds them. Cells unable

Figure 7 A Cell's Response to a Hypertonic Environment

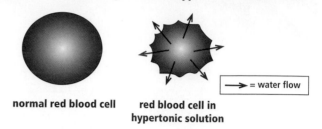

normal red blood cell   red blood cell in hypertonic solution

⟶ = water flow

to do so would swell and burst. Some single-celled freshwater organisms have an organelle that pumps water out of the cell. More complex multicellular organisms, like freshwater fish, have kidneys that pump water out of the body but retain solutes such as salt and glucose.

Land-dwelling animals are not surrounded by water. They maintain chemical homeostasis primarily by preventing water loss. Humans lose water by evaporation from the lungs and skin and by excretion in the urine. If too little water is taken in, or too much water lost, the chemical balance of body cells is disturbed. The imbalance may result in fever and can interfere with the nervous system and other body functions. Human kidneys help maintain chemical homeostasis by filtering waste products out of the blood. The kidneys produce urine that contains a higher concentration of dissolved substances than the blood. This process, which requires active transport in the kidney cells, accounts for a significant amount of the body's overall energy requirement.

There are also other mechanisms that contribute to the maintenance of osmotic balance in the human body. For example, when the body becomes dehydrated, cells in the mouth lose water and produce a sensation we call thirst. This leads us to drink more water to replenish the body's supply.

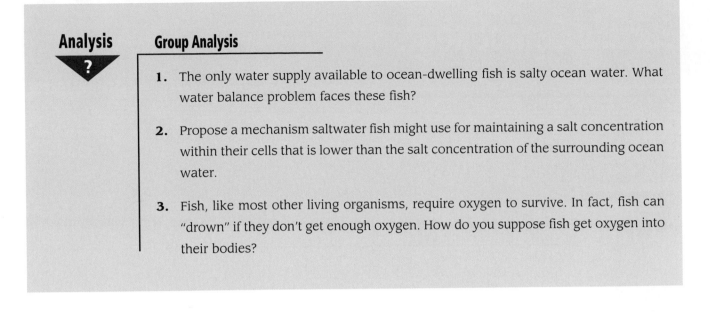

Analysis

?

Group Analysis

1. The only water supply available to ocean-dwelling fish is salty ocean water. What water balance problem faces these fish?

2. Propose a mechanism saltwater fish might use for maintaining a salt concentration within their cells that is lower than the salt concentration of the surrounding ocean water.

3. Fish, like most other living organisms, require oxygen to survive. In fact, fish can "drown" if they don't get enough oxygen. How do you suppose fish get oxygen into their bodies?

Earth's Components

14

14.1 What's It Made Of?

Purpose ▶ **R**eview the structure of atoms and molecules and investigate the structure of nutrients and other substances.

Introduction

The physical world is made up of elements, which consist of atoms. An **element** is a pure substance that is made of only one kind of atom and cannot be broken down into a simpler substance. An **atom** is the smallest particle of an element that has the properties of that element. A **compound** is a pure substance made up of atoms of two or more elements that are attached, or bonded, together. In some compounds, adjacent atoms are bonded strongly together to form **molecules**. For example, a water molecule consists of two hydrogen atoms and one oxygen atom bonded together. Alternatively, atoms in a **non-molecular substance** can be bound more or less equally to all adjacent atoms in a repeating crystalline pattern. An example of a non-molecular substance is table salt. It consists of many atoms of sodium and many atoms of chlorine held together in a repeating pattern.

The processes of life that take place within cells involve interactions among atoms, molecules, and non-molecular substances. While air, water, food, and most other organic materials consist of substances with molecular structures, remember that many inorganic substances are not molecular in nature. Many of the molecules that are important to living organisms, such as proteins, are very large and can contain millions of atoms. An atom, molecule, or non-molecular substance is identified by its name and by a chemical formula that indicates which atoms it is made of. All of these—atoms, molecules, and non-molecular substances—can be referred to as particles. To understand how these tiny particles interact, it helps to understand the structure of atoms.

The Physical Structure of an Atom

Figure 1 provides a summary of the terms used to describe atoms. The components of an atom include the **nucleus**, which usually contains protons and neutrons. **Protons** have a positive charge; **neutrons** are neutral because they have no electrical charge. All atoms with the same number of protons are atoms of the same element. The number of protons defines an element's atomic number.

The densely packed nucleus is surrounded by a relatively large area, or cloud, containing rapidly moving, negatively charged **electrons**. In a neutral atom, the total number of electrons is equal to the total number of protons. For instance, each atom of hydrogen, which has an atomic number of 1, has one proton and one electron. Carbon, which has an atomic number of 6, has six protons and six electrons. Electrons move so quickly that the location of any electron within the cloud cannot be pinpointed. However, the probable region in which a specific number of electrons will be found can be defined and is called an orbital. In any atom, the volume of the region occupied by the electrons is much larger than the volume of the nucleus, although the electrons themselves are much smaller than the protons or neutrons. Because of this, most of the region occupied by an atom is actually empty space!

An atom's mass comes almost entirely from the protons and neutrons in its nucleus. For example, an atom of hydrogen, with an atomic mass of 2, has one proton and one neutron. Carbon, which has an atomic mass of 12, has six protons and six neutrons. Protons and neutrons have almost the same mass, with the neutron being slightly more massive. The mass of a proton is defined as 1 atomic mass unit (amu). The mass of a proton is approximately 1,200 times that of an electron.

The number of neutrons in an atom of a given element can vary; this variability affects the mass of the atom, but not the type of element. Atoms that have the same atomic number but different numbers of neutrons are called isotopes. The most common carbon isotope, carbon-12, has six protons and six neutrons; a less common isotope, carbon-14, has six protons and eight neutrons. The atomic mass given on the Periodic Table for each element is not a whole number because it is an average of the mass of all naturally occurring isotopes of that element.

Many of the less common isotopes are radioactive, meaning that they emit energy from the nucleus. Some radioactive isotopes have important applications in science, medicine, energy production, and other fields, which you will investigate in Part 4 of this course. For example, isotopes of carbon and some other elements can be used to determine the age of ancient objects. However, uncontrolled exposure to radioactive substances can have negative health effects, a problem that raises significant safety issues.

Figure 1 The Parts of an Atom

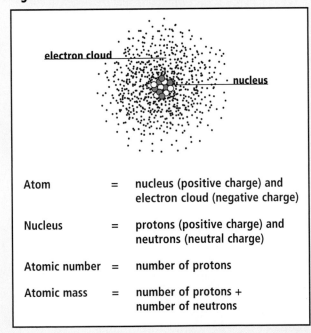

Atom	=	nucleus (positive charge) and electron cloud (negative charge)
Nucleus	=	protons (positive charge) and neutrons (neutral charge)
Atomic number	=	number of protons
Atomic mass	=	number of protons + number of neutrons

Materials

■ **For each student**

1   Periodic Table of the Elements

Procedure

1. Use the Periodic Table of the Elements to find out what atoms make up a molecule of each of the following molecular substances. The first one has been completed for you.

Molecular Substance		Chemical Formula	Atoms That Make Up the Molecule
	Water	H_2O	2 hydrogen atoms + 1 oxygen atom
Air	Nitrogen	N_2	
	Oxygen	O_2	
	Carbon dioxide	CO_2	
Food	Glucose (sugar)	$C_6H_{12}O_6$	
	Alanine (amino acid)	$C_3H_7O_2N$	
	Oleic acid (fat)	$C_{12}H_{34}O_2$	
	Amylose (starch)	$C_{2,000}H_{10,000}O_{5,000}$	

Analysis **?**

Group Analysis

1. How does the number of atoms in the molecules that are found in air and water compare with the number of atoms in molecules that are found in food?

2. How do the elements that make up the molecules found in air and water compare with the elements that make up the molecules found in food?

3. Explain why you think the molecules found in air, water, and food have the similarities and differences you noticed when answering Analysis Questions 1 and 2.

4. A healthy human diet includes foods that contain many elements, such as Ca, Fe, Zn, K, P, and S. Name each of these elements and propose a source for them. (**Hint:** They're not found in air or water.)

Individual Analysis

5. How might the size of a molecule affect the way in which an organism obtains the nutrients in that molecule? Be as specific as you can.

Exploring the Physical Properties of Elements

Purpose ▶ **I**nvestigate the physical properties of five elements and discover their similarities and differences.

Introduction

What happens to food consumed by an organism? The food is broken down into its components, and these substances eventually make their way into the organism's cells. Inside the cell, they are used for cellular processes that are essential to life. The specific use for each substance depends upon the chemical and physical properties of that substance. These properties are related to the properties of the element(s) from which the substance is made. There are currently 115 known elements, each with a distinct set of chemical and physical properties. Some elements share properties, and some are not at all similar. In this activity, you will observe, describe, and compare samples of the five elements shown in Figure 2.

Figure 2   Samples of Five Solid Elements

Materials

For the class

supply of water

For each group of four students

1   carbon rod

1   lump of sulfur

1   piece of silicon

1   aluminum cylinder

1   iron cylinder

1   magnet

1   conductivity apparatus

1   50-mL graduated cylinder

1   250-mL beaker

access to a balance

For each team of two students

1   magnifier

access to a Periodic Table

Procedure

1. Examine the samples provided for each of the five elements—aluminum, carbon, iron, silicon, and sulfur. Create a data table to record your observations for each element with regard to the following properties, as well as any others you think are important:

color	odor
texture (roughness)	luster (shininess)
magnetism	hardness (resistance to scratching)
electrical conductivity (see Figure 3)	density $D = \frac{m}{v}$ (See Figure 4)

2. Make your observations and record them in your data table.

Figure 3   Using the Conductivity Apparatus

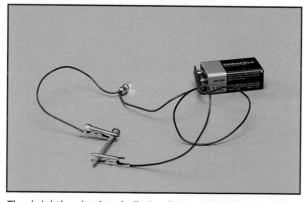

The brightly glowing bulb in the conductivity apparatus indicates that this metal nail is a good conductor of electricity.

Analysis

?

Group Analysis

1. Which of your samples appear to be in a natural state and which appear to have been processed? Explain your reasoning for each answer.

2. Do you think the processing affected any of the properties you tested? Explain why or why not.

3. Divide the five elements into two groups, with each group containing the elements you think are the most similar. Use evidence from the observations you made during this activity to explain your classification system.

4. Which two samples were most difficult to categorize? Why?

5. Describe the general location of each of the five elements on the Periodic Table of the Elements.

Individual Analysis

6. You are given a sample of an unknown element and are asked to perform three of the tests from this activity to help you decide in which group to place it. Describe the procedure you would follow and the data you would collect to accomplish this task.

Figure 4   Determining Volume of an Object

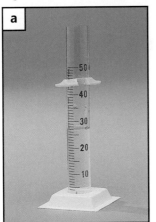
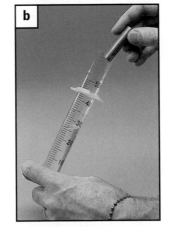
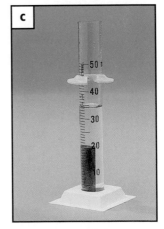

To determine an object's volume by water displacement, (a) record the initial volume of the water; (b) carefully submerge the object in the water; and (c) record the final volume.

14.3 The Origins of the Periodic Table

Purpose ▶ **I**dentify the patterns underlying the arrangement of the Periodic Table of the Elements.

Introduction

Early scientists thought that all of the matter in the physical world was derived from what they called "the four elements"—air, water, earth, and fire. We now know that air, water, and earth are made up of chemical elements, and that fire is a form of energy. Every substance on Earth is made up of one or more of the chemical elements listed in the **Periodic Table of the Elements**. The Periodic Table is arranged in an orderly pattern that makes it easy to obtain important information about an element's properties (see Figure 5) and its relationship to other elements. The modern periodic table evolved from several earlier attempts to categorize and organize the elements.

Figure 5   Key to Information Provided in the Periodic Table

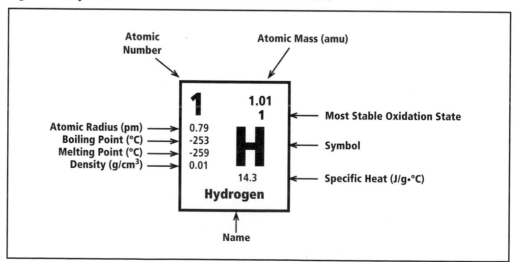

Finding Patterns Among the Elements

Imagine you have a collection of items that you often look through, like a collection of photographs. Imagine that the photographs are not in a photo album and are in no particular order—they are all jumbled together in a shoebox. Every time you want to look at a particular photograph, you must go through the entire collection to find what you want. How could you organize them so that each photograph is easy to locate? Different individuals might choose to organize them differently. Some people might organize them by who is pictured, others by date or by place. Is one organizational scheme better than another?

In the 1800s, chemists found themselves in this position as they attempted to arrange the elements based upon their known physical and chemical properties. Many elements had been discovered and data were available on them, but they had not yet been organized into any sensible order.

In 1829, a German chemist named Johann Dobereiner attempted to make sense of the elements by grouping them according to similar chemical properties. He found that many of the known elements could be grouped into threes, or triads. Lithium, sodium, and potassium formed one triad; calcium, strontium, and barium formed another. He then noticed that there was a pattern to the atomic masses of the elements in each of his triads.

In 1864, John Newlands, an Englishman, tried arranging the known elements in order of increasing atomic mass. Starting with lithium, he observed that the first and eighth elements, the second and ninth elements, and so on along the row, had similar chemical properties. Because this pattern reminded him of the division of musical notes on a piano keyboard, he referred to it as the "law of octaves." Unfortunately, this pattern was inconsistent and did not always work; what chemists did not know at the time was that there were elements as yet undiscovered that would have filled in the gaps in Newlands' law of octaves.

It was only a few years later, in 1869, that Dmitri Mendeleev, a Russian chemist, carried the work of his predecessors to a new level. First, Mendeleev placed the known elements, of which there were only 63 at the time, in order of increasing atomic mass. He then created a grid by placing elements with similar chemical properties in the same column, one underneath the other, while still maintaining the rows that placed the elements in order of increasing atomic mass. Consistent with Newlands' work, his grid had seven elements in each horizontal row; the eighth element is placed in the second row, under the first element, the ninth element under the second element, and so on, as shown in Table 1. Mendeleev had discovered that the properties of elements are "periodic functions of their atomic weights"—that is, their properties repeat periodically after each seven elements.

What really set Mendeleev's work apart, though, was what he did from that point on. As he was constructing his table, he noticed that there were places for which no known element had the appropriate set of properties. Mendeleev boldly predicted that there were still-undiscovered elements that would fill in these gaps. He not only anticipated the discovery of three new elements, which he named eka-boron, eka-aluminum, and eka-silicon, but also predicted their properties, based on the position of the gaps in his periodic table. Amazingly, his predictions were proven correct by the discovery in 1875 of gallium (eka-aluminum) with an atomic mass of 70. Scandium (eka-boron), with an atomic mass of 45, was discovered in 1879, and germanium (eka-silicon), with an atomic mass of 73, was discovered in 1886.

Although Mendeleev is credited as the "father" of the modern periodic table, an interesting historical note is the almost simultaneous development of this idea by Julius Lothar Meyer, a German chemist. Meyer independently published his table of the elements in 1870, and it was almost identical to that developed by Mendeleev.

The entire Periodic Table is now organized into the basic pattern suggested by Mendeleev: vertical columns, known as groups or **chemical families**; and rows, known as periods. The members of each chemical family have similar chemical properties. The arrangement of the modern periodic table allows us to extrapolate information about one element to other elements within the same chemical family. What an ingenious time saver!

Today, the Periodic Table of the Elements includes 115 elements. The elements with atomic numbers higher than 92 (uranium) are known as the transuranium elements. These elements are not usually found on Earth. They can be created in high-energy colliders, which smash atoms together with so much force that they combine to form larger atoms. Most of the transuranium elements are radioactive, and exist for only a short time before breaking down into more common elements.

Table 1   Mendeleev's 1872 Periodic Table

	R_2O	RO	R_2O_3	RH_4 RO_2	RH_3 R_2O_5	RH_2 RO_3	RH R_2O_7	RO_4			
1	H 1										
2	Li 7	Be 9.4	B 11	C 12	N 14	O 16	F 19				
3	Na 23	Mg 24	Al 27.3	Si 28	P 31	S 32	Cl 35.5				
4	K 39	Ca 40	(44)	Ti 48	V 51	Cr 52	Mn 55	Fe 56	Co 59	Ni 59	{Cu 63}
5	{Cu 63}	Zn 65	(68)	(72)	As 75	Se 78	Br 80				
6	Rb 85	Sr 87	?Yt 88	Zr 90	Nb 94	Mo 96	(100)	Ru 104	Rh 104	Pd 106	{Ag 108}
7	{Ag 108}	Cd 112	In 113	Sn 118	Sb 122	Te 125	J 127				
8	Cs 133	Ba 137	?Di 138	?Ce 140							
9											
10			?Er 178	?La 180	Ta 182	W 184		Os 195	Ir 197	Pt 198	{Au 199}
11	{Au 199}	Hg 200	Tl 204	Pb 207	Bi 208						
12				Th 231		U 240					

KEY

RH_4 = ability of element to combine with hydrogen

R_2O = ability of element to combine with oxygen

H = element symbol

1 = atomic mass

(44) = predicted atomic mass of undiscovered element

{Cu 63} = element appears in two places

Analysis ▼?

Group Analysis

1. What elements do you think have chemical properties similar to those of fluorine (F)?

2. Johann Dobereiner saw a pattern in the atomic masses of groups of elements with similar properties. He called these groups "triads." What pattern do you observe within each triad shown in the table below?

   Hints: Do some math. Remember, patterns don't have to be perfect.

Triad 1		Triad 2	
Element	Atomic Mass	Element	Atomic Mass
Lithium	7	Calcium	40
Sodium	23	Strontium	88
Potassium	39	Barium	137

3. Working independently of each other, and at about the same time, Mendeleev and Meyer developed similar periodic tables. How would you explain this?

Individual Analysis

4. Mendeleev's periodic table was arranged in order of increasing atomic mass. Is the modern periodic table arranged according to atomic mass? If not, how is it arranged? Provide evidence to back up your claims.

14.4 Specific Heat and Atomic Structure

Purpose ▶ **M**ake and use a scatterplot to help determine the underlying cause of a substance's specific heat.

Introduction

Earlier in this course, you discovered that specific heat describes how energy is transferred through a material. Now that you know a little more about the elements that make up all materials, you can explore how specific heat and atomic structure are related.

Imagine that you have samples of two materials. The two samples are of equal mass. If you heat both samples in an identical manner, the sample with the higher specific heat will change temperature less rapidly—in other words, its particles will absorb more heat— than the sample with the lower specific heat.

Many properties of a material, including specific heat, are related to the material's atomic structure. Since all atoms in an element are identical, the study of the properties of each element provides useful information about properties associated with atomic structure.

Figure 6   Representation of Equal Mass Samples of Two Elements

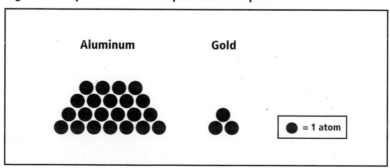

Materials

■ **For each student**

1 sheet of graph paper

Procedure

1. Set up and label the axes for a graph of atomic mass vs. specific heat. Use Table 2, which provides data on the atomic mass of each elemental solid, as a guide.

2. On your graph, make a scatterplot of the specific heat and atomic mass of each of the substances listed in Table 2.

Table 2   Atomic Mass and Specific Heat for Select Elements

Element	Atomic Mass (amu)	Specific Heat (J/g°C)
Sodium (Na)	23.0	1.23
Magnesium (Mg)	24.3	1.02
Aluminum (Al)	27.0	0.90
Silicon (Si)	28.1	0.70
Iron (Fe)	55.8	0.45
Copper (Cu)	63.5	0.39
Zinc (Zn)	65.4	0.39
Silver (Ag)	107.9	0.24
Tin (Sn)	118.7	0.23
Gold (Au)	197.0	0.13
Lead (Pb)	207.2	0.13

Analysis

?

Individual Analysis

1. Describe any relationship(s) you see between atomic mass and specific heat.

2. Could you make a reasonably accurate prediction of the specific heat of an element if all you knew about it was its atomic mass? Explain why or why not.

3. Propose a theory that could explain why elements with a higher atomic mass have a lower specific heat. (**Hint:** See Figure 6.)

Classifying Elements

15

15.1 Modeling Molecules

Purpose ▶ **U**se physical models to explore how atoms combine to form molecules, and discover how the structure of an atom influences molecule formation.

Introduction ▼

Most of the substances found on Earth are not elements made up of a single type of atom. Most substances consist of several types of molecules. Atoms can be held together to form molecular or non-molecular substances by "energy connections" called **chemical bonds**.

A compound is a substance made up of atoms of more than one element in specific, regular proportions. For example, ammonia (NH_3) is a compound because its molecules are made from atoms of nitrogen and hydrogen in exact proportions, as indicated by its formula. Compounds necessary for life include the sugars, proteins, and lipids found in foods. Common, everyday materials—from gasoline to cotton to aspirin—are also made up of compounds. Many of these compounds have molecular structures; many others are non-molecular in nature. In this activity you will concentrate on a few molecular compounds.

Materials **For each team of two students**

    1   molecular model set
    1   set of colored pencils (optional)

Procedure

1. Check to see if you and your partner have the following pieces in your molecular model set:

> 32 yellow "atoms"
>
> 14 blue "atoms"
>
> 4 red "atoms"
>
> 18 black "atoms"
>
> 54 white "bonds"

Note: During this activity, you will be constructing and sketching molecules. After sketching each molecule, pull it apart before proceeding.

2. **a.** Construct all the molecules you can according to the following two rules:

   - Each molecule must contain only two atoms.

   - All the bonding sites (the protruding "sticks") of an atom must be connected to those of another atom.

   Note: The holes in the nuclei of the atom models are not to be used as bonding sites. They are a result of the manufacturing process.

   b. Draw each molecule that you construct.

3. **a.** Use the following rules to construct four more molecules:

   - Each molecule must contain between three and five atoms.

   - All the bonding sites must be connected to those of another atom.

   b. Draw each molecule that you construct.

4. **a.** Construct two more molecules following only one rule:

   - All the bonding sites must be connected to those of another atom.

   b. Draw each molecule that you construct.

5. Make sure all the molecules have been taken apart and return all the pieces to the set. Compare your drawings with those made by the other team in your group.

Analysis

?

Group Analysis

1. Besides the rules you were given, what characteristics of the models limited the types of molecules you could construct?

2. Based on your experience with the models, what factors influence the kinds of bonds an atom can make?

15.2 ___Trends in the Periodic Table___

Purpose ▶ **I**dentify how the arrangement of the Periodic Table assists in predicting chemical properties of elements.

Introduction ▼ The Periodic Table is an orderly arrangement, based on chemical and physical properties, of all the known elements. Elements found in the same vertical column are considered members of the same chemical family. In this activity you will begin to explore how elements from different chemical families bond to form molecules. You will also become more familiar with the information provided in the Periodic Table.

Table 1   Predicting Chemical Formulas

Element Name	Element Symbol	# of Hydrogen Atoms With Which It Bonded	Sketch of Molecule	Molecular Mass (amu)	Standard Chemical Formula	Common Name
Oxygen	O	2	O H H	18	H_2O	water
Nitrogen						
Carbon						
Hydrogen						

Materials

■■ **For each team of two students**
1   molecular model set
1   set of colored pencils (optional)

■ **For each student**
access to a Periodic Table

Procedure

1. **a.** The pieces of the molecular model set can be used to represent atoms of the following elements:

black	=	carbon
yellow	=	hydrogen
red	=	nitrogen
blue	=	oxygen

For each of the four elements, make a molecule using only one atom of that element and as many hydrogen atoms as you need.

Remember: All the bonding sites (the protruding "sticks") of an atom must be connected to those of another atom.

   b. Draw the four molecules that you construct.

2. Create a table similar to Table 1 and complete the first four columns (Element Name, Element Symbol, # of Hydrogen Atoms..., and Sketch of Molecule).

3. Using data from the Periodic Table, calculate molecular mass by adding up the masses of each atom in the molecule you sketched. Enter your results in the column titled Molecular Mass.

Note: The final two columns should remain blank for now. You will fill them in during the class discussion that follows this activity.

4. Place a model atom of each of the four elements on its symbol in the Periodic Table. Look at the relationship between the number of bonding sites on each element and the column it is in. How many bonding sites do you predict for the following elements?

   a. fluoride

   b. beryllium

Hint: Boron has three bonding sites.

Analysis

Group Analysis

1. You may have realized that the molecule you made by bonding oxygen and hydrogen is water, which is commonly represented by the chemical formula H_2O. (The subscript "2" indicates that there are two atoms of hydrogen in the molecule.) Write a possible chemical formula for the three other molecules you constructed.

2. How many hydrogen atoms do you think an atom of the following elements would bond with?

   a. silicon

   b. phosphorus

   c. sulfur

   d. sodium

   Hint: Recall that columns of the Periodic Table represent families of elements that have similar properties.

Individual Analysis

3. What generalization can you make about the bonding properties of an element and its location on the Periodic Table?

4. Which model provides more information—a chemical formula or a sketch of the molecule?

15.3 Patterns in the Properties

Purpose ▶ **G**ain a more complete understanding of the Periodic Table by exploring the properties that are used to classify and categorize the chemical elements.

Introduction

As you learned in Activity 14.3, "The Origins of the Periodic Table," early versions of the Periodic Table organized the known elements according to their chemical and physical properties. Today, even though we know much more about the properties of the elements, we still use a periodic table very similar to the one that was developed over 100 years ago. (When Mendeleev designed his periodic table, he placed some elements in a different order than that of the modern periodic table.) As scientists continue to learn more about the elements—and even create and observe new elements in the laboratory—these properties still tend to fit the organizational trends of the Periodic Table. In this activity, you will explore more thoroughly the relationship between the position of an element on the Periodic Table and the properties of that element, as well as the properties of neighboring elements.

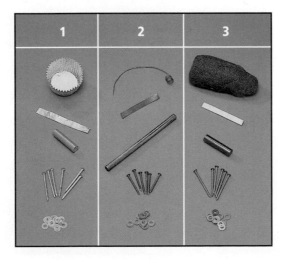

Column 1 contains aluminum items, Column 2 contains copper items, and Column 3 contains iron items. What characteristics determine which row each item is in?

Materials For each group of four students

    1   set of 24 element cards

    1   piece of poster board or paper

    tape, glue, or a stapler

Procedure

1. Use one of the first three properties listed on your element cards to divide all the elements you were given into categories. Make a chart showing the names of your categories and the elements that belong in each category.

2. Within each of your categories, place the elements in sequential order according to one of the other properties listed on the element cards.

   Hint: Use one of the quantitative properties.

3. Arrange all the elements into a grid so that

   a. most of the elements in the same category are grouped together.

   b. most of the elements are organized according to your chosen quantitative property.

   Hint: If you are having trouble doing this, try organizing the elements into different categories or using a different quantitative property.

4. Sketch your arrangement of the elements. (Use the property you chose in Step 2 to identify each element.)

There are many different ways to sort these buttons into categories.

Analysis

?

Group Analysis

1. Explain the reasoning behind your choice of categories in Procedure Step 1. Do you think there were other categories that would have given you a similar table?

2. What property did you choose in Procedure Step 2? Why did you choose it? Do you think another property would have worked as well or better?

3. Is there anything in your table that does not seem to fit the patterns you observe for most of the elements?

4. Describe what you think is the most useful feature of your table.

5. Compare your table to the standard periodic table. What similarities do you notice? What differences are there?

Individual Analysis

6. Invent a new element that does not appear on the table you made. Where will it be located in your table? Predict one chemical property and one physical property of this element. Explain how you came up with each of your predictions.

7. Make a graph of atomic mass vs. atomic radius. Describe the relationship shown by your graph.

8. Compare the graph you made for Analysis Question 7 to the graph of atomic mass vs. specific heat that you made in Activity 14.4, "Specific Heat and Atomic Structure." How are they similar? How are they different?

Extension

Select a group of common objects, such as the buttons shown on the previous page, that can be classified into groups based on at least two characteristics, and construct a periodic table that organizes these objects. Other possibilities include shoes, automobiles, and playing cards.

15.4   Building Blocks of Chemical Change

Purpose ▶ **O**bserve patterns in bonding between reactants and products.

Introduction

You have just explored how atoms combine to form molecules. When atoms are combined into molecules, or when atoms in existing molecules are rearranged to form different molecules, a chemical reaction has taken place. All chemical reactions begin with one or more elements, molecules, or non-molecular substances, which are called the **reactants**. The reaction proceeds when, under the right conditions, the reactants break apart and the atoms recombine to make one or more different compounds, called the **products**. Chemical reactions occur during all life processes, including digestion and photosynthesis, and when energy is used or transferred, as in the operation of a car engine or the burning of wood.

Chemical equations are commonly used to describe chemical reactions. Each atom that participates in a chemical reaction is represented by its chemical symbol, and each molecule or non-molecular substance is represented by its chemical formula. The chemical symbol or formula for each reactant appears on the left side of the equation. The chemical symbol or formula for each product appears on the right side of the equation. The chemical equation for the reaction that describes a water molecule breaking apart to form hydrogen and oxygen gas is:

$$2H_2O \rightarrow 2H_2 + O_2$$

reactants   products

This equation shows that when two molecules of water break apart, two molecules of hydrogen and one molecule of oxygen are formed.

Materials

■■ **For each team of two students**

1   molecular model set

access to a Periodic Table

Procedure

1. Use the molecular model set to construct two water molecules. Using only the atoms contained in these two molecules, rearrange the pieces to make two hydrogen molecules and one oxygen molecule. Write the chemical equation for the reaction you just modeled.

   Remember: All the bonding sites (the protruding "sticks") of an atom must be connected to those of another atom.

2. Construct one molecule of nitrogen (N_2), and three molecules of hydrogen (H_2). Using only these atoms, construct as many molecules of ammonia (NH_3), as you can. Draw the structural formula for a molecule of nitrogen, hydrogen, and ammonia. Write the chemical equation for the reaction you just modeled.

3. Hydrogen peroxide readily breaks down into water and oxygen gas. Start with two molecules of hydrogen peroxide (H_2O_2), and rearrange the atoms and bonds to make molecules of water and oxygen. Draw structural formulas for all molecules involved and write the chemical equation for the reaction you just modeled.

4. When carbon dioxide (CO_2) dissolves in water, it makes an acidic solution because carbon dioxide combines with water to form carbonic acid (H_2CO_3). Use the molecular modeling set to model this reaction. Draw structural formulas for each molecule and write the chemical equation for the reaction you just modeled.

Analysis
?

Group Analysis

1. The hydrogen, oxygen, and nitrogen molecules you constructed are called diatomic elements. Why isn't carbon a diatomic element?
   Hint: Use your molecular model set to try to construct C_2.

2. Why do you think atoms naturally join together to form molecules rather than remain as individual atoms? Explain your reasoning.

Individual Analysis

3. Use the Periodic Table to predict the chemical formulas for each of the following molecules:
   a. a compound of oxygen and silicon
   b. a compound of phosphorus and hydrogen
   c. a compound of hydrogen and sulfur

4. Draw the predicted structural formula for each compound in Analysis Question 3.

5. For each chemical reaction you have been asked to model, the products were assembled from the same pieces used to make the reactants. Do you think that all chemical reactions are like that, or just those discussed here? Explain.

Photosynthesis

16

16.1 Do Plants Pass Gas?

Purpose   **C**onduct an experiment to determine the rate of gas production and consumption in plants.

Introduction   **Photosynthesis** is the chemical process that produces sugar molecules from carbon dioxide gas (CO_2) and sunlight. It takes place in cells of green plants and other producer organisms. **Respiration** is the chemical process all living cells use to release energy from sugar molecules. Cellular respiration that involves oxygen is called aerobic respiration. During aerobic respiration, sugar produced by the digestion of food reacts with oxygen to produce water, CO_2, and energy.

Simplified chemical equations for respiration and photosynthesis are shown in Figure 1. As you can see, these two equations are very similar—both involve energy, oxygen, CO_2, water, and sugar. In fact, these two reactions are opposites of each other.

Do plants respire? How does light affect respiration and photosynthesis? This investigation will help you explore these questions. In this activity, you will observe the production and consumption of CO_2 that occurs during respiration and photosynthesis. This observation is reasonably easy to accomplish because when CO_2 dissolves in water it produces an acid that changes the pH of the solution. You made models of the production of this acid, H_2CO_3, in Activity 15.4, "Building Blocks of Chemical Change."

Figure 1   Chemical Equations for Respiration and Photosynthesis

Aerobic Respiration

$$C_6H_{12}O_6 + 6O_2 \rightarrow 6H_2O + 6CO_2 + energy$$

One molecule of glucose (a type of sugar) reacts with six molecules of oxygen to form six molecules of water, six molecules of carbon dioxide, and energy.

Photosynthesis

$$6H_2O + 6CO_2 + energy \rightarrow C_6H_{12}O_6 + 6O_2$$

Six molecules of water react with six molecules of carbon dioxide and energy from sunlight to form one molecule of glucose and six molecules of oxygen.

Do all these plants respire and photosynthesize?

Photosynthesis

Materials

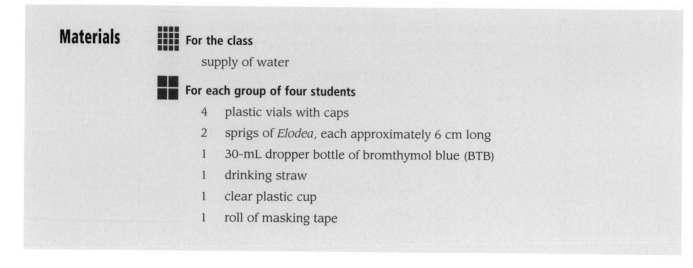

For the class

supply of water

For each group of four students

4	plastic vials with caps
2	sprigs of *Elodea*, each approximately 6 cm long
1	30-mL dropper bottle of bromthymol blue (BTB)
1	drinking straw
1	clear plastic cup
1	roll of masking tape

Procedure

1. Fill one plastic cup approximately half-way with water. Add 15 drops of bromthymol blue (BTB) to the water and swirl the cup until it is thoroughly mixed.

2. Have one person from your group exhale slowly through a straw into one of the cups until the solution turns green. Be careful not to blow so much air through the straw that the solution turns yellow. Do not drink the solution!

3. Label your vials as Vials 1–4 and add one sprig of *Elodea* to Vials 1 and 4.

4. Fill all four vials with the green solution.

5. Cap all four vials tightly and tape down the lids so that no gas can escape. Create a data table and record the colors of the solutions in the tubes.

6. Place Vials 1 and 2 in the sun or under fluorescent lights. Place Vials 3 and 4 in a dark location that is designated by your teacher.

7. Reflect: What was the purpose of preparing the green solution (of BTB, water, and a small amount of CO_2)?

8. Predict: Do you think any of the vials will show a color change over the next few minutes or hours? Describe and explain your prediction(s).

9. Check the vials every 15 minutes until the class period ends. Record your observations in your data table.

10. Make a final observation of the vials when you next return to class.

Analysis

?

Group Analysis

1. What, if any, color changes did you observe? Describe them.

2. How would you interpret the change(s) in color you observed? That is, what gas was present in each tube at the start and at the end of the experiment?

3. What was the purpose of Vials 2 and 3?

4. Did the color changes you observed occur at the same rate? What explanation can you provide as to why the changes would occur at the same or different rates?

Individual Analysis

5. What evidence do you have from this activity that plants photosynthesize? Explain.

6. What evidence do you have from this activity that plants respire? Explain.

7. Does light affect respiration and/or photosynthesis? Use evidence from this experiment to support your answer.

Highlights From the History of Botany

Purpose ▶ **E**valuate historical experimental evidence that led to the modern understanding of photosynthesis.

Introduction ▼

Botany is the study of plants.

Modern scientific knowledge is based on a long history of reproducible experiments and scientific conclusions. Many scientists have studied the question of how plants grow. Over the centuries, the combined results of their investigations have developed into the current scientific description of the chemical process of photosynthesis.

Read the following descriptions of the work of four scientists. While reading each passage, take notes. Before proceeding to the next scientist, discuss the Group Analysis Questions with your group members.

Aristotle (384–322 B.C.)

Aristotle, a naturalist and teacher who lived in ancient Greece, taught his students that plants feed on the soil. He told them that plant roots absorb materials from the soil in a manner similar to the way people suck juice from a slice of orange. He also said that plants use these absorbed soil materials for growth, just as animals use the food they eat.

Analysis

Group Analysis

1. How accurate do you think Aristotle's ideas were? Which of his ideas, if any, do you think are correct and which, if any, do you think are incorrect? Explain.

2. Develop a theory that explains the growth of plants by addressing questions like those listed below. Provide evidence, either from the reading or from your own background knowledge, to support your theory.
   - How do plants grow?
   - What do plants need to grow?
   - What do plants produce as they grow?

Johann Baptista van Helmont (1579–1644)

J. B. van Helmont, a Belgian physician, carried out scientific experiments on plants during the early 1640s. In one experiment, he took a large pot and placed in it exactly 200 lbs. of soil that had been dried in a furnace and carefully weighed. To this he added a 5-lb. branch from a willow tree, which soon took root and started to grow. For five years, van Helmont protected the tree from dust and added nothing but pure rainwater to the soil. At the end of that time, he separated the tree from the soil and dried the soil. The tree now weighed 164 lbs., and the soil weighed 199 lbs. 2 oz.

Analysis

Group Analysis

3. Describe a hypothesis van Helmont could have been testing with his experiment.

4. What conclusion(s) can you draw from the results of van Helmont's experiment? Support your conclusion(s) with experimental evidence.

5. Review your answer to Analysis Question 2. Would you modify or change your theory based on van Helmont's experiment? Explain.

Jan Ingenhousz (1730–1799)

In 1779, Jan Ingenhousz, another Belgian physician, published *Experiments Upon Vegetables*, a summary of his work in plant research. He hypothesized that plants in the presence of heat or sunlight change the air. In his experiments, Ingenhousz placed the leaves of a plant in a closed container completely filled with water. The gas produced by the plant collected inside the container, displacing the water and rising to the top. Ingenhousz thought to test this gas by seeing if a candle would burn in it. If the candle burned, he called the gas "good" air; if not, it was "bad" or "foul" air. He wrote:

> *A jar full of walnut tree leaves was placed under the shade of other plants, and near a wall, so that no rays of the sun could reach it. It stood there the whole day, so that the water in the jar had received there about the same degree of warmth as the surrounding air; the air obtained was worse than common air, whereas the air obtained from other jars kept in the sun-shine during such a little time that the water had by no means received a degree of warmth approaching that of the atmosphere, was [good] air.*
>
> *I placed some [other] leaves in a jar and kept it near the fire to receive a moderate warmth. [I placed] a similar jar, filled with leaves of the same plant, in the open air in the sun [so that it too received a moderate warmth]. The result was, that the air obtained by the fire was very bad, and that obtained in the sun was [good] air.*

Ingenhousz struggled with the question of where the "good air" comes from, and came to the following conclusion:

> *The air obtained from the leaves, as soon as they are put in the water, is by no means air from the water, but air continuing to be produced by a special operation carried on in a living leaf exposed to the daylight, and forming bubbles, because the surrounding water prevents this air from being diffused through the atmosphere.*

Analysis ▼ ?

Group Analysis

6. Why is the air described in the second paragraph "worse than common air?"

7. What can you conclude from Ingenhousz's experiments? Explain.

8. Do you think that Ingenhousz's experiment was well designed? Explain why or why not. If not, describe how you would set up a similar, but better designed, experiment.

9. Review your answers to Analysis Questions 2 and 5. Would you modify or change your theory based on Ingenhousz's experiments? Explain.

Nicolas Theodore de Saussure (1767–1845)

In 1804, Nicolas Theodore de Saussure found that if he put plants with thin leaves in an atmosphere of only oxygen and nitrogen, no oxygen was produced. However, if he added carbon dioxide, the plants produced oxygen. After more careful study, he found a direct relationship between the amount of carbon dioxide consumed and the amount of oxygen produced. In other words, twice as much added carbon dioxide led to the production of twice as much oxygen. He concluded that the process of photosynthesis requires both water and carbon dioxide. He also concluded that the carbon found in sugar, starch, and other plant molecules comes from carbon dioxide.

Analysis
?

Group Analysis

10. According to Saussure, what is the "good" air described by Ingenhousz?

11. Based on Saussure's experiments, what gas is required by plants for photosynthesis?

12. Review your answers to Analysis Questions 2, 5, and 9. Would you modify or change your theory based on Saussure's experiments?

16.3   Where's It Happening?

Purpose ▶ **E**xamine the organelles responsible for respiration and photosynthesis in cells.

Introduction

In the last two activities you investigated respiration and photosynthesis. The cells of all eukaryotic organisms, including green plants, have a cell membrane, a nucleus, and other organelles, including mitochondria and, in some cases, chloroplasts. **Mitochondria** are the organelles in which respiration takes place. **Chloroplasts** are the organelles in which photosynthesis occurs. They are present only in cells that carry out photosynthesis. Chloroplasts contain a photosynthetic pigment, usually **chlorophyll**, that absorbs the light needed for photosynthesis. It is chlorophyll that gives green plants their characteristic green color. Figure 2 is a diagram of a leaf cell from a corn plant, with the chloroplasts and a few other organelles labeled. The nucleus is on the right side of the cell. Note that, like all cells, plant cells can come in many shapes. In this activity you will observe some of these structures under a microscope.

Figure 2   Cell From the Leaf of a Corn Plant

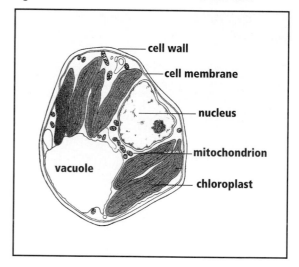

cell wall
cell membrane
nucleus
mitochondrion
vacuole
chloroplast

Materials

■■ **For each team of two students**

1   microscope slide
1   coverslip
1   30-mL dropper bottle of water
1   microscope
several leaves of *Elodea*

Procedure

1. Place several leaves of *Elodea* on the center surface of a microscope slide. Make sure the leaves lie flat.

2. Place 3–4 drops of water directly onto the leaves.

3. Carefully touch one edge of the coverslip to the water on the slide. Holding the coverslip at a 45° angle, slowly lower it into place over the leaves.

4. Center the slide on the microscope stage so that the leaves are directly over the light aperture. Adjust the microscope settings as necessary.

5. Begin by observing the leaves on the lowest objective. Center the tip of a leaf in the microscope field.

6. Increase the magnification to 40x and draw what you observe. Label your drawing.

Analysis

?

Group Analysis

1. Does a typical *Elodea* leaf cell contain a single chloroplast or more than one? Why do you suppose that this type of cell has this number of chloroplasts?

2. Predict the relative number of chloroplasts that you think would be found in a typical cell from the parts of a plant listed below. Explain your reasoning.

    a. the stem

    b. the root

    c. the flower

Plant Genetics and the Green Revolution

17

17.1 Modeling Inheritance

Purpose ▶ **M**odel how genes are passed from one generation to another.

Introduction

Why do many children resemble one or both of their parents?

Genetics is the study of the way physical traits are passed from parents to offspring. The scientists who developed the higher-yielding grains that have contributed to the Green Revolution could not have done so without a good understanding of genetics.

Prediction

?

Write two competing theories describing how genes are passed from one generation to another.

Scenario

An eight-year-old boy has blue eyes. His mother has blue eyes and his father has brown eyes. You have been asked to explain to the boy why he has blue eyes.

Materials ■■ **For each team of two students**

At least:

4   colored objects, 3 of which are the same color

3   paper tasting cups

Procedure

1. Using only the materials provided (but not necessarily all of them) and the guidelines listed below, develop a model to explain how the physical trait of eye color is passed from parents to child.

   • Establish what the different objects, including the cups, will represent.

   • Determine the exact number of each object required for your model.

   • Make sure your explanation is clear enough to be understood by an eight-year-old.

2. Gather your materials and set up your model.

3. Explain your model to another pair of students. Be sure to explain exactly how the physical trait of eye color is passed from the parents to the child.

Analysis

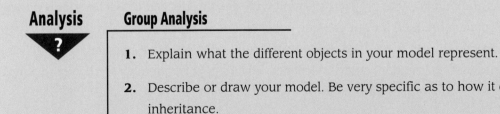

Group Analysis

1. Explain what the different objects in your model represent.

2. Describe or draw your model. Be very specific as to how it explains the process of inheritance.

3. A young girl has type AB blood. Her father has type B blood and her mother has type A blood. Can your model explain how blood type is passed from parents to child? How? Be specific.

17.2 Genes and Traits

Purpose ▶ **B**egin the study of genetics.

Introduction

Humans have lived among plants and animals for thousands of years, using them for food, to help with work, for companionship or decoration, and many other purposes. It is easy to see that when two parent organisms—whether plant or animal—reproduce, their offspring will have some characteristics of one parent and some of the other parent. Through careful selection of the parents, humans have been able to breed varieties of plants and animals that produce offspring with specific, desired characteristics. One way to produce more food for our growing population is to breed plants and animals that are healthier and yield more food. Early human efforts at selective breeding were the first steps toward the modern study of genetics, which, in part, attempts to answer the question, "How do physical traits pass from parent to child?"

Many flowers grown today have been selectively bred so that they have specific characteristics such as color or height.

The Basics of Genetics

Gregor Mendel was a 19th-century monk who set out to explore this question scientifically. He chose to work with pea plants, which were relatively easy to use for his experiments. He studied how certain traits were passed from one generation of pea plants to the next. The traits he analyzed included seed color (yellow vs. green) and stem length (long vs. short). Over several years, he conducted carefully controlled experiments that involved selecting parent plants and keeping detailed records of the traits inherited by their offspring. Mendel found that there appeared to be units of information that were passed from parent to offspring. He also noticed that these units behaved as though each one was made up of two parts.

Since the time of Mendel, many scientific studies have provided huge amounts of evidence that support and expand upon his basic ideas about inheritance. Today, Mendel's "units" are called **genes**, and each gene is present in an individual in two copies, called **alleles**. We now know that each parent provides one allele for every gene of every offspring. In any organism, some traits are determined by the alleles of only one gene, while others are determined by many genes. Some traits are determined by a combination of genes and environmental conditions.

In the simplest cases, a gene has only two possible traits, one of which is dominant over the other. An example of this is a gene that determines whether corn kernels are smooth or wrinkled. The scientific evidence tells us that a corn plant having two smooth alleles will always have smooth kernels and a plant with two wrinkled alleles will always have wrinkled kernels. However, the evidence also shows that a plant with one smooth and one wrinkled allele will always have smooth kernels. This evidence leads to the conclusion that the smooth trait is **dominant** and the wrinkled trait is **recessive**. The dominant trait is defined as the trait that is seen in an organism with one of each type of allele. The observed feature, such as smooth or wrinkled corn kernels, is known as the organism's **phenotype**.

An organism has many genes, or allele pairs. Biologists use an uppercase letter to represent the dominant allele and a lowercase letter to represent the recessive allele. (To make it easier to distinguish between uppercase and lowercase letters, we underline the uppercase letters.) For the corn kernel alleles, we can use <u>S</u> for smooth and s for wrinkled. Any allele pair, such as <u>SS</u>, <u>S</u>s, or ss, is known as the organism's **genotype**. Genotypes that have two identical alleles, such as <u>SS</u> or ss, are called **homozygous**. The prefix *homo-* means "same." Genotypes with two different alleles, such as <u>S</u>s, are referred to as **heterozygous**. The prefix *hetero-* means "different." Organisms with homozygous recessive alleles will express the recessive phenotype (e.g., wrinkled kernels), organisms with homozygous dominant alleles will express the dominant phenotype (e.g., smooth kernels), and organisms with heterozygous alleles will also express the dominant phenotype (e.g., smooth kernels).

Analysis ▼?

1. Explain the difference between genotype and phenotype.

2. A plant with red flowers is bred with a plant with white flowers.

   a. If all offspring have red flowers, what does this tell you?

   b. If all offspring have pink flowers, what does this tell you?

3. Organisms that reproduce asexually have only one parent. How would you expect the genotype and phenotype of asexually produced offspring to compare to the genotype and phenotype of the parent?

17.3 Rearranging Rice Genes

Purpose ▶ **E**xplore the patterns of genetic inheritance.

Introduction

In Activity 17.1, "Modeling Inheritance," you created your own model to describe how one trait is passed from parents to offspring. In the simplest situations, a trait is determined by one gene. In organisms that reproduce sexually, each parent contributes one allele for each gene. Organisms that reproduce asexually have only one parent; both alleles for each gene come from that one parent.

One trait that is very important for the production of high-yield rice is plant height. This is determined by a single gene, called the "semi-dwarf" gene. Semi-dwarf rice plants are shorter and sturdier than tall rice plants, so they survive better and thus produce more rice. The semi-dwarf trait is recessive, and the tall trait is dominant; plant breeders use the symbols **sd1** and <u>**SD1**</u> for the alleles associated with the semi-dwarf and tall traits, respectively. (For this activity, an uppercase <u>T</u> will represent <u>**SD1**</u> and a lowercase t will represent **sd1**.) In this activity you will use a model to investigate what happens when a homozygous, tall rice plant (genotype <u>TT</u>) is crossed with a homozygous, semi-dwarf rice plant (genotype tt).

This picture shows twelve different types of rice, all with different genetic traits.

Materials

■■ **For each team of two students**

2   blue semi-dwarf rice allele cards (t)

2   blue tall rice allele cards (T̲)

Procedure

1. Working with your partner, place the four rice allele cards face up on the desk or table. One partner, who will represent a homozygous tall plant, should take the two cards with the tall symbol, T̲. The other partner, who will represent a homozygous semi-dwarf plant, should take the other two cards, each with the symbol t.

2. Each partner should close their eyes, shuffle their two cards, place one of the cards on the desk, and then open their eyes.

3. The two cards on the desk (one from you and one from your partner) represent the pair of alleles inherited by the offspring of a cross between the semi-dwarf plant and the tall plant. Record the genotype of the offspring.

4. Repeat Steps 1–3 nine more times.

5. Make a data table to record the genotype and the phenotype of each of your 10 offspring.

6. Report your data to your teacher. Data from all groups will be used to determine the total number of each genotype and phenotype in the first-generation offspring.

Analysis

?

Group Analysis

Use the class data to answer these questions.

1. What type of reproduction—sexual or asexual—was modeled in the Procedure? Explain.

2. List the genotype(s) of the offspring produced during this simulation.

3. List the phenotype(s) of the offspring produced during this simulation.

4. Explain why only these genotypes and phenotypes were possible.

17.4 Generation Next: Crossing the Offspring

Purpose ▶ **S**imulate a cross between heterozygous plants.

Introduction

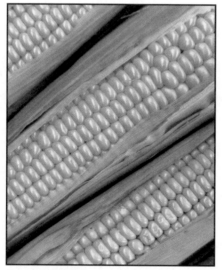

Do the kernels on these ears of corn exhibit the same phenotype?

In Part A of this activity you will investigate the genetics of corn plants. Each ear of corn has hundreds of kernels; each kernel is a really an embryo, an offspring of two adult plants. One gene determines the color of the corn kernels, which may be yellow or purple, depending on the alleles contributed by the parents. You will examine and compare pictures of offspring produced by two different sets of parents and use evidence from these pictures to extend your investigations on the inheritance of dominant and recessive traits.

In Part B, you will model the inheritance of the semi-dwarf gene in a rice plant. As you learned in the last activity, this gene determines the plant's height. The tall trait is dominant, and the tall allele is represented by T̲. The semi-dwarf trait is recessive, and the semi-dwarf allele is represented by t. What type of offspring are possible when you breed a heterozygous rice plant (T̲t) with another heterozygous rice plant (T̲t)?

Part A   Counting Corn

Materials

 For each group of four students

1   Corn Ear A
1   Corn Ear B

Procedure

1. Count the number of purple kernels and the number of yellow kernels on each of your two corn ears. Record your results in a data table.

Corn Ear A

Corn Ear B

Analysis

?

Group Analysis

1. When a corn plant with all purple kernels is crossed with a plant with all yellow kernels, the offspring are all purple kernels. For this species of corn, which kernel color is dominant? Explain your evidence.

2. Given this information, determine

   a. the genotype of each parent of Corn Ear A.

   b. the genotype of each parent of Corn Ear B.

   Explain how you reached your conclusions.

Part B   Rice Ratios

Materials   ■■ **For each team of two students**

   2   blue semi-dwarf rice allele cards (t)

   2   blue tall rice allele cards (T)

Procedure

1. Working with your partner, place the four rice allele cards face up on the desk or table. This time, each person will represent a heterozygous tall plant and should take one tall allele card (T) and one semi-dwarf allele card (t).

2. Each partner should close their eyes, shuffle their two cards, place one of the cards on the desk, and then open their eyes.

3. The two cards on the desk (one from you and one from your partner) represent the pair of alleles inherited by the offspring of a cross between two heterozygous tall plants. Record the genotype of the offspring.

4. Repeat Steps 1–3 nine more times.

5. Make a data table to record the genotype and the phenotype of each of your 10 offspring.

6. Report your data to your teacher. Data from all groups will be used to determine the total number of each genotype and phenotype in the first-generation offspring.

Analysis

Group Analysis

Use the class data to answer these questions.

1. For this rice breeding simulation,

   a. list all the possible genotypes in the offspring.

   b. determine what percent of the offspring possessed each genotype.

   c. list the possible phenotypes for the offspring.

   d. determine what percent of the offspring displayed each phenotype.

   e. calculate which ratio best expresses the ratio of the tall to the semi-dwarf phenotype: 1:1, 2:1, 3:1, or 4:1.

2. Do you see any pattern or relationship between the parent genotypes and the frequency of certain genotypes in the offspring? Explain.

Analysis
?

Individual Analysis

3. Imagine that you have two corn plants, and one of them (Parent 1) produces offspring with only brown corn kernels; the other (Parent 2) produces offspring with only white corn kernels. You breed the two plants and find that all of the first-generation offspring have brown kernels. You then breed two first-generation offspring and discover that the second-generation plants produce both brown and white kernels.

   a. Which trait is dominant—white or brown? How do you know?

   b. List the genotypes and phenotypes you would predict to be present in
      • Parent 1.
      • Parent 2.
      • the first-generation offspring.
      • the second-generation offspring.

   c. For a second-generation corn ear with 800 kernels, what is the approximate number of kernels likely to have each phenotype? Explain and show your work.

Breeding Improved Crops

18.1 Double Crossing Corn

Purpose ▶ **I**nvestigate genetic crosses involving two different traits in corn.

Introduction

In the last activity, you investigated how traits that are determined by a single gene are passed from one generation to another. The traits you modeled included height in rice plants and kernel color in corn. Breeding new crop varieties that will improve the food supply usually involves manipulating many genes to produce an organism that expresses the phenotype for many desirable traits.

In this activity you will examine the inheritance of two traits in corn: the color of the corn kernels (purple vs. yellow) and the taste of the kernels (sweet vs. starchy). Sweet kernels are wrinkled in appearance; starchy kernels are smooth. Purple is dominant over yellow, and starchy is dominant over sweet. The genotype of a homozygous purple and starchy variety can be written **PPSS**; the genotype of a homozygous yellow and sweet variety can be written **ppss**.

In many cases, the inheritance of multiple genes follows a pattern first noticed by Mendel. According to this pattern, known as the **Law of Independent Assortment**, genes for different

Introduction
(cont.)

Figure 1 Possible Allele Donations From Each Parent

Figure 2 Punnett Square for PPSS x ppss Cross

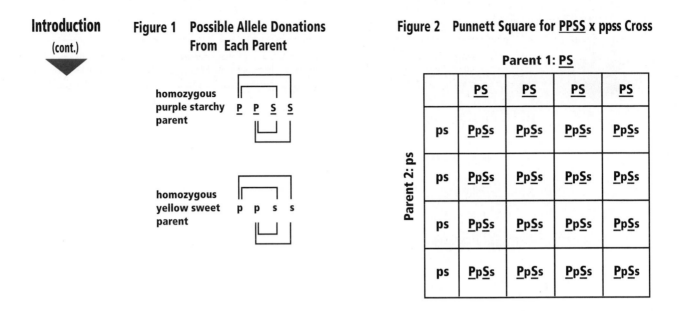

traits are inherited independently of each other. In other words, the inheritance of alleles for starchy or sweet kernels has no effect on which alleles for color will be inherited by the same offspring. Not all genes follow the Law of Independent Assortment.

We will refer to the offspring of a cross between two homozygous plants as the first generation. What type of corn will we get if we cross a homozygous purple and starchy plant (PPSS) with a homozygous yellow and sweet one (ppss)? All of the first-generation offspring will have the genotype PpSs because one parent can donate only P and S alleles and the other can donate only p and s alleles. The possible combinations of parent donations are shown in Figure 1. As you can see, each parent can donate four possible combinations, and each of the four possibilities are identical. The Punnett square in Figure 2 shows the genotypes of the first-generation offspring resulting from this cross.

Suppose we crossbreed two of the first-generation offspring from Figure 2. We will refer to their offspring as the second generation. What genes will the second generation have?

Materials

For each group of four students

1 Corn Ear C
1 Corn Ear D

Procedure

1. Prepare a data table with the following headings:

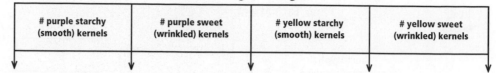

# purple starchy (smooth) kernels	# purple sweet (wrinkled) kernels	# yellow starchy (smooth) kernels	# yellow sweet (wrinkled) kernels

2. Corn Ear C represents the second generation, which is the result of a cross between two first-generation PpSs plants. Count, then record in your data table, the number of each of the four kernel types found on Corn Ear C.

3. Complete a Punnett square showing the genetic possibilities for the offspring of a cross between two PpSs parents.

   Hint: Each PpSs parent can potentially donate one of the following combinations: PS, Ps, pS, ps.

4. Make a table that shows the relative numbers of each genotype and phenotype predicted by the Punnett square.

Analysis **?**

Group Analysis

1. According to your Punnett square, how many different genotypes should be present in the kernels of second-generation Corn Ear C?

2. According to your Punnett square, how many different phenotypes should be observed in the kernels of Corn Ear C?

3. Use your Punnett square to predict the ratio of the different phenotypes produced by this cross. Express your ratio in this form:

   # purple starchy: # purple sweet: # yellow starchy: # yellow sweet

Individual Analysis

4. How do your results from Procedure Step 2 relate to your results from Analysis Question 3? Explain.

5. Imagine you have several corn plants that have produced only purple starchy kernels. You want to determine whether these plants will always "breed true"—meaning that they will always produce offspring with only purple starchy kernels. Design an experiment to find out which ones will breed true. In addition to providing a procedure to follow, describe the data you will collect and how you will analyze it.

Extension Count the kernels on Corn Ear D, then create a data table and Punnett square(s) you can use to determine the possible genotypes for the parents of Corn Ear D.

18.2 Breeding Rice

Purpose ▶ **A**pply what you have learned about genetics to the breeding of a desirable strain of rice.

Introduction

Different strains of rice plants grow better and produce larger yields in different parts of the world. For example, rice plants that grow well in the cool, dry highlands of Bhutan perform poorly in semi-tropical areas prone to flooding, such as parts of southeast Asia. Different strains of rice also have different tastes and textures. The best rice to grow in any particular region is a rice that is resistant to com-mon local diseases, grows rapidly to give the highest yield in the shortest time, is suited to the local climate, and has a taste and texture enjoyed by the residents of the region.

These grains come from many different strains of rice plants.

In this activity, you will investigate the breeding of a hybrid rice from parent rice plants that have different desired characteristics.

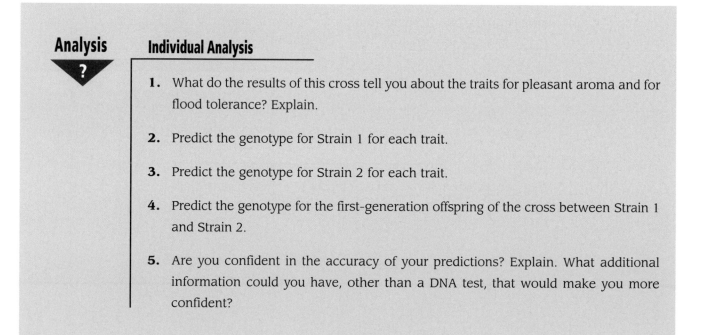

Scenario

Imagine you have discovered two different strains of rice. Strain 1 has an pleasant aroma and flavor (aromatic) but is easily damaged by flooding (flood intolerant). Strain 2 can be submerged under water for up to two weeks, then recover and grow once the flooding is over (flood tolerant), but has an unpleasant aroma and flavor (non-aromatic). You would like to develop a strain of plants that has the interesting aroma and flavor of Strain 1 and the flood tolerance of Strain 2.

Part A   First Generation

Procedure

1. You breed plants from Strain 1 with plants from Strain 2 and find that 100% of the offspring are flood tolerant, but none of them have a pleasant aroma and flavor. Assume that

   - <u>A</u> stands for the allele for the dominant, non-aromatic trait.
   - a stands for the allele for the recessive, aromatic trait.
   - <u>F</u> stands for the allele for the dominant, flood-tolerant trait.
   - f stands for the allele for the recessive, flood-intolerant trait.

Analysis ?

Individual Analysis

1. What do the results of this cross tell you about the traits for pleasant aroma and for flood tolerance? Explain.

2. Predict the genotype for Strain 1 for each trait.

3. Predict the genotype for Strain 2 for each trait.

4. Predict the genotype for the first-generation offspring of the cross between Strain 1 and Strain 2.

5. Are you confident in the accuracy of your predictions? Explain. What additional information could you have, other than a DNA test, that would make you more confident?

Part B   Second Generation

Materials   ■■ **For each team of two students**

  2   green flood-tolerant allele cards (<u>F</u>)
  2   green flood-intolerant allele cards (f)
  2   yellow non-aromatic allele cards (<u>A</u>)
  2   yellow aromatic allele cards (a)

Procedure

Now that you have determined that the genotype of all the first-generation offspring is A<u>a</u>F<u>f</u>, you will use cards to represent the different alleles for the flood-tolerant and aromatic genes to simulate a cross that produces second-generation offspring.

1. You and your partner should each take one green card with an uppercase **<u>F</u>**, one green card with a lowercase **f**, one yellow card with an uppercase **<u>A</u>** and one yellow card with a lowercase **a**. Both you and your partner now have a set of four alleles for a first-generation plant.

2. To simulate a cross between two first-generation plants, place your four cards face down on the table and mix them up while your partner does the same with his or her cards. Turn one of your partner's green cards and one of your partner's yellow cards face up while your partner does the same with your cards.

3. The four cards that are face up represent the genotype of a second-generation plant. Record the genotype.

4. Repeat Steps 1–3 until you have genotypes for 16 second-generation plants. Report your 16 genotypes to your teacher.

5. Prepare a table to record the following data on the second generation:
   • number of each genotype your team produced
   • total number of each genotype produced by the class
   • phenotype expressed by each genotype

Analysis

Group Analysis

Use the class data to answer these questions.

1. What is the ratio of the different phenotypes expressed by the second-generation offspring?

2. What fraction of the offspring have the desired phenotype?

3. What fraction of the offspring have the desired phenotype and will breed true for this phenotype?

4. Imagine that you will be doing crosses for three different genes. How will this affect your breeding effort?
   Hint: What will it do to the percentage of plants that will breed true for the desired phenotype? How will it affect the number of crosses you will have to do to get exactly what you want?

Individual Analysis

5. Prepare a Punnett square to show all of the possible second-generation genotypes that could have resulted from crossbreeding two <u>Aa</u><u>Ff</u> parent plants.

6. Compare the ratio of genotypes produced by your team with the ratio produced by the class and with the ratio predicted by the Punnett square. Describe and explain any similarities and differences.

7. Imagine that you will be crossing two plants in an attempt to obtain offspring with 10 desirable traits. What effect will this have on your breeding efforts?

18.3 Breeding Crops With Desirable Traits

Purpose ▶ **U**nderstand some of the methods used to breed desirable crops.

Introduction Many crops have been bred selectively for different traits, such as the apples shown below. In the last activity, you simulated the breeding of a new variety of aromatic, flood-tolerant rice. You learned that it would take several generations to develop strains of rice that breed true for these two traits. Plant breeders often attempt to breed for many different traits at once. Some of these traits depend on several genes, rather than just one, which complicates the process. Using conventional breeding techniques to develop a new strain of a crop like corn, rice, or tomatoes usually takes 8–12 years.

Apples come in many colors. The color of an apple gives consumers clues to other traits such as sweetness and texture.

Breeding for Blight Resistance

Gains achieved through selective-breeding techniques do not come without trade-offs. Although these techniques may produce crop strains that yield larger harvests, taste better, or contain more nutrients, the new varieties may also be less resistant to disease. In addition, too much emphasis on a few desirable varieties of a particular crop can lower biodiversity by limiting the number of varieties that are grown. A reduction in biodiversity could seriously limit our ability to breed new varieties, because many genes for desirable traits come from wild species or from older cultivated strains.

Consider a recent case that involved improving the disease-resistance of high-yield rice plants. There are a number of diseases that affect rice. One of these is blight, which causes infected leaves to turn yellow and wilt. Blight can cause a severe reduction in the rice harvest.

A blight-resistant strain of rice grows wild in Mali. It contains genes that help prevent infection by the bacteria that cause rice blight. Unfortunately, the Mali strain of rice is not a good food crop because it doesn't taste good and yields relatively few rice grains. A group of scientists at the International Rice Research Institute (IRRI) in the Philippines set out to attempt to transfer blight resistance from the Mali strain of rice to a different strain that has a higher yield, and a good flavor, but is not blight resistant. It took them twelve years to develop just one blight-resistant variety. And although this resistant variety might grow well in the Philippines, it will not necessarily be successful in other rice-growing climates.

The Selective-Breeding Process

To produce a blight-resistant, high-yield plant with good-tasting grains, pollen from a blight-resistant rice plant is used to pollinate the flowers of a rice plant that produces high yields of good-tasting grains. In this case, the blight-resistant strain has just one desirable trait. The other parent plant, which is already the result of extensive breeding, has many desirable traits. When the blight-resistant wild rice is crossed with the high-yield, good-tasting rice, the offspring inherit many traits, some of which may not be desirable. So it is necessary to breed the first generation of blight-resistant offspring with the original high-yield strain. Breeders then select offspring from this second-generation cross that show high yield and good taste as well as blight resistance. These second-generation offspring are again crossed with the original high-yield strain. This process is repeated many times. Each time, offspring that show blight resistance are selected and crossed again with the high-yield parent, so that eventually a variety is produced that has lost most of the wild parent's genes, except for blight resistance.

Twelve years may sound like a long time for breeding just one variety of a plant, but the breeding process that developed today's large-eared, sweet-kerneled corn has an even longer history. The ancestor of our familiar corn-on-the-cob was a species of grass, probably similar to a grass that grows in Mexico today. Corn breeding was first carried out over a period of hundreds of years by Native American peoples. They selectively planted the varieties of this grass that were most desirable as food—plants that produced large and numerous seeds, had a more pleasant taste, and so on. Over the past century, more modern breeding techniques have been used to make further improvements to cultivated corn. The corn we purchase in the market today would do poorly in the wild. One reason is that its seeds, or kernels, stay on the cob when ripe. This trait is desirable to humans who want to harvest and store the corn for food, but it is not an advantage to corn species living in the wild, where seed dispersal helps the plants to spread.

Evolution via natural selection produces adaptations that enable a plant to grow and reproduce. Although evolution has produced literally millions of plant species with traits that enable them to thrive in the wild, these traits are not necessarily ideal for human agriculture. Most of today's crop varieties are the product of extensive breeding efforts, many of which artificially select plants with characteristics that aren't particularly advantageous to wild species, including high yield, longer shelf life, and better taste, color, and texture.

Using Technology to Produce Crops With Desirable Traits

A variety of new practices for developing desirable crop strains have emerged over the past several decades. Known collectively as **biotechnology** and **genetic engineering**, these new techniques can produce the same results as selective breeding, but in a much shorter time and with much more predictable results. Genetic engineers can develop detailed knowledge about the genetic makeup of a particular plant, then use that information to choose which traits will be passed on to the next generation. Using biotechnology, desirable genes from one variety or species can be inserted directly into another, greatly accelerating any natural or selective breeding process. These techniques have even made it possible to take genes from an entirely different organism, such as a bacteria, and insert them into a plant! For example, genes from bacteria have been inserted into strawberries to produce a fruit that resists frost damage. Some crop varieties have been given genes that resist the effect of herbicides. This trait allows the farmer to spray herbicides on a field to get rid of weeds without harming the crop. An organism that has been modified through the insertion of genes from a different species is called a **transgenic** organism.

Some scientists, in this country and globally, are concerned about transgenic plants for three reasons. First, genes for tolerance to herbicide or disease can be transferred to wild plants, causing an increase in hardiness of weed species and potentially making it more difficult to control weed growth. Second, when a plant carries genes for resistance to a virus, some of the genes for that virus are included in the plant's DNA. These viral genes help the plant protect itself from infection by causing it to produce antibodies to the virus. However, the viral genes can also be transferred to other viruses that may infect the plant, potentially creating new, more powerful viruses. Finally, new proteins produced by a transgenic plant can change the way the plant uses energy and disposes of waste products. These changes can cause the plants to build up compounds not normally present in that plant species. These compounds might turn out to be toxic to animals, including humans, that eat the plant.

Scientific evidence suggests that these kinds of issues should be considered by scientists who create transgenic plants and by policy makers who are involved in decisions about their use. In addition to scientific evidence, social and political concerns must be taken into account when determining the appropriate role of genetic engineering in food production. However, more evidence must be collected to accurately assess the known and perceived risks involved in creating transgenic organisms. Have the potential effects described here been adequately studied? Can they cause significant problems or will they be limited to rare occurrences?

Analysis
?

Group Analysis

1. What is rice blight and what causes it?

2. Why are some rice strains resistant to blight?

3. Why did it take so long to breed a blight-resistant, high-yield rice plant?

Individual Analysis

4. Is it important to maintain wild varieties of plants like rice? Explain.

5. Would you recommend the use of biotechnology to reduce the time required to produce a blight-resistant plant? Explain.

18.4 Cattle Calls

Purpose ▶ **M**ake choices concerning breeding and genetically engineering cattle for use in a specific country.

Introduction

Remains of domesticated cattle dating back to 6500 B.C. have been found in Turkey. According to some researchers, cattle were domesticated as early as 8000 B.C. There is evidence of even earlier dates for the domestication of sheep, goats, pigs, and dogs. Modern domestic cattle probably evolved from ancestral animals called aurochs. These animals are the subject of many prehistoric paintings. Aurochs survived until relatively modern times; it is believed that the last surviving member of the species was killed by a poacher in 1627 in Poland. Historically, people have used domesticated cattle to provide meat and milk and to perform labor as draft animals. In more modern times, cattle have been replaced as a labor source, first by horses and then by machinery. Most of today's cattle have been selectively bred for either dairy or beef production, or both.

As you have learned, selective breeding to obtain desired characteristics requires many years of successive crossbreeding of offspring. Today's technology allows genetic engineers to identify portions of the DNA strand that code for certain desired traits. Once these DNA segments have been identified, they can be removed from the cells of one breed and inserted into the DNA of another, thus eliminating the need for years of selective crossbreeding.

Each cattle breed has unique traits that make it suited to particular environments and uses.

Materials  **For each group of four students**
1   set of 12 Cattle Cards

For each team of two students
1   copy of *Material World*

Procedure

1. Cattle can serve three main purposes—meat, milk, and labor. Use the information provided in *Material World* to determine which use(s) for cattle would be most desirable to meet the needs of people living in your chosen country.

2. If you could genetically engineer a new breed of cattle for your country, which traits from which existing breeds would you want it to have?

3. Use the information provided on the Cattle Cards to choose two breeds of cattle that, when mated, could produce offspring with traits appropriate for your country. Record the name of each breed and the desirable traits each one has to offer.

4. Ask your teacher for a "Breeding Results" card to find out how many generations of crossbreeding it will take for you to achieve offspring with all your desired traits.

5. Repeat Step 3. This time, choose a different combination of parent breeds.

6. Repeat Step 4 to find the results of crossbreeding the cattle breeds you chose in Step 5.

Analysis

Group Analysis

1. Why did you choose the traits you selected in Procedure Step 2? Describe any trade-offs you had to consider.

2. Why did you choose the two breeds you selected in Procedure Step 3?

3. Why did you choose the two breeds you selected in Procedure Step 5?

4. Which set of parent cattle—those from Procedure Step 3 or those from Procedure Step 5—do you think was the better choice for crossbreeding? Explain.

5. Would you prefer to use crossbreeding or genetic engineering to develop a new breed of cattle with desirable traits? Describe any trade-offs you would have to consider.

Genetically Engineering Food

19

19.1　Genes, Chromosomes, and DNA

Purpose **L**earn about the physical structure and location of genes within cells.

Introduction

You have been learning about genetics and its role in developing new and improved agricultural products. So far, you have explored the use of selective breeding to control inheritance and to produce crops and domestic animals with desirable traits. Since the early 1950s, when researchers discovered that genes are made up of **deoxyribonucleic acid (DNA)** molecules, our understanding of the genetic code has grown tremendously. This knowledge has increased our ability to alter the characteristics of agricultural products because it has made it easier for us to control which genes are passed from one generation to the next. As a result, we are able to produce more varieties with more desirable traits at a faster pace than is possible with conventional selective-breeding techniques.

In this activity, you will read about the molecular nature of the genetic code and learn how modern technology can be used to alter these molecules. You will also explore the potential benefits and trade-offs of using DNA-altering technology.

DNA: Carrier of the Genetic Code

You may recall that eukaryotic organisms are made up of cells containing nuclei. The cell's genetic material, or DNA, forms chromosomes that are located inside the nuclei. DNA molecules stain differently than the molecules of the cytoplasm; it was stained DNA that allowed you to observe nuclei in the cells you examined in Activity 13.1, "In and Out Nutrients." Under normal conditions, individual strands of DNA cannot be seen with a light microscope. However, they do become visible during the process of cell division.

Chromosomes are long strands of DNA that contain information for directing all functions of the cell. A **gene** is a specific segment of a chromosome containing instructions for the synthesis of a single protein. Because cells make and use hundreds or even thousands of proteins, most of which are enzymes crucial to cell processes, each chromosome contains many genes. Different organisms have different numbers of genes and chromosomes. All the genes of a species, taken together, comprise that species' **genome**.

As Figure 1 shows, human cells contain 23 pairs of chromosomes, making a total of 46. Chromosomes in a pair contain genes for the same traits. Scientists are currently collaborating in an effort called the Human Genome Project to develop a map of the 46 human chromosomes that will indicate where each gene is located.

Normally, each individual inherits 23 single chromosomes from each parent: one chromosome in each pair is contributed by the mother; the other is contributed by the father. The fact that chromosomes are paired explains why a gene is made up of two alleles (one on each chromosome in a pair). It also explains how offspring receive one allele from each parent.

The Behavior of Chromosomes During Cell Division

According to cell theory, which you studied in Activity 13.2, "Inside the Membrane," all new cells come from pre-existing cells. How are new cells formed? After a period of growth, every cell goes through the process of cell division. Normal cell division, which occurs constantly within your body, produces two identical cells containing a full complement of chromosomes. This is called **mitosis**. When a cell prepares to divide by mitosis, it first makes a copy of the DNA that forms its chromosomes.

Figure 1   Chromosome Pairs of a Human Male

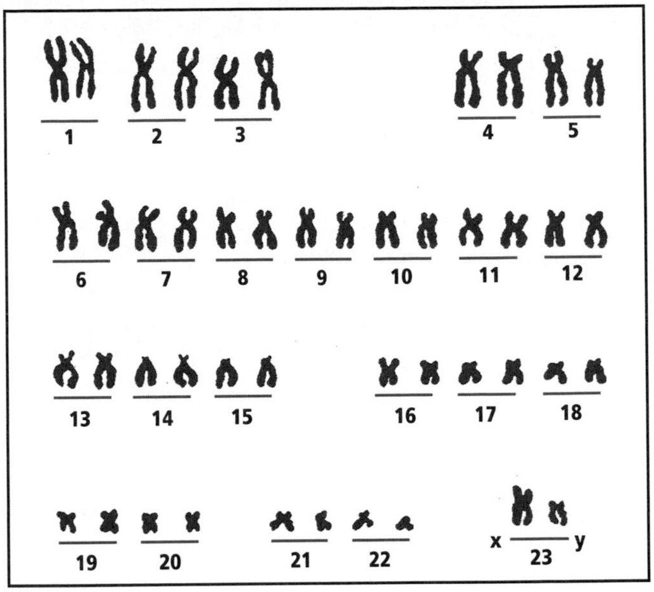

Most of the time, a cell's chromosomes are extremely thin threadlike strands of DNA that cannot be observed through a light microscope. As cell division begins, these thin strands condense into tight coils that become visible. The condensed, visible forms of all 23 pairs of human chromosomes are shown in Figure 1. Before the cell divides, an exact duplicate of each chromosome is made by a process known as **replication**. During this process, the nuclear membrane begins to break down.

Once the chromosomes have been replicated, they line up along the center of the cell. Duplicate chromosomes line up side by side. When the cell divides, two new cells, called **daughter cells**, are formed, with each daughter cell receiving one copy of each chromosome. Figure 2, on the next page, illustrates mitosis in an organism with two pairs of chromosomes. As soon as the new cells are formed, the nuclear membrane reforms and the chromosomes once again become thin and threadlike.

During sexual reproduction, a new organism is created when a male sex cell (sperm) fertilizes a female sex cell (egg). Both sex cells contribute genes to the new organism. If the sperm cell and the egg cell each contained the organism's full complement of chromosomes, the offspring would have twice as many chromosomes as the parents. This does not happen! Instead, egg and sperm cells are formed through a different process of cell division, called **meiosis**. Meiosis results in the formation of four sex cells (sperm or egg), each having one complete set of single chromosomes. For example, while a typical human body cell contains 23 pairs, or 46 total chromosomes, a typical human sex cell contains only 23 chromosomes. Thus, when a sperm cell fertilizes an egg, a cell with a full set of 46 chromosomes is usually pro-

Figure 2   Mitosis in an Organism With Two Pairs of Chromosomes

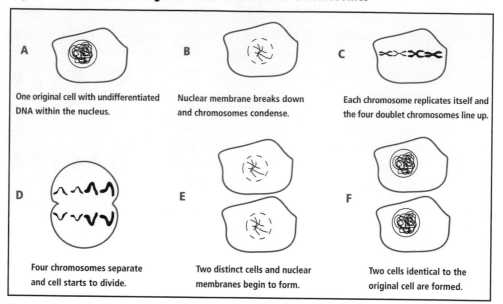

A One original cell with undifferentiated DNA within the nucleus.

B Nuclear membrane breaks down and chromosomes condense.

C Each chromosome replicates itself and the four doublet chromosomes line up.

D Four chromosomes separate and cell starts to divide.

E Two distinct cells and nuclear membranes begin to form.

F Two cells identical to the original cell are formed.

duced. This newly fertilized cell then divides by mitosis over and over again to form the embryo and eventually the adult organism. Every time mitosis takes place, the genetic material of the cell is duplicated. This duplication is not always perfect—occasionally it results in errors, called **mutations**, that are inherited by successive generations of cells. If a mutation occurs in the DNA of an egg or sperm cell, that mutation can be inherited by all cells of the offspring.

In Activities 17 and 18 you modeled the selective breeding of traits, such as purple/yellow and sweet/starchy corn kernels, that show independent assortment. Independent assortment means that a plant with purple starchy kernels and a plant with yellow sweet kernels can produce offspring in which the purple trait appears with the sweet trait and the yellow trait appears with the starchy trait. The frequency at which these new combinations of traits occur is what you would expect if there were no connection between the two traits. The Law of Independent Assortment applies to unlinked traits, which are controlled by genes located on separate chromosomes. Traits that do not follow this law are controlled by genes located on the same chromosome and are called linked traits. Linked traits produce offspring with phenotype ratios that are significantly different from the ratios characteristic of unlinked traits.

Analysis ▼?

Group Analysis

1. Explain how the terms chromosome, gene, and DNA are related.

2. Where is the genetic material of a eukaryotic organism located? Is this the same for prokaryotic organisms?

3. At what point in the process of cell division might mutations occur? Explain.

4. How does knowing where a gene is located help explain how traits are passed from parents to offspring?

5. Why might there be concern about using biotechnology to replace or remove a gene that is linked with others on the same chromosome?

19.2 Modeling DNA Structure

Purpose ▶ **M**odel the subunits that make up a DNA molecule and model the double-helix structure of DNA.

Introduction ▼

You have learned that it can take many years to breed desirable traits into crops or livestock via selective breeding. The same goal can be accomplished much more quickly by directly changing an organism's genetic information. However, doing so requires some knowledge about the chemical and physical nature of that genetic information.

Experiments conducted in the 1940s and 1950s provided evidence that DNA molecules are incredibly long. In fact, if all the DNA from just one of your cells were stretched out, it would be two meters long! Nonetheless, all this DNA easily fits into every cell because it is extremely thin and threadlike.

DNA is built of three kinds of smaller molecules: 1) a five-carbon sugar called deoxyribose, 2) a phosphate group made up of one phosphorous atom and four oxygen atoms, and 3) a nitrogen base. Each **nitrogen base** is a single- or double-ring structure that contains nitrogen, carbon, hydrogen, and oxygen. Three of these smaller molecules—one deoxyribose molecule, one phosphate group, and one nitrogen base—join together to form a simple **nucleotide**. Thousands of nucleotides link together in a chain to form a DNA molecule. An organism's genetic makeup is determined by the sequence of nucleotides in its DNA.

During the 1950s, determining the structure of DNA became a subject of intense scientific interest and rivalry. Several research groups were competing to be the first to find the answer. These groups included the laboratory of Linus Pauling (who had earlier discovered the structure of proteins) at the California Institute of Technology; scientists at Cambridge University in England; and another group of scientists at King's College in London.

▶

Introduction
(cont.)

The Cambridge University group, led by James Watson and Francis Crick, won the race. In 1953, after careful analyses of structural data—much of which was provided by Maurice Wilkins and Rosalind Franklin from King's College—Watson and Crick proposed that the DNA molecule is shaped like a **double helix.** The double-helix model for DNA is now accepted by scientists worldwide. This structure not only accounts for the chemical makeup of DNA, but also correctly predicts how DNA is replicated when a cell divides. Watson, Crick, and Wilkins were awarded a Nobel Prize in 1962 for their breakthrough. The story of these scientists and their work is the subject of a number of fascinating books and articles.

In Part A of this activity, you will use molecular model sets to illustrate the structure of each of the molecular subunits of DNA. In Part B, you will use a different modeling set to represent the double-helix structure of a DNA strand. The structures of deoxyribose, phosphate, and the four nitrogen bases are shown in Figures 3 and 4.

Figure 3   DNA Backbone Components

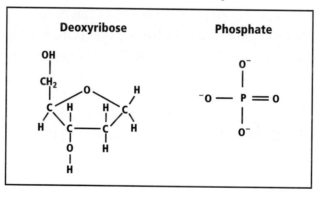

Figure 4   Nitrogen Bases in DNA

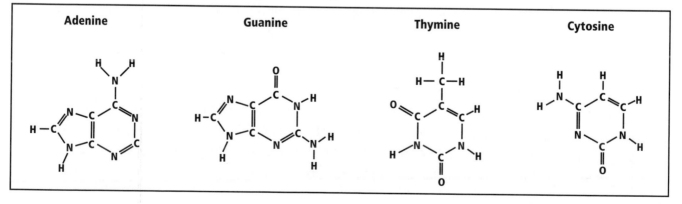

Part A Modeling a Single Nucleotide

Materials ■■ **For each team of two students**

1   molecular model set

Procedure

1. One team of students in your group should make the deoxyribose sugar; the other team should make the nitrogen base thymine. See the diagrams in Figures 3 and 4 for the structures of these molecules.

2. Attach the sugar and the base together to form a partial subunit of DNA. Figure 5 shows how deoxyribose bonds to thymine. A phosphate would complete this structure to form a nucleotide.

Figure 5   Deoxyribose Sugar and Base

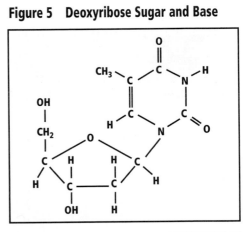

Note: Phosphate is difficult to model with your molecular model set, because the phosphorus in the phosphate group forms five bonds in this structure. This is an exception to the pattern of bonding rules you learned about in Activity 15.1, "Modeling Molecules."

Part B Modeling the DNA Double Helix

DNA molecules store the codes that direct all life processes. The genes on a strand of DNA code for proteins; these proteins carry out specific tasks or are used in the construction of the various parts of an organism. Understanding the chemical structure of DNA allows us to understand how proteins are produced. We can use that information to modify the proteins produced by an organism's DNA, which in turn modifies the characteristics of the organism itself.

The sugar and phosphate groups of a DNA molecule form two long chains, like the sides of a ladder. In each chain, the sugar and phosphate alternate. The nitrogen bases are arranged in pairs to form the steps, or rungs, of the ladder. Each base extends inward, so that two bases meet at the center of each step of the ladder. As you can see in Figure 6, the two bases, often called a base pair, are joined together by a hydrogen bond.

One way to build a model of a DNA molecule is to represent each molecular subunit—sugar, base, and phosphate—as a single piece, rather than as a group of connected atoms.

Figure 6 "Ladder" Arrangement of DNA

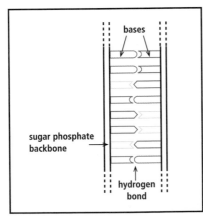

Materials	**For each team of two students**
	1 DNA model set

Procedure

1. Review your DNA model packet to identify the following pieces:

   - a black five-sided piece that represents a deoxyribose sugar

   - a white tube that represents a phosphate group

   - colored tubes that represent nitrogen bases

orange tube	adenine (A)
green tube	thymine (T)
yellow tube	guanine (G)
blue tube	cytosine (C)

   - a small white rod that represents the hydrogen bonds between each pair of nitrogen bases

Figure 7   Modeling a Single Nucleotide

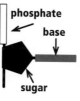

Figure 8   Joining Two DNA Nucleotides

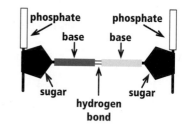

2. Put together one sugar, one phosphate, and one green base to make a complete thymine nucleotide subunit of DNA, as shown in Figure 7.

3. Now, use an orange base to build a complete adenine nucleotide.

4. Next, use a white hydrogen bond to attach your nucleotides to each other, as shown in Figure 8.

Procedure
(cont.)

5. Study the diagram of segments of DNA in Figure 9. Decide which of the segments you will make and which one your partner will make.

6. Put together the four base pairs of your chosen segment. Note that the guanine (yellow tube) must always link to cytosine (blue tube) and thymine (green tube) must always link to adenine (orange tube).

7. While only certain pairs of nitrogen-based nucleotides can link to each other, these pairs can be arranged in any order. Get together with your partner and attach your two segments to form a ladder-like structure with eight base-pair rungs.

8. Twist your ladder-like structure into a helix (like a spiral staircase). You now have a model of a DNA molecule with the basic characteristics that Watson and Crick proposed.

Figure 9 Two Segments of a DNA Strand

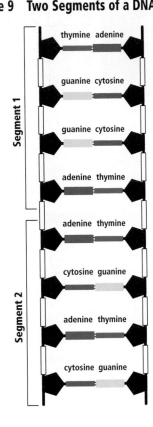

9. Starting at one end, gently pull apart the two sides of your DNA ladder. Describe what happens.

Analysis

Group Analysis

1. Every time a cell divides, an exact copy of its DNA must be made, with all the bases in the gene in the correct order. Propose a mechanism that would explain how the DNA might be copied and how the structure of the DNA helps to prevent errors.

Individual Analysis

2. You used two kinds of models in this activity. What differences did you see between the molecular models you used to make the deoxyribose sugar and the thymine base, and the DNA models you used to construct the double helix?

3. What were the advantages and disadvantages of each of the models you used?

Extension

For more information about the discovery of the structure of DNA, read James Watson's account of the process in his book *The Double Helix*.

19.3 Fight the Blight

Purpose ▶ **I**nvestigate the use of recombinant DNA technology to produce new varieties of rice.

Introduction

In Activity 18.3, "Breeding Crops With Desirable Traits," you were introduced to the idea of using genetic engineering to choose which genes a plant will carry. In this reading, you will learn more about how that process is accomplished. Figure 11 on page 240 gives a historical perspective on the development of food technologies and shows how selective breeding and genetic engineering fit in with other practices.

The selective breeding of blight-resistant rice took over ten years and produced only one new blight-resistant variety. Breeding blight resistance into other varieties of rice would require separate, long-term breeding efforts for each variety. With genetic engineering, it is possible to reduce the time needed to transfer a specific gene, such as blight resistance, from one variety of plant to another. It is also possible to transfer genes from one species to another. Researchers have even succeeded in transferring animal genes into plants: a gene from a firefly has been used to create tobacco plants that glow!

One key step in the process of genetic engineering is to make copies of the DNA strand that codes for the desired trait. A schematic diagram of the DNA replication process is shown in Figure 10.

Figure 10   DNA Strand Unzipping and Replicating

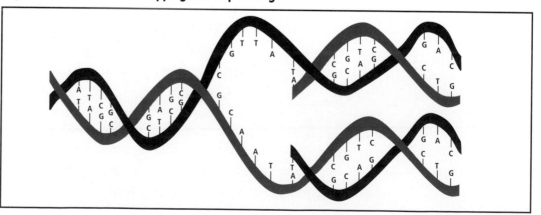

Genetic Resistance to Disease

The blight-resistance gene was the first gene to be used to create a disease-resistant variety of rice using genetic-engineering techniques. How does genetic engineering differ from traditional selective breeding? In genetic engineering, the gene of interest is isolated from the cells of one organism, copied, and transferred directly into the cells of another organism.

According to Pamela Ronald, who, along with her colleagues, successfully copied the blight-resistance gene, the obstacles involved in making disease-resistant rice fell into two basic categories: first, scientists had to find the genes; then they had to figure out how to move them around.

Finding the gene for blight resistance took several years and required complex biological and biochemical techniques. The search depended on a map of rice genes that had been developed by another group of researchers. This kind of map is created by finding out which genes are linked—in other words, which genes are located on the same chromosome. Genetic crosses like the ones you investigated in Activity 18 can be used to show which genes exhibit ratios typical of independent assortment and which do not.

For example, when organisms with different traits for genes A and B are crossed, a 9:3:3:1 ratio of phenotypes in the second-generation offspring demonstrates that these genes show independent assortment—they are not linked on the same chromosome. However, if offspring do not show as many new phenotypes as expected, the genes must be linked; they must be located on the same chromosome.

Once the desired gene is located, biochemical techniques can be used to cut the chromosomes into small pieces. The piece containing the blight-resistance gene is then isolated from the rest of the chromosome; now the gene can be cloned. **Cloning** is a general term that refers to making identical copies of a cell or an organism. Gene cloning refers to making identical copies of a gene. Gene cloning is performed by artificially inserting a gene from another organism into bacteria, and then allowing the bacteria to make many copies of the gene. The bacterial DNA that contains a piece of DNA from another organism (in this case, rice) is called **recombinant DNA**, because the DNA of one organism is combined with DNA from a different one. Cloning the gene for disease resistance was less of a barrier than finding the gene, because the techniques for cloning, first developed in 1973, are fairly straightforward. Still, cloning the blight-resistance gene took about a year.

The next major obstacle that had to be overcome to produce recombinant, blight-resistant rice plants was to insert the copied gene into the cells of rice plants. It is much more difficult to get DNA into eukaryotic cells than it is to get it into bacteria. A gun-like device was used to shoot microscopic pieces of gold coated with rice genes into rice cells. These cells were used to grow 1,500 rice plants that were then exposed to the bacteria that cause blight. Fifty of the 1,500 plants were resistant to blight and could pass the blight resistance to their offspring.

It took about six years to find and clone the blight-resistance gene and then insert it into rice cells to obtain disease-resistant plants. But now that the gene has been cloned, it can be inserted into other varieties of rice in a process that takes just a few months. Another advantage of genetic-engineering techniques is that the gene can be inserted into rice species that cannot breed with the blight-resistant rice. It can even be inserted into other species of crops that suffer from blight.

Bacteria have been used to clone genes from many species. Even genes from humans can be cloned in this way. The genes responsible for the production of human insulin have been cloned to produce this hormone for diabetics who cannot produce their own. Cloning from such organisms raises ethical questions, such as whether people know enough about its effects on the gene pool to choose which genes, and which organisms, should survive. For this reason, decisions about the role genetic-engineering technologies should play in producing food for Earth's human population involve not only the scientific principles you have read about here, but also ecological, ethical, environmental, and social considerations.

Figure 11 A Brief History of Food Technology

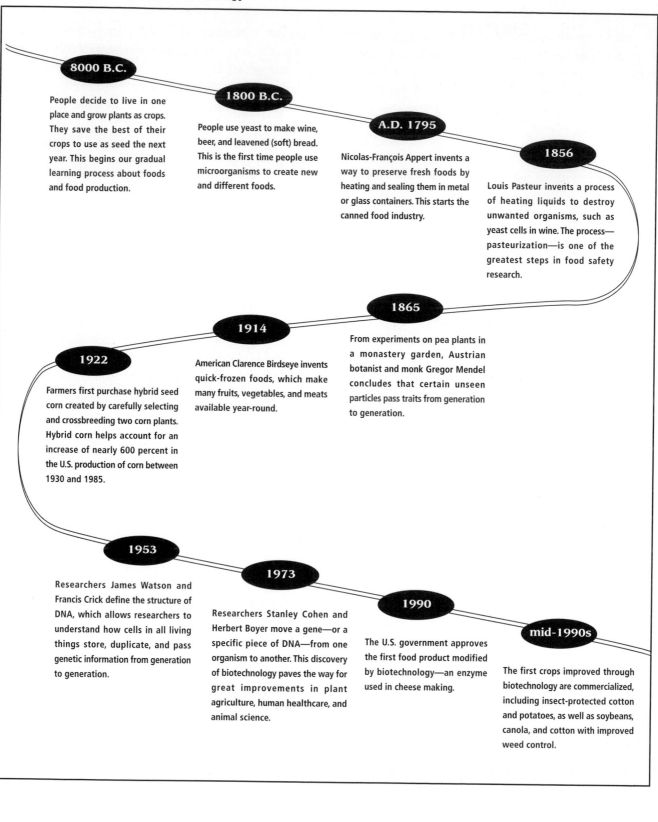

Analysis

?

Group Analysis

1. Describe how, in the effort to produce blight-resistant rice, researchers overcame each technical barrier to genetic engineering that you identified in your class discussion.

2. Developing a process to find, isolate, and clone the blight-resistance gene took a long time—considerably longer than the time it took to put the process to practical use. Explain how the process of science makes it possible to quickly use procedures that took years to develop.

3. Describe the trade-offs of genetic engineering vs. conventional selective-breeding techniques.

4. Think back to earlier activities related to feeding the world's human population. In what ways can science and technology help provide an adequate food supply?

5. What factors other than science and technology have an impact on our ability to adequately feed the world?

6. Choose one scientific advance described in this activity and explain how the scientists involved depended upon the research of other scientists.

The Role of Cloning in Food Production

20.1   The Clone Zone

Purpose ▶ **C**onsider evidence concerning whether genetic engineering and cloning should or should not be researched in an effort to enhance food development and production efforts worldwide.

Introduction

Currently, the world produces enough food for everyone. Continuing to do so in the future means that food production must increase as long as the population continues to grow. In addition, there is an increasing worldwide demand for high-quality, fresh food during every season of the year. Agricultural engineers are developing methods of using genetic engineering and cloning technologies to help meet the demand for more food and for high-quality fresh food.

A **clone** is an exact genetic replica. A clone can be an exact copy of a single gene or an entire organism. Cloning occurs in nature whenever a cell divides or an organism reproduces asexually. During cell division and sexual reproduction, genes are copied when DNA is replicated. Asexual reproduction produces an entire cloned organism. For some organisms, that is the only means of reproduction; other organisms can reproduce both sexually and asexually. For example, plants often reproduce sexually, yet cloning in plant reproduction is long established and commonplace, both in nature and by humans. It can be as simple as taking a leaf or stem from a plant and allowing it to grow roots. The result is a new plant that is genetically identical to the original. In the past few decades, scientists have been developing technologies for cloning individual genes and entire organisms. Cloning of genetically complex animals, including mammals, has recently become possible.

Introduction
(cont.)

Methods of controlling reproduction, including cloning, can lead to increased food production. But they can also give rise to environmental concerns, such as loss of biodiversity, lowered soil fertility, and decreased resistance to new strains of disease. The possibility of cloning farm animals on a global scale could increase the risk for widespread disease: decreasing the number of genes in the gene pool might eliminate genes that could confer resistance to a new disease-causing organism that might emerge in the future.

In this activity you will use what you have learned about genetics and DNA, together with information collected during additional research, in preparation for a debate on the role of genetic engineering and cloning in our efforts to increase food production. When your research is complete, your teacher will schedule a debate. With your group, you will decide if the evidence you have gathered leads you to decide yes or no on the following question:

Should humans continue to research genetic engineering and cloning for the purpose of increasing food production?

Procedure

This activity has three parts, each of which will be conducted on a different date. You will start Part A today. Make sure you record the dates your teacher assigns for Parts B and C.

Part A   Researching the Topic

1. Complete the following lists in your group of four.

   a. List facts you already know about genetic engineering, cloning, and other topics that might inform your decision about whether humans should continue to research these technologies in an effort to increase food production.

   b. Widespread use of genetic engineering and cloning would require the assessment of many issues that lie beyond science and technology. Make a list of some of the environmental, economic, social, political, and/or ethical issues that you think are relevant in deciding whether humans should continue to research genetic engineering and cloning in an effort to increase food production.

   c. List any questions you have about genetic engineering and cloning, particularly those for which the answers might be helpful in making a decision about whether humans should continue to research these technologies in an effort to increase food production.

2. At the direction of your teacher, share your lists with the class.

Procedure
(cont.)

3. With your teacher's help, divide the questions that the class has listed into eight general topics. While the topics your class comes up with may vary, they may include some of the ones displayed by your teacher.

4. As directed by your teacher, meet with your group and select at least two questions that you will research in preparation for the debate.

5. Begin conducting your research.

> You may wish to find information on the Internet. Go to the *Science and Sustainability* page on the SEPUP website to find updated sources on cloning and genetic engineering.

6. Periodically report to your group on the progress of your research.

> **Note:** During the research stage of this activity, you may be asked by your teacher to share your progress with the rest of the class. These sharing sessions can help provide answers to some of the questions previously generated by the class or generate new lists of questions to guide further research.

Part B   Preparing for the Debate

7. With your group, decide if the evidence you have gathered leads you to answer yes or no to the question that will be debated: Should humans continue to research genetic engineering and cloning for the purpose of increasing food production?

8. During the debate, you will present to the entire class the evidence your group has gathered. Each student should participate in the group presentation in some way. Use your teacher's guidelines to organize your presentation.

Part C   Debating the Continuation of Research on Genetic Engineering and Cloning

9. Your teacher will give each group an opportunity to present its research findings and recommendations about the future of research on genetic engineering and cloning.

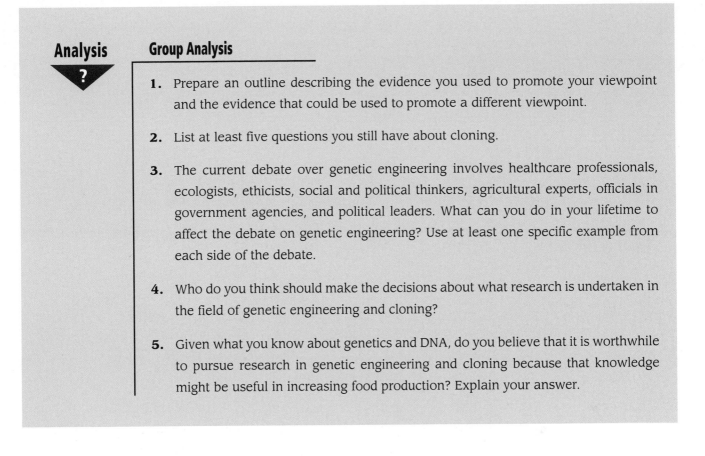

Analysis

?

Group Analysis

1. Prepare an outline describing the evidence you used to promote your viewpoint and the evidence that could be used to promote a different viewpoint.

2. List at least five questions you still have about cloning.

3. The current debate over genetic engineering involves healthcare professionals, ecologists, ethicists, social and political thinkers, agricultural experts, officials in government agencies, and political leaders. What can you do in your lifetime to affect the debate on genetic engineering? Use at least one specific example from each side of the debate.

4. Who do you think should make the decisions about what research is undertaken in the field of genetic engineering and cloning?

5. Given what you know about genetics and DNA, do you believe that it is worthwhile to pursue research in genetic engineering and cloning because that knowledge might be useful in increasing food production? Explain your answer.

Using Earth's Resources

Manufactured products, like living organisms, have a life cycle. They are made of the same elements, atoms, and molecules as plants or animals; energy must be used to make them; and they may also consume energy when they are functioning. Like a living organism, once a product ceases to be functional, it is laid to rest. Many living organisms are laid to rest in their natural environment. Their component parts are recycled through one or more of Earth's many natural cycles. Although some material goods are recycled, many more are placed in dumps, landfills, and other disposal sites. As a result, their component parts are no longer available for use.

The goal of a new field of science, called industrial ecology, is to reduce the amount of materials and energy that are "wasted" because they are disposed of in a fashion that makes them unavailable for reuse. There is no longer a clearly defined boundary between wastes and resources. Whether a material is considered useful or not depends on knowledge, technology, customs, and values. A major objective of industrial ecology is to develop and promote the use of processes to produce goods that are equivalent, or better than, those we use now, require fewer resources, create less waste, and are easy to recycle. A second objective is to encourage companies to determine how the "wastes" they produce could be used as resources for another process or product.

This type of thinking, in part spurred on by increasingly stringent environmental regulations, has led to the development of "industrial ecoparks." One of these parks is located in Kalundberg, Denmark. In this community, a network of companies has been exchanging by-products for over 25 years. For instance, waste gas from an oil refinery is sold to other factories for use as a heating fuel. Under ordinary circumstances, the waste gas would be burned off and its energy would not serve any useful purpose. Another example is a power plant that produces the pollutant sulfur dioxide (SO_2). In this plant, the sulfur dioxide is trapped before it can enter the atmosphere and converted to gypsum ($CaSO_4$). The gypsum is sold to a drywall factory. Taken together, Kalundberg industries generate much less waste and consume many fewer resources than similar groups of factories that do not swap waste products.

If industrial ecoparks are to become commonplace, public policies that embrace and subsidize waste disposal practices, such as landfills, need to be limited. In their place, policies that provide incentives for industries and individuals to engage in resource conservation and waste reduction practices need to be enacted.

Part 3

Using Earth's Resources

In Part 2 of *Science and Sustainability*, "Feeding the World," you investigated a variety of scientific and societal issues surrounding food production. In addition to food, many of Earth's natural resources are essential to the survival of humans and other species. The activities in Part 3, "Using Earth's Resources," focus on scientific concepts that can enhance our understanding of these resources and help us use them more wisely. Today, many people, especially those living in more developed countries, use far more material goods than they need to survive. They also produce huge amounts of waste. In this part of *Science and Sustainability,* you will investigate techniques for measuring and analyzing how our resource use affects Earth and its inhabitants. You will also consider alternatives to present-day patterns of resource use and explore a variety of questions:

- How has our growing understanding of the elements and of chemical reactions affected the development of technology?

- How do we obtain the materials we need?

- What are the properties that make these materials suitable for our use?

- How do our efforts to obtain and utilize resources affect the environment?

- What types of wastes are produced during the mining and processing of natural resources, such as copper or petroleum, and how do we dispose of them?

- What lessons about resource use can we learn from the past, and what precautionary measures should we take to ensure the safety of new products and processes?

- How can we best use our scientific knowledge to maintain or improve the quality of life for people living around the world today, and for future generations?

Identifying and Separating Hydrocarbons

21.1 Differentiating Liquids

Purpose ▶ **D**iscover how to use physical and chemical properties to differentiate liquids.

Introduction

Part 3 of this course focuses on how humans use natural resources. The study of liquids and their properties is important to society because liquids, especially water, are critical to the existence of life. In addition, many of the fuels that humans rely upon to enhance their survival are liquids.

In Part 2, you identified properties that can be used to characterize the elements on the Periodic Table. Most of the elements you investigated are solids at room temperature. Liquids, like solids, have distinct properties that can be used to identify them. Some of these properties, such as color, odor, and density, also apply to solids. Properties specific to liquids include **viscosity** and the ability to dissolve other substances.

Water, the most important liquid on Earth, has many distinctive properties.

Materials

  For each group of four students

1	30-mL dropper bottle of water
1	60-mL dropper bottle of methanol
1	60-mL dropper bottle of glycerin (glycerol)
1	10-mL graduated cylinder

■■ **For each team of two students**

1	SEPUP tray
1	dropper
1	stir stick
1	salt packet
1	sugar packet
1	small piece of effervescent tablet
	access to a balance

Procedure

1. Prepare a data table similar to the one below.

Table 1   Differentiating Liquids

Substance	Viscosity (low to high)	Density (g/mL)	Reaction With Effervescent Tablet	Dissolves Sugar?	Dissolves Salt?
Water					
Methanol					
Glycerin					

2. Work with your partner, but share your materials with the other team in your group as you perform the following tests. Record all results in your data table.

a. **Viscosity**

Viscosity describes a liquid's ability to flow. Honey, which flows relatively slowly, is an example of a liquid that has a high viscosity. You can examine each liquid's viscosity as you determine its density in Step 2b. Characterize each liquid's viscosity as low, medium, or high and record this in your data table.

b. **Density**

Density is calculated using this formula: $density = \dfrac{mass}{volume}$ or $D = \dfrac{m}{v}$

To determine the density of each liquid, you must first detemine the mass and volume of a sample of each liquid by following the procedure below,

(1) Find the mass of an empty graduated cylinder.

(2) Add 10 mL of liquid to the graduated cylinder.

Procedure
(cont.)

(3) Find the combined mass of the graduated cylinder and the 10 mL of liquid.

(4) Use the following equation (g.c. stands for graduated cylinder):

mass of liquid = mass of g.c. and liquid – mass of empty g.c.

                       (Step 3)   (Step 1)

(5) Carefully pour the liquid back into its original container.

(6) Thoroughly clean and dry the graudated cylinder before finding the density of the next liquid.

> **Hint:** Find the density of the glycerin last.

c. Reaction with effervescent tablets

In your SEPUP tray, add 20 drops of water to Cup 1, 20 drops of methanol to Cup 2, and 20 drops of glycerin to Cup 3. Add a small piece of effervescent tablet to each sample. Observe any bubbling that may occur and record your results under "Reaction With Effervescent Tablet" in your data table.

d. Solubility of sugar

In your SEPUP tray, add 20 drops of water to Cup 4, 20 drops of methanol to Cup 5, and 20 drops of glycerin to Cup 6. Using the scooped end of your stir stick, add one scoop of sugar to each cup. Stir each sample vigorously for one minute and record how well the sugar dissolved.

e. Solubility of salt

In your SEPUP tray, add 20 drops of water to Cup 7, 20 drops of methanol to Cup 8, and 20 drops of glycerin to Cup 9. Using the scooped end of your stir stick, add one scoop of salt to each cup. Stir each sample vigorously for one minute and record how well the salt dissolved. Clean up your equipment.

Analysis
?

Individual Analysis

1. Could you confidently identify a sample of an unknown liquid using only one of the properties you measured or observed during this activity? Provide evidence to support your answer.

21.2 Origins and Uses of Petroleum

Purpose ▶ **D**iscover how petroleum is formed and how its uses have evolved.

Introduction Some of the most common fuels we use are liquids derived from petroleum. The word **petroleum**, which comes from Greek words that mean "rock oil," refers to liquid hydrocarbons formed beneath Earth's surface. A **hydrocarbon** is an organic substance made only of hydrogen and carbon atoms. Unlike many other organic compounds, hydrocarbons do not contain oxygen. Petroleum is a mixture of many different hydrocarbon compounds and a variety of impurities.

Some of the many different petroleum products are stored in these refinery tanks.

Texas Tea

The Origins of Petroleum

Where do the hydrogen and carbon atoms that make up petroleum come from? How does petroleum form? Where is petroleum found and how did it get there?

To find the answers to these questions, we must go back millions of years in time. During certain periods in Earth's history, geological and environmental conditions created oceans teeming with billions upon billions of microscopic plants and animals called plankton. Plankton, like all other living organisms, are made of molecules formed mostly of carbon, hydrogen, and oxygen atoms. Many of the ancient plankton became food for other animals, but many died and sank to the ocean floor. Over thousands of years, thick layers of dead plankton gradually covered huge areas of the ocean floor. As conditions changed, these layers of dead plankton were buried beneath thousands of meters of mud, sand, and rock. The weight of the overlying mud and sand created high temperatures and pressures in the layers of dead plankton below. These conditions caused the chemical rearrangement of the oxygen, hydrogen, and carbon atoms of the dead plankton, releasing the oxygen and forming the hydrocarbon mixtures we now call petroleum.

Petroleum in its natural form—as it is collected from the ground—is often called **crude oil**. Its chemical constituents and physical properties can vary widely depending on where it is found. For example, it can flow like water or be as viscous as peanut butter; crude oil from some parts of the world has a much higher density and viscosity than crude oil from other parts. Crude oil can be yellow, red, green, brown, or black. In addition to hydrogen and carbon, crude oil also contains small amounts of other elements. Although its composition can be quite varied, "average" crude oil contains 84% carbon, 14% hydrogen, 1%–2% sulfur, and less than 1% each of nitrogen, oxygen, metals, and salts.

Early Uses of Petroleum

When geological conditions are right, small amounts of crude oil can naturally ooze to Earth's surface. As a result, humans have known about and used crude oil throughout history. The Egyptians coated mummies and sealed pyramids with it. The Babylonians, Assyrians, and Persians used it to pave streets and hold walls and buildings together. Boats along the Euphrates River were constructed with woven reeds and sealed with crude oil to make them watertight. The Chinese used it for heating. Native Americans used it for paint, fuel, and medicine. Desert nomads used it to treat camels for mange, and the Emperor of the Holy Roman Empire, Charles V, used it to treat his gout. The belief that crude oil had medicinal value remained popular through the 19th century, when jars of petroleum "snake oil" were advertised as miracle tonics "able to cure whatever ails you."

Although raw, unprocessed crude oil was used for a variety of purposes in the past, it is valuable today because it can be turned into hundreds of useful products, including gasoline, motor oil, waxes, dyes, plastics, and even fibers that can be woven into cloth. The process used to convert crude oil into these other products is called **refining**. An early type of refinery, called a "tea kettle" still, consisted of

Oceans are the primary source of the plankton from which petroleum is derived. How did arid places such as Texas and Saudi Arabia come to contain huge deposits of petroleum?

a large iron kettle with a long metal tube coming from the top, very similar to stills used to make "moonshine" whiskey. The crude oil was poured in the kettle and heated over a coal fire. These early refineries were only used to produce **kerosene,** a flammable liquid similar to gasoline. The other components, or fractions, of crude oil were considered valueless and were usually dumped on the ground.

Most of the kerosene made in tea kettle stills was burned in lanterns to produce light. Before kerosene became available, whale oil was the most widely used lamp oil. During the 19th century, whales were overhunted and became scarce. The resulting shortage of whale oil caused most families in the U.S. to start using kerosene in their lamps. In the rush to make as much kerosene as possible, some unscrupulous refiners failed to distill all the impurities from the kerosene. The impure kerosene caused many lamp explosions and fires.

Petroleum Use Today

Petroleum has become so valuable that teams of geologists and other experts continue to search the world for oil deposits located beneath Earth's surface. Approximately 90% of the crude oil pumped from these deposits is converted into some type of fuel. Fuels made from petroleum are considered fossil fuels, because petroleum, like other fossil fuels, is formed from the ancient remains of plants and animals. Of the remaining 10%, about 5% of crude oil is used to produce plastics. The other 5% is used to make dyes, inks, household detergents, pharmaceuticals (medicines), and many other compounds suitable for a wide variety of applications.

When crude oil reaches a modern refinery, it is first put through a process called **fractional distillation**. This process separates the crude oil into its main components, or fractions. Each **fraction** is a complex mixture of chemical compounds that have a similar boiling point. Fractional distillation is carried out in large towers, where the crude oil is heated to about 350°C. The vapors released from the hot crude oil rise up the tower, cool, and are removed as they condense back to liquids. The liquids with high boiling points condense and are removed from the lower sections of the tower; those with lower boiling points reach the upper part of the tower. The lightest fraction, the gases, are collected at the very top. In this way, different

Figure 1 Diagram of a Fractional Distillation Tower

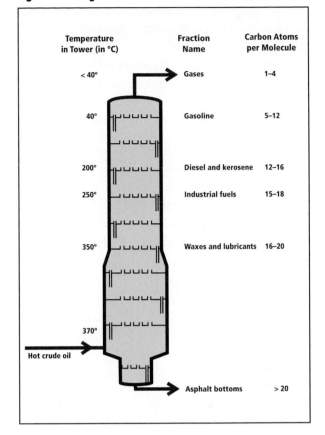

fractions are separated. Figure 1 is a diagram of a typical fractional distillation tower.

The complete refining process uses heat, pressure, chemicals, and catalysts to rearrange the structures and bonding patterns of each fraction to produce different hydrocarbon molecules and compounds. For example, the amount of gasoline produced by distillation alone is far less than that needed to meet the demand for fuel for automobiles. Additional gasoline is produced by a process called **cracking**, in which larger hydrocarbon molecules, generally from the kerosene fraction, are broken into smaller ones. Cracking is accomplished by heating the kerosene fraction, usually under high pressure or in the presence of catalysts. Catalysts are molecules that decrease the amount of energy necessary to start a chemical reaction. You will investigate catalysts in Activity 26, "Catalysts, Enzymes, and Reaction Rates." Catalytic cracking is used to produce hydrocarbon compounds that increase the octane rating and improve the anti-knock quality of gasolines.

Identifying and Separating Hydrocarbons

The simplest hydrocarbon molecule, methane, contains one carbon atom and four hydrogen atoms. Although methane gas makes up only a small fraction of petroleum, it is the main component of another fossil fuel, natural gas. In many countries, natural gas is piped into homes for heating and cooking. Hydrocarbons containing up to four carbon atoms are usually gases at temperatures and pressures normally encountered at Earth's surface.

Hydrocarbons with 5 to 19 carbon atoms are usually liquids at normal temperatures and pressures. Important liquid components of crude oil that are used as fuels include gasoline, diesel fuel, and kerosene, much of which is now used as jet fuel. Gasoline is often called "gas," even though it is a liquid. This confusing use of terms developed because the first internal combustion engines ran on "town gas," which is a mixture of carbon monoxide (CO) and hydrogen (H_2) gases. These engines were therefore called "gas engines." When gasoline replaced "town gas," people still called the motors "gas engines" and started calling gasoline "gas." Today, the average U.S. resident uses about 1,750 liters (450 gallons) of gasoline a year.

Hydrocarbons with 20 or more carbon atoms are solids, the most common of which is asphalt. The cracking process can be used to break down virtually all of the larger hydrocarbon molecules. Molecules formed by cracking are often used as chemical reactants for the synthesis of other molecules. These reactants are called **petrochemicals** or principal feedstocks, and are used to make a variety of valuable products, such as pharmaceuticals (medicines) and plastics.

The Future of Petroleum

Petroleum takes millions of years to form. The demand for plastic products and petroleum-based fuels is steadily increasing throughout the world. In energy terms, petroleum makes the single largest contribution to world energy supply, at 38%, followed by coal, at 26%, and natural gas at about 22%. Earth's supply of petroleum is finite and ultimately exhaustible: it is a non-renewable resource. In the future, as oil supplies diminish and production costs rise, petroleum use will decline as we rely more upon other energy sources and materials. How long our supplies of petroleum will last depends on how fast we use them and whether or not we continue to find more petroleum deposits that can be obtained economically. Current and future scientific, political, economic, and social circumstances will influence how long society continues to rely on petroleum products.

Analysis

?

1. How is petroleum formed?

2. Why is there concern about our current rate of petroleum use? (**Hint:** Consider sustainability.)

3. If you had to choose between using petroleum only to make products or only for fuel, which would you choose? Explain.

Extension

Research the history of human use of petroleum and create a timeline showing how it has changed over time.

21.3 Distillation of Simulated Crude Oil

Purpose ▶ **D**istill a sample of simulated crude oil into three fractions and identify each fraction by its physical and chemical properties.

Introduction

Oil refineries begin the process of converting crude oil to useful products. The refining process consumes energy and creates waste products.

In this activity, you will separate the components of a mixture of liquids by the process of fractional distillation. Each of these liquids has distinct physical properties. During fractional distillation, a liquid mixture is boiled. The gases produced during boiling are collected, then cooled, so that they condense back into liquid form. The various components of the original liquid mixture boil (change from a liquid to a gas) at different temperatures. This temperature difference makes it possible to separate the original components by collecting gases at different boiling points, then cooling and condensing them. The end result is a series of liquids, called fractions, each of which is a relatively pure sample of one of the substances that made up the original mixture.

Part A Fractional Distillation

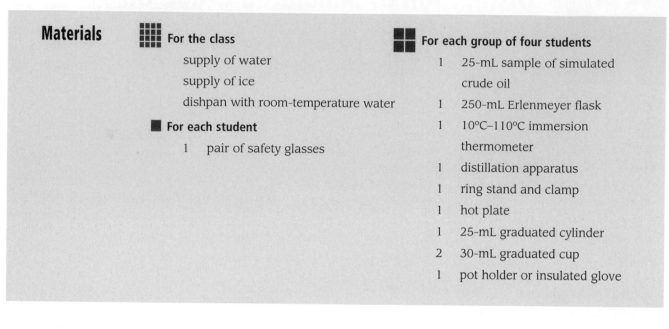

Materials

For the class

supply of water

supply of ice

dishpan with room-temperature water

For each student

1 pair of safety glasses

For each group of four students

1 25-mL sample of simulated crude oil

1 250-mL Erlenmeyer flask

1 10°C–110°C immersion thermometer

1 distillation apparatus

1 ring stand and clamp

1 hot plate

1 25-mL graduated cylinder

2 30-mL graduated cup

1 pot holder or insulated glove

Safety Note

Wear safety glasses! Keep your hands away from the hot plate and use insulated gloves or pot holders to move the flask as you finish. One of the liquid fractions is highly flammable. Do not perform this experiment near open flames! Know the safety procedures in case of a fire.

Procedure

1. Prepare a data table similar to the one on the next page.

2. Measure 25 mL of "crude oil" and pour it into your flask. Observe some of the physical properties of this substance and record your observations. Be sure to clean and dry the graduated cylinder after use.

Figure 2 Distillation Apparatus

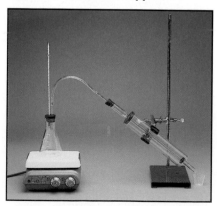

3. Set up your distillation apparatus as directed by your teacher. Use Figure 2 as an example. Be sure that the tubing hoses are properly connected and that the thermometer can be easily read. You will be recording observations of what is happening in the flask and the amount of liquid (distillate) you collect over time.

4. Fill the outer sleeve of your distillation apparatus with ice water, then turn on the hot plate to the medium setting.

Table 2   Products of Distillation

	Fraction Collected at 50–92°C	Fraction Collected at 92–102°C	Fraction Remaining in Flask
Observations During Collection			
Mass of 25-mL Graduated Cylinder (g)			
Viscosity (low to high)			
Volume of Liquid Fraction (mL)			
Mass of Cylinder with Liquid Fraction (g)			
Mass of Liquid Fraction (g)			
Density of Liquid Fraction (g/mL)			
Reaction With Effervescent Tablet			
Dissolves sugar?			
Dissolves salt?			

Procedure

(cont.)

Collecting the First Fraction

5. Use the graduated cup to collect the material that condenses and drips from the distillation apparatus. When the first drop falls into the cup, record the time, and continue to make observations of time, boiling behavior, and volume of distillate obtained, every minute until the temperature inside the flask of boiling crude oil reaches 92°C.

> **Note:** Keep the liquid in the flask boiling slowly at all times. To accomplish this, you may have to adjust the heat setting on your hot plate.

Collecting the Second Fraction

6. When the temperature reaches 92°C, remove the first graduated cup and replace it with another one. Continue to make observations of time, boiling behavior, and volume of distillate obtained every minute until the temperature reaches 102°C.

7. When the temperature reaches 102°C, turn off the hot plate, but continue to collect liquid in the second graduated cup until liquid stops dripping into the cup.

Collecting the Third Fraction

8. Use insulated gloves or a pot holder to carefully remove the flask containing the residue from the hot plate and allow it to cool away from any student work areas.

Part B   Identifying and Testing the Fractions

Materials    **For each group of four students**

3	distillate fractions from Part A
1	25-mL graduated cylinder
1	SEPUP tray
1	dropper
1	stir stick
1	salt packet
1	sugar packet
1	small piece of effervescent tablet
	access to a balance

Procedure

Follow the procedure below to determine some properties of each of the three liquid fractions you collected. Allow the flask to cool before determining the properties of the residue fraction.

1. Determine the mass of your clean, dry graduated cylinder.

2. Carefully pour the entire contents of one of your collected fractions into the graduated cylinder and record the viscosity and volume of the liquid.

3. Determine the combined mass of the graduated cylinder and the fraction it contains.

4. Determine the mass of the fraction and then calculate its density. Record all results in your data table.

5. Pour the fraction back into the graduated cup.

6. Using the effervescent tablet, sugar, and salt, test each fraction as you did in Activity 21.1, "Differentiating Liquids," and record the results in your data table.

7. Clean and dry the graduated cylinder and repeat Steps 1–6 for the other fractions you collected.

8. Clean up as directed by your teacher.

Analysis

?

Individual Analysis

1. From your tests, identify each of the liquid fractions obtained from distilling the "crude oil." For each fraction, give evidence to support why you identified it as glycerin, methanol, or water.

2. Why do you think the properties of the liquid fractions were not identical to the properties of any of the liquids you examined in Activity 21.1?

3. From this activity, what can you conclude about the boiling point of the residual fraction (the liquid remaining in the flask)?

4. How might the amount of energy used to distill a mixture determine whether or not the process is economical?

5. When petroleum is distilled, some residual waste materials are left behind. What concerns and recommendations would you have about disposal of these wastes?

The Chemistry of Hydrocarbons

22

22.1 The Role of Carbon

Purpose **D**iscover why carbon is an important element.

Introduction

Although carbon makes up much less than 1% of Earth's mass, it is in more different compounds in and around the planet's surface than almost any other element. Carbon is an essential component of all living organisms. Virtually all of the molecules that make up any organism—including DNA, proteins, fats, and sugars—contain carbon. Partly because carbon plays such an important role in life processes, there are more compounds on Earth that contain carbon than compounds that do not. The branch of chemistry that deals exclusively with the study of carbon-based molecules is called **organic chemistry**.

Common Carbon Compounds

Carbon Chemistry

Atoms of carbon are the foundation of the molecular structure of petroleum, paper, peanuts, and people, as well as many other things. In fact, because of its unique bonding ability, carbon serves as the basis for a wide variety of compounds, from the relatively simple structure of carbon dioxide to the more complex structure of leucine (Figure 1).

As the number of carbon atoms in a molecule increases, more molecular structures become possible. Carbon's capacity to form a total of four bonds with up to four other atoms, including other carbon atoms, leads to this structural flexibility. Because carbon atoms are able to bond with each other, they can join together to form long chains. These chains may be straight, branched, or interconnected to form networks. The atoms of elements in the same chemical family as carbon on the Periodic Table, such as silicon or germanium, can also form four bonds, but are too large to form long chains.

Carbon in Life

Large and small carbon-based molecules play an important role in the composition and growth of living organisms. You are already familiar with carbon dioxide, which is one of the simpler carbon-based molecules. Carbon dioxide is a major component of the atmosphere. It is produced whenever fuels are burned; it is also a reactant in photosynthesis. **Biochemistry** is the study of the compounds that make glycerin up living organisms and the chemical reactions that take place within cells. The vast majority of the large molecules studied in biochemistry contain carbon. Biochemical compounds include proteins, nucleic acids, carbohydrates, and lipids. The molecules that form each of these groups of compounds are based on carbon atoms.

Amino acids are biological molecules that join together to form very large molecules known as **proteins**. Leucine, shown in Figure 1, is an amino acid. Notice the N–C–C structure at the top of the molecule. This structure is common to all amino acids and forms the basis for all proteins. Proteins may be composed of 100 or more amino acids and have a variety of complex shapes, one of which is shown in Figure 2. Proteins play an important part in a variety of life functions. For example, they form muscle tissue, serve as hormones, and control chemical reactions in cells. One of the proteins essential to humans is hemoglobin, which is found in red blood cells. **Hemoglobin** is the molecule that transports oxygen through the body via the bloodstream.

Nucleic acids, though quite complex, are composed exclusively of atoms of carbon, hydrogen, oxygen, nitrogen, and phosphorus. There are two kinds of nucleic acids. As you learned in Activity 19.1, "Genes, Chromosomes, and DNA," DNA, or deoxyribonucleic acid, contains genetic information and is found in the nucleus of the cell. The structural formula for DNA is shown in Figure 3, on the next page. RNA, or ribonucleic acid, transfers genetic information from the nucleus to other parts of the cell. It is also involved in protein synthesis. The only difference between DNA and RNA is that DNA has one less oxygen atom in each ring structure.

Figure 1   Structural Formulas for Two Carbon Compounds

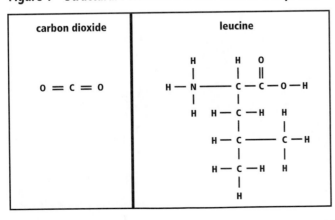

Figure 2   Diagram of Folded Protein Molecule

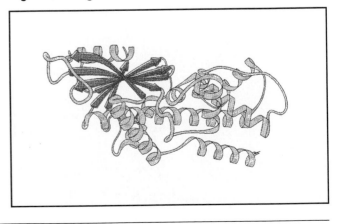

Figure 3   Structural Formula for DNA

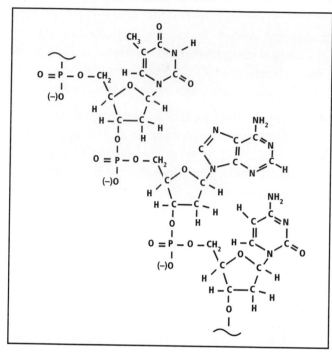

Figure 4   Structural Formula for Glucose

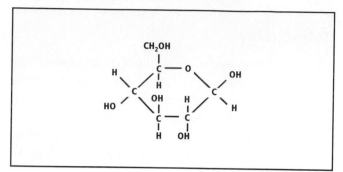

Carbohydrates are compounds that have the general chemical formula $(CH_2O)_n$, where **n** is a number ranging from a single digit to many thousands. Sugars and starches are examples of carbohydrates. Glucose, a sugar with the formula $C_6H_{12}O_6$, is the most common carbohydrate used for energy by living cells. Its basic structure consists of a six-atom ring made of five carbon atoms and one oxygen atom, as shown in Figure 4. This is a glucose monomer unit. Glucose can exist in the monomer form, or many glucose monomers can join together to form a long chain. Long chains of molecules are called polymers. A polymer

made up of a chain of sugar monomers is called a polysaccharide. One example of a polysaccharide is cellulose, the compound that forms the cell walls of plants. Its structure is shown in Figure 1 in Activity 23.1, "Polymers for Clothing" (page 271). The human digestive system must break down carbohydrate polymers into glucose before these carbohydrates can be used to provide energy to the cells.

Like carbohydrates, proteins and nucleic acids are also capable of forming long polymer chains. Lipids, however, do not form long chains of molecules, even though they are made up of the same atoms as carbohydrates—carbon, hydrogen, and oxygen. **Lipids** are a class of molecules that includes fats, waxes, and steroids. One example of a steroid molecule is cholesterol, which is a precursor material for many human hormones. Its structural formula is shown in Figure 5. Although cholesterol is necessary for the healthy functioning of the human body, the consumption of high amounts of certain types of cholesterol in the diet can result in an increased risk of heart disease.

Figure 5   Structural Formula for Cholesterol

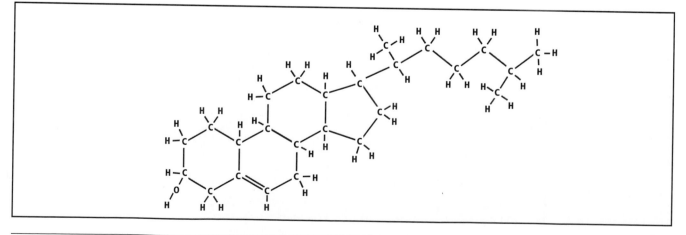

As part of the processes of metabolism and growth, plants and animals also synthesize many other carbon compounds, including cellulose, vitamins, and fibers. Because scientists originally thought that carbon compounds could be produced only by living organisms, the chemistry of carbon is called organic chemistry. This name came from the belief, now discarded, that living materials are organized in a unique way or contain some special ingredient not present in non-living matter.

Carbon on Earth

The organic compounds present in the tissues of living organisms and derived from the decay of dead organisms account for thousands of tons of carbon-rich material. However, most of the carbon on Earth is not found in these forms: Earth's largest store of carbon exists in rock and soil in the form of carbonate minerals. These minerals include the chemical group $-CO_3$. Calcite, $CaCO_3$, is one example of a carbonate mineral; it is found in limestone rock. It has been estimated that the carbon dioxide in Earth's atmosphere contains about 40 times as much carbon as all fossil fuels and forests combined—in spite of the fact that only 0.04% of the atmosphere is carbon dioxide.

Both organic and inorganic carbon contribute to the carbon cycle. As illustrated in Figure 6, the **carbon cycle**

Figure 6 The Carbon Cycle

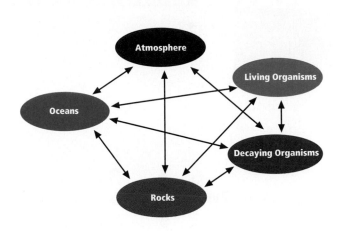

involves the exchange of carbon among all components of the environment. Carbon is used by living creatures, dissolved in oceans, stored in rock, or released into the atmosphere. It is not only necessary for the very structure of life, but is also found in virtually all non-living parts of the environment as well. Since carbon is an element, the amount present on Earth never changes. Carbon does change form, however. The chemical reactions it takes part in, and the chemical bonds it forms at various points in the carbon cycle, have a significant effect on our planet and on all organisms living on Earth.

Analysis

?

1. What role does carbon play in life on Earth? How does the chemistry of carbon relate to its role in life on Earth?

2. What role does life on Earth play in the cycling of carbon?

3. How can knowing about the chemical properties of different elements help a society become more sustainable?

Constructing Models of Hydrocarbons

Purpose ▶ **B**ecome familiar with the chemical structure of hydrocarbon molecules and learn how the number of carbon atoms in a hydrocarbon relates to the number of possible molecular structures it can form.

Introduction

In Activity 21, "Identifying and Separating Hydrocarbons," you were introduced to hydrocarbons, compounds composed only of carbon and hydrogen atoms. They are the major component of the fossil fuels that are used to meet many of the globe's energy and material needs. Organic chemistry—the chemistry of carbon and its compounds—explains how hydrocarbons can be used to produce energy or be synthesized into modern materials, such as plastics or pharmaceuticals. Organic chemistry also provides a basic understanding of how fats, proteins, carbohydrates, and other carbon compounds are used by the human body.

Energy is stored in the chemical bonds that hold atoms together to make compounds. Because each carbon atom has four bonding sites, organic molecules can contain three types of bonds—**single bonds**, **double bonds**, and **triple bonds** (Figure 7). Double bonds store more energy than single bonds; triple bonds store more energy than double bonds. The properties of a compound are determined by the type and number of elements in the compound, the geometrical arrangement of these elements, and the types of bonds formed between the elements.

Figure 7   Three Common Bonds

A single bond:	H — Br	(hydrogen bromide)
A double bond (and two single bonds):	O $\parallel$ H — C — H	(formaldehyde)
A triple bond (and a single bond):	H — C $\equiv$ N	(hydrogen cyanide)

Materials   ■■ **For each team of two students**

1   molecular model set

Procedure

During this activity, you will construct molecules. Remember, structures that contain the same atoms but have different geometric arrangements are different molecules.

1. **a.** Construct as many different molecules as you can according to the following rules:

   • Use only two carbon atoms and any number of hydrogen atoms.

   • All bonding sites of an atom must connect to those of another atom.

   Hint: Remember that carbon can form double and triple bonds.

   b. Draw the structural formulas of the molecules that you construct.

2. **a.** Construct as many different molecules as you can in 10 minutes according to the following rules:

   • Use only four carbon atoms and any number of hydrogen atoms.

   • All bonding sites of an atom must connect to those of another atom.

   b. Draw the structural formulas of the molecules that you construct.

3. Construct a molecule of benzene, which has the chemical formula C_6H_6.

4. **a.** Construct as many different molecules as you can in 10 minutes according to the following rules:

   • Use only six carbon atoms and any number of hydrogen atoms.

   • All bonding sites of an atom must connect to those of another atom.

   Hint: Molecules can form closed geometric figures such as triangles, squares, pentagons, etc.

   b. Draw the structural formulas of the molecules that you construct.

5. Write the appropriate chemical formula next to each of the structural formulas you drew.

Analysis

?

Individual Analysis

1. Describe the relationship between the number of available carbon atoms and the number of different molecules that can be constructed.

2. Explain the main reasons that this relationship exists.

22.3 Molecular Mysteries

Purpose ▶ **G**limpse a chemist's fascination with the struggle to determine the physical structure of a chemical compound. Investigate how the discovery of the structure of benzene changed the study of chemistry.

Introduction ▼

Benzene (C_6H_6), a hydrocarbon containing six carbon atoms and six hydrogen atoms, is derived from the fractional distillation of coal tar and petroleum. The 19th-century discovery that benzene molecules have a ring structure had an enormous impact on the field of organic chemistry. Today, thousands of organic chemicals are known to have ring structures. Carbon compounds that have this structure are called "closed chains"; those that do not are called "open chains." Closed-chain hydrocarbons play a central role in the manufacturing of many products. Benzene is an excellent solvent and is widely used in the production of medicines, plastics, dyes, and explosives.

A Dream Come True

In September of 1850, Friedrich August Kekulé, a young student of architecture, appeared before a grand jury in Giessen, Germany, to help clear up the circumstances surrounding the death of his neighbor Countess Görlitz. Her charred body had been found in an otherwise undamaged room a few weeks earlier, and the cause of death was assumed to be spontaneous combustion, brought on by excessive drinking.

Among the other witnesses called was the organic chemist Justus von Liebig, who had testified that spontaneous combustion of human tissue is physically impossible—that alcohol would poison a person long before it could raise the flammability of the body by an appreciable amount. Kekulé was asked about Görlitz's servants, in particular, about a man named Stauff, who

had been caught selling stolen goods, including a gold ring consisting of two intertwined snakes biting their own tails—the alchemy symbol for the unity and variability of matter. The ring, Kekulé told the court, had been the countess' talisman. This and other evidence convinced the jury that Stauff was guilty, and he was convicted of murder.

The trial changed Kekulé's life, as well. Deeply impressed by von Liebig's testimony, the young student switched from architecture to chemistry, with von Liebig as his mentor, and went on to make outstanding contributions to the field. Even the incriminating ring left a lasting impression, for it was the serpent symbol that bubbled up from Kekulé's subconscious fifteen years later, while he was trying to unravel the structure of the benzene molecule.

Judging from the weight and chemical properties of benzene, Kekulé had deduced that the molecule consists of six carbon atoms and six hydrogen atoms, but he couldn't figure out how these were arranged. In the midst of his puzzlement, he had an illuminating dream. "One night," he later wrote, "I turned my chair to the fire and sank into a doze." There appeared before his eyes a vision of atoms that danced and gamboled, occasionally joining together and turning and twisting in snakelike motion. Then, suddenly, "one of the serpents caught its own tail and the ring thus formed whirled exasperatingly before my eyes. I awoke as by lightning, and spent the rest of the night working out the logical consequences of the hypothesis."

Kekulé proposed that the benzene molecule is a hexagonal ring of carbon atoms from which hydrogen atoms dangle like charms from a bracelet.

Although it took some time for Kekulé's hypothesis to gain acceptance, it eventually inspired the entirely new field of **structural chemistry**, which relates the chemical properties of molecules to their geometric structures. For more than a hundred years now, Kekulé's ring model has been considered an established fact, which is remarkable, given that, until recently, no one had so much as glimpsed an actual benzene molecule.

Excerpted from "A Dream Come True" by Hans Christian van Baeyer, *The Sciences*, January/February 1989.

Figure 8   Three Versions of the Structural Formula for Benzene

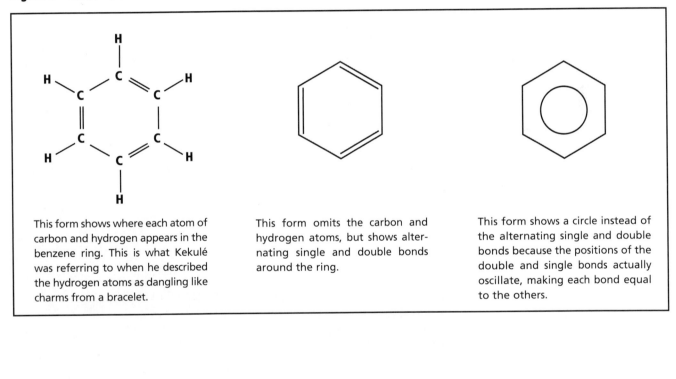

This form shows where each atom of carbon and hydrogen appears in the benzene ring. This is what Kekulé was referring to when he described the hydrogen atoms as dangling like charms from a bracelet.

This form omits the carbon and hydrogen atoms, but shows alternating single and double bonds around the ring.

This form shows a circle instead of the alternating single and double bonds because the positions of the double and single bonds actually oscillate, making each bond equal to the others.

Analysis
?

1. Why was Kekulé's discovery of the ring structure for benzene an important milestone in organic chemistry?

2. Benzene is now known to be a **carcinogen** (cancer-causing chemical). Discuss whether you think the production and use of benzene should be banned.

23.1 Polymers for Clothing

Purpose ▶ **E**xplore the use of polymers in the manufacture of clothing and the benefits and drawbacks of natural and synthetic fabrics.

Each type of fabric has unique properties that make it suitable for specific uses.

Introduction

Many small hydrocarbon molecules can combine to form a larger molecule called a **polymer**. DNA is a polymer you have previously studied; some of the molecules you modeled in Activity 22, "The Chemistry of Hydrocarbons," are also polymers. Natural polymers are extracted from plant or animal products. Synthetic polymers are made from hydrocarbon molecules that have been derived from petroleum. One of the many uses for both types of polymers is to make fibers that can be woven into cloth. Cotton fibers are made from the polymer cellulose, a natural part of the cotton plant; wool fibers are made from a natural polymer called keratin. The structural formulas for cellulose and keratin are shown in Figure 1. Synthetic polymers include nylon and polyester, shown in Figure 2. Polymers make good clothing materials because their long chains form strong fibers that can be spun into thread, which can then be woven or knitted into cloth.

Figure 1   Natural Polymers

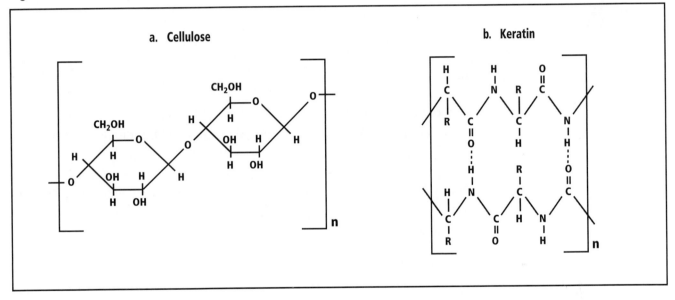

a. Cellulose   b. Keratin

Figure 2   Synthetic Polymers

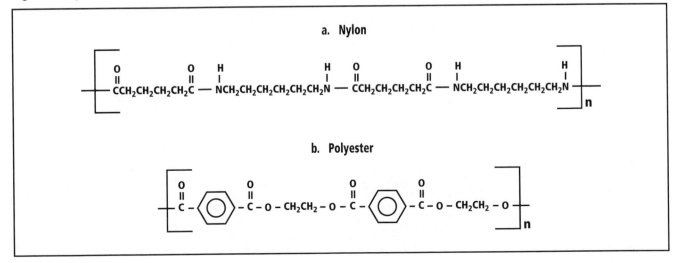

a. Nylon

b. Polyester

Materials

⬛ For each group of four students

1	sample of polyester
1	sample of cotton
1	sample of wool
1	sample of nylon

other materials as needed for experimentation

⬛ For each team of two students

1	magnifier
1	copy of *Material World*

Procedure

1. People choose clothes based in part on the local climate. Use the photographs in *Material World* to describe the clothing people wear in countries with the following climates:

   a. hot

   b. cold

   c. rainy

2. Make and record observations about the four cloth samples your teacher has provided. Focus on characteristics that are important for clothing, such as texture, thickness, durability, and so on.

3. Think about the clothing you wear. What properties are most important in determining whether you will buy and wear a new piece of clothing? Could you reliably test any of these properties?

4. Design an investigation to test the four cloth samples for properties that are important for clothing. In addition to the properties you identified in Step 3, you may want to test for flammability, water absorption, drying speed, or insulation.

5. Have your procedure approved by your teacher.

6. Make a data table to record all your data.

7. Conduct the investigation you designed. Record your data and observations.

What makes these people's clothing appropriate for their environment?

Analysis

?

Group Analysis

1. What advantages do you think are provided by the clothes you described in Procedure Step 1?

2. Of the four materials you tested, which would make the best clothing for someone living in a very hot climate? A very cold climate? A very rainy climate?

Individual Analysis

3. Choose one of three climates (hot, cold, rainy) and design an outfit that would be suitable for a person living in that climate. The outfit should be made of one of the four materials you have tested. Explain why the material you selected is appropriate for the climate.

4. Recall the clothing you saw people wearing in *Material World*.

    a. Is there anyone shown living in the climate you selected who is wearing an outfit similar to the one you designed?

    b. Why do you think the people in the country you selected are (or are not) shown wearing clothes similar to the outfit you designed?

    c. Describe any factor(s) that might be important to them that you did not take into consideration.

Extension

Prepare an advertisement or television commercial for the outfit you designed for Analysis Question 3. In your advertisement, use evidence from your investigation about the properties of different fabrics.

23.2 Modeling a Simple Polymer

Purpose ▶ **M**odel the formation of a large molecule from many small molecules.

Introduction

You have learned that polymers can be used to make clothing materials. The simplest polymers are **chain polymers**. Like a chain made from many identical links, chain polymers are made from many identical molecules. Each of the smaller molecules that join together to make the polymer is called a **monomer**. Chain polymers are especially useful in making fibers for fabrics and are also components of many plastics. Polyethylene, a common type of plastic, is a particularly long chain polymer. A single polyethylene molecule is made of thousands of linked ethylene monomers.

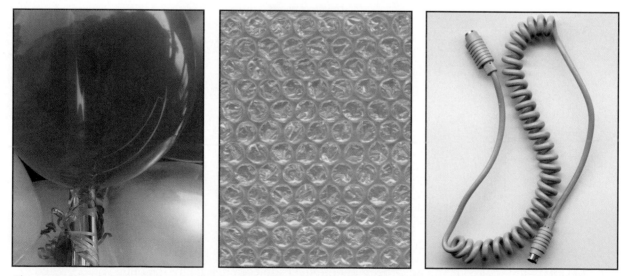

The properties of plastics that make them so versatile and useful are related to the fact that plastics are polymers. Three of the many plastic products in use today—balloons, bubble wrap, and wire insulation—are pictured here.

Materials

■■ **For each team of two students**

1   molecular model set

Procedure

1. Ethane, a gas, is a simple hydrocarbon. It has two carbon atoms and six hydrogen atoms.

    a. Write the chemical formula for ethane.

    b. Make a model of ethane and draw its structural formula.

2. Ethene, also known as ethylene, is a gas with the chemical formula C_2H_4. Make two models of ethene and draw its structural formula.

3. Link your two models of ethylene molecules together. Draw the structural formula for your model.

4. Draw the structural formula of a polyethylene molecule made of 10 linked ethylene monomers.

Figure 3   General Structural Formula for Polyethylene

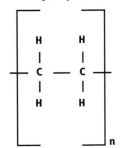

Note: Because a drawing like the one you just made is time-consuming to produce, polyethylene is often represented like the diagram shown in Figure 3. The brackets enclose an ethylene monomer. The subscript **n** means that the ethylene monomer is bonded to many other ethylene monomers. If you knew the exact number of monomers, you would replace the **n** with that number.

Analysis
?

Group Analysis

1. Why is there no double bond in polyethylene, when there is one in ethylene?

2. Write the chemical equation for the reaction you modeled in Procedure Step 3, involving

    a. the polymerization of two ethylene monomers.

    b. the polymerization of **n** ethylene monomers.

Individual Analysis

3. In the chemical equation you wrote for Analysis Question 2a, how are the atoms in the reactant molecules (on the left) related to the atoms of the product molecule? Are there more reactant atoms? More product atoms?

23.3 Researching Oil Production and Use

Purpose ▶ **C**ollect and present information about petroleum production and use.

Introduction

Through previous activities, you have explored several aspects of petroleum production and use, and learned that petroleum is a non-renewable resource. After a considerable amount of processing, petroleum can be used to manufacture a wide variety of valuable products, including fuels, plastics, and medicines, as shown in Figure 4. In this activity you will have the opportunity to pose a question about the production, processing, and use of petroleum, and propose an answer based on published data.

Figure 4   Major Products of Crude Oil

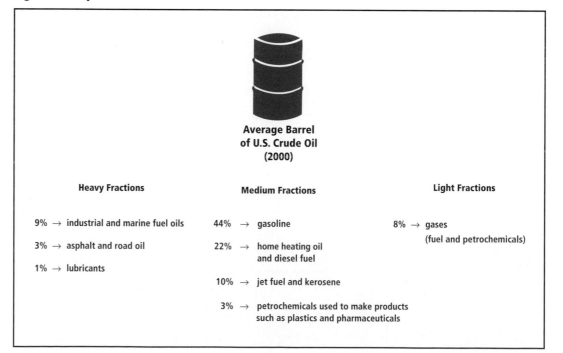

Average Barrel of U.S. Crude Oil (2000)

Heavy Fractions	Medium Fractions	Light Fractions
9% → industrial and marine fuel oils	44% → gasoline	8% → gases (fuel and petrochemicals)
3% → asphalt and road oil	22% → home heating oil and diesel fuel	
1% → lubricants	10% → jet fuel and kerosene	
	3% → petrochemicals used to make products such as plastics and pharmaceuticals	

Procedure

1. With your group, prepare a list of at least 10 questions that relate to past and present trends in the use, extraction, and production of petroleum. Think about what kinds of factors could change or sustain those trends in the future. Try to make sure that your questions address both factual and ethical issues. Below are some examples of good questions on similar topics:

   • How does the use of recycled paper affect the health of rain forests?

   • Have increased fishing activities in Alaska led to changes in the populations of species other than those sought by fishing crews?

   • How is the production of solar panels similar to the production of computer chips, and what effects do these production processes have on water quality?

   • How has the increasing use of refrigeration affected the lives of individuals, of urban populations, of the global population?

2. From the questions you listed for Procedure Step 1, select one or two that are especially interesting to you. Consider the following factors when making your selection:

   • Which questions do you already have information about?

   • Which questions would you like to learn more about?

   • Which questions do you think are most likely to affect you in your lifetime?

   • Which questions are most likely to affect future generations?

3. With the approval of your teacher, choose one question to research. The question you choose should have some factual components that you can use as evidence and some ethical components that will enable you to make a statement about how the facts and research findings affect society. It may also include the need to address projected future trends.

4. Begin your research, using the materials described by your teacher. You will eventually need to report all your sources in a bibliography, so you may want to use index cards to keep track of each source of information.

   You may wish to gather more information about the production and use of petroleum on the Internet. Go to the *Science and Sustainability* page on the SEPUP website to find updated sources.

5. Before the date specified by your teacher, prepare your project. Your teacher will give you more specific directions about how your project and bibliography should be prepared and presented.

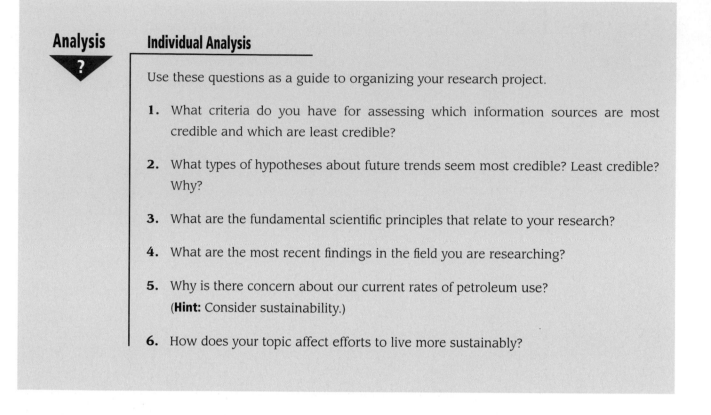

Analysis

?

Individual Analysis

Use these questions as a guide to organizing your research project.

1. What criteria do you have for assessing which information sources are most credible and which are least credible?

2. What types of hypotheses about future trends seem most credible? Least credible? Why?

3. What are the fundamental scientific principles that relate to your research?

4. What are the most recent findings in the field you are researching?

5. Why is there concern about our current rates of petroleum use?
    (**Hint:** Consider sustainability.)

6. How does your topic affect efforts to live more sustainably?

23.4 Modeling Cross-Linked Polymers

Purpose ▶ **C**onstruct a model of cross-linked polymers and show two different cross-linking mechanisms.

Introduction

In this activity, you will model polymer formation using paper clips to represent monomers. Keep in mind that a scientific model does not have to look like the thing it represents; it just has to act like it in important ways. Paper clips, like monomers, can link together to form a larger structure—a paper-clip polymer. Long chain polymers that are lined up side by side can be bonded together, or **cross-linked**, to form mesh-like structures. Kevlar, an extremely strong polymer used to make cables, bulletproof vests, and windsurfing sails, is one example of a cross-linked polymer.

Although Kevlar was developed in the late 1960s, scientists still don't completely understand why it is so strong. However, Kevlar does have some structural features that contribute to its strength. The long chains of Kevlar molecules don't get tangled up like many other polymer chains. Instead, they stay lined up parallel to each other. Each strand is attached to other parallel strands by

Figure 5   Chemical Structure for Kevlar®

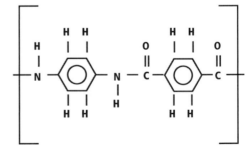

hydrogen bonds, which are formed by the four hydrogen atoms located on either side of each monomer. This structural arrangement is shown in Figure 5. A hydrogen bond is not as strong as the bond that links the end of one Kevlar monomer to another to form a chain. However, because there are so many hydrogen bonds along the entire length of each polymer chain, the combined effect is very strong. The numerous hydrogen bonds that connect one Kevlar polymer strand to another produce a fabric so strong it can resist bullets and 60-mile-per-hour winds.

Materials **For each group of four students**

    6   colored paper clips

    1   sample of Kevlar®

    transparent tape

For each team of two students

  24 paper clips in a clear plastic cup

    1 60-mL wide-mouth bottle

    1 plastic spoon

    1 clear plastic cup

Procedure

Prepare a data table to record your observations. You will need four rows, one for each of your models: "Monomer," "Linear Chain Polymer," "Cross-Linked Polymer #1," and "Cross-Linked Polymer #2." You will also need three columns, one for each of the properties tested: "Stirrability," "Pourability," and "Pullability." Be sure to label the rows and columns of your table.

Part A   Examining the Properties of a Monomer

1. Put 24 unconnected paper clips into the bottle. Each unconnected paper clip represents one monomer. Pour them from the bottle into a plastic cup. Describe how easily the paper clips flow out of the bottle. If necessary, shake the bottle. Record your observations in your data table in the column labeled "Pourability."

2. Use the plastic spoon to stir the 24 unconnected paper clips in the plastic cup. Record your observations in the column labeled "Stirrability."

3. Reach into the cup and pull on a single paper clip "monomer." Record your observations in the column labeled "Pullability."

Part B   Forming a Straight Chain Polymer

4. Each member of your group should link six paper clips together to form a straight chain and then link all four chains together to make one long chain of 24 paper clips.

5. Put the "polymer chain" into a plastic cup and stir it with the spoon. Record your "stirrability" observations.

6. Reach into the cup and pull on a single paper clip. Record your "pullability" observations.

7. Now put your paper clip "polymer chain" in the bottle, making sure to leave one clip hanging out of the top. Pour the chain into a plastic cup and record your "pourability" observations.

Procedure
(cont.)

Part C Cross-Linking Polymers

8. Separate the 24-paper-clip chain into four 6-paper-clip chains. Place the chains in four parallel lines and use the colored paper clips to connect, or cross-link, the four chains together as shown in Figure 6.

9. Perform the tests for "stirrability," "pullability," and "pourability" and record your observations in the row labeled "Cross-Linked Polymer #1" in your data table.

Figure 6   Modeling a Cross-Linked Polymer

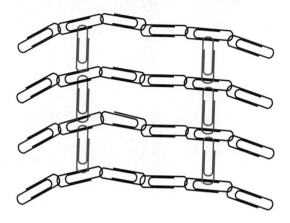

Part D Examining and Modeling Kevlar®

10. Describe the properties of your sample of Kevlar fabric.

11. Remove the cross-linking colored paper clips, and once again place the four 6-paper-clip chains in four parallel lines.

12. Use 18 small pieces of transparent tape to cross-link each paper clip from one chain to the corresponding paper clip of the adjacent chain as shown in Figure 7.

Figure 7   Modeling Kevlar®

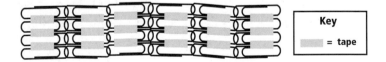

Key
▬ = tape

13. Repeat Step 9, but record your observations under "Cross-Linked Polymer #2."

14. Remove all the tape and separate all of the paper clips.

Analysis

?

Group Analysis

1. Describe the differences you observed in the properties of your paper-clip models of monomers, chain polymers, and cross-linked polymers.

2. Explain what causes the difference(s) in the properties of these models.

3. Would you expect actual long chain polymer molecules to have different physical properties from single monomer molecules? Explain.

Individual Analysis

4. Viscosity is a term that describes how easily a liquid flows. Highly viscous liquids do not flow easily, while liquids of low viscosity are runny. How would you expect the viscosity of a monomer to compare to the viscosity of a polymer made from that monomer? Explain.

5. Describe what happens to the physical properties of a substance as more and more molecules of that substance are polymerized (linked together).

6. Describe the advantages and disadvantages of the paper-clip model as compared to the molecular models you have used in previous activities.

7. How could you make a better model of a cross-linked polymer?

23.5 Creating a Cool Cross-Linked Polymer

Purpose ▶ **C**ompare the properties and structures of chain polymers and cross-linked polymers.

Introduction

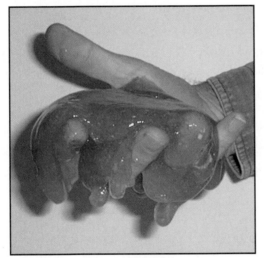

This gooey oozing polymer was made by combining two low-viscosity liquids.

In the last activity you used paper clips to model the formation and properties of different polymer structures. In this activity you will take liquid polyvinyl alcohol, a simple chain polymer, and cross-link it to make a slimy, squishy polymer. To make the cross-linked polymer, you will use sodium borate molecules to connect the polyvinyl alcohol chains together, just as the colored paper clips did in your model polymer.

Materials

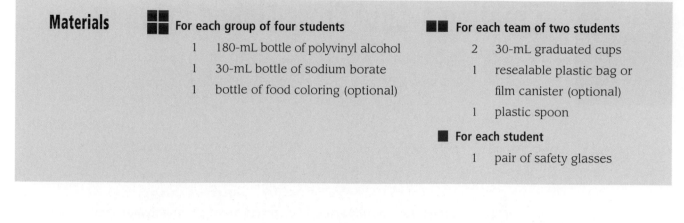

For each group of four students

- 1   180-mL bottle of polyvinyl alcohol
- 1   30-mL bottle of sodium borate
- 1   bottle of food coloring (optional)

For each team of two students

- 2   30-mL graduated cups
- 1   resealable plastic bag or film canister (optional)
- 1   plastic spoon

For each student

- 1   pair of safety glasses

Safety Note

Wear safety glasses. Wash your hands before and after this activity. Avoid getting your polymer on the table, the floor, or your clothes.

Procedure

1. Place 10 mL of polyvinyl alcohol (PVA) in one graduated cup and 2.5 mL of sodium borate in the other graduated cup. Observe the properties of each liquid and record them in your notebook.

2. Add 1 drop of food coloring (optional) to the 10 mL of PVA. While stirring the PVA with a plastic spoon, add the 2.5 mL of sodium borate. Closely observe what happens. Continue to stir until nothing further happens. Record your observations in your notebook.

3. Spoon the slimy product from your cup onto a paper towel.

4. You have made a form of Slime, a product sold in toy stores. Try pulling slowly on your slime. What happens?

5. Pull sharply on your slime. What happens?

6. What happens when you press down on your slime?

7. Try to bounce a small piece of your slime on your paper towel. What happens?

8. Wash your hands, graduated cups, and spoons thoroughly. Your teacher will instruct you on what to do with your slime.

Analysis

Group Analysis

1. Summarize what you learned about the properties of your slime.

2. Compare the properties of your slime with those of PVA and sodium borate.

3. Thoroughly describe how the behavior of your paper-clip model relates to the behavior of PVA, sodium borate, and your slime.

Individual Analysis

4. How is viscosity affected by cross-linking? Explain.

Material Resources: Metals

24

24.1 Extracting Metal From a Rock

Purpose ▶ **C**hemically refine malachite to produce copper.

Introduction

Malachite is a common mineral ore of copper. An ore is a rock that contains economically valuable concentrations of metallic elements or compounds and a number of other minerals. Malachite contains copper in the form of copper carbonate ($CuCO_3$). Once ores are mined, they are typically crushed and then processed, or refined, to extract the desired metals. The remaining components of the ore are often discarded as wastes. Processing of an ore often involves treatment with acid or heat, or both. These treatments not only contribute to the ultimate cost of the refined metal, but also produce additional waste products. In this activity, you will use a chemical process to refine malachite ore. The first step in the refining process involves treating the ore with sulfuric acid. During this step, you will observe the formation of copper sulfate solution ($CuSO_4$) by the chemical reaction shown in the following equation.

$$CuCO_3 \text{ (s)} + H_2SO_4 \text{ (aq)} \rightarrow CuSO_4 \text{ (aq)} + CO_2 \text{ (g)} + H_2O \text{ (l)}$$

In the second step of the refining process you will obtain solid copper as a product of the reaction between the copper sulfate solution and solid iron, as shown below.

$$3CuSO_4 \text{ (aq)} + 2Fe \text{ (s)} \rightarrow 3Cu \text{ (s)} + Fe_2(SO_4)_3 \text{ (aq)}$$

Designations for Physical States

(aq) = dissolved in water

(s) = solid

(l) = liquid

(g) = gas

Safety Note

The materials used in this investigation are toxic. They can irritate and burn the skin and eyes. Always wear safety glasses. If any of the solid malachite comes in contact with your skin, wash it off immediately. If any acid comes in contact with your skin, especially your hands or face, immediately wash with plenty of water and inform your teacher! If any acid gets on clothing, immediately remove the clothing and rinse in plenty of water.

Part A Dissolving the Copper

Materials

For each group of four students

1 piece of malachite
1 sample of crushed malachite ore
1 60-mL bottle of 1M sulfuric acid
1 30-mL dropper bottle of water
access to a balance

For each team of two students

1 SEPUP tray
1 filter funnel
1 coarse filter-paper circle
2 30-mL graduated cup
1 stir stick
access to a balance

For each student

1 pair of safety glasses

Procedure

1. Examine the malachite rock and the crushed malachite and describe the physical properties of each.

2. Place a 1-gram sample of crushed malachite in a 30-mL graduated cup.

3. Carefully and slowly add 1M sulfuric acid solution up to the 10-mL mark on your graduated cup.

4. Use your stir stick to gently stir the contents of the graduated cup for about 5 minutes. Describe your observations.

5. Determine the mass of your piece of filter paper. Prepare a filter set-up as shown in Figure 1. Position your filter so that it drips into one of the large cups in your SEPUP tray.

Figure 1 Single Filter Set-up

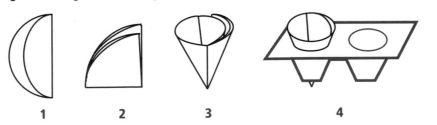

1   2   3   4

Procedure
(cont.)

6. Carefully pour the liquid contents of your graduated cup into the filter.

7. When all the liquid has dripped through the filter, use the stir stick to scrape any solids left in the cup into the filter. (If necessary, add a small amount of water to help get all of the solids into the filter. To avoid an overflow caused by the use of additional water, you may need to move the filter so that it drips into a different cup of the SEPUP tray.)

8. Use your dropper to transfer all the filtered liquid that dripped into the large cup of your SEPUP tray into a clean graduated cup. Describe the liquid.

9. Write your team members' names on a paper towel. Place your filter paper containing the solid wastes on the paper towel and then bring this and your graduated cup of liquid to your teacher to store for use in Part B of this activity.

10. Thoroughly clean your SEPUP tray for use in Part B.

Part B   Retrieving the Copper

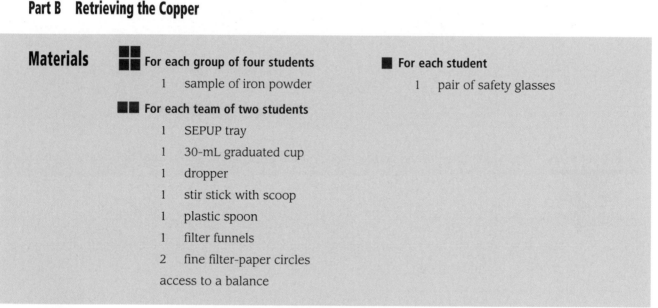

Materials

For each group of four students
 1   sample of iron powder

For each team of two students
 1   SEPUP tray
 1   30-mL graduated cup
 1   dropper
 1   stir stick with scoop
 1   plastic spoon
 1   filter funnels
 2   fine filter-paper circles
 access to a balance

For each student
 1   pair of safety glasses

Procedure

1. Retrieve the dried filter paper and solids that you set aside in Part A. Describe and find the mass of the solid waste material.

2. Use a graduated cup to obtain from your teacher about 5 mL of the liquid from Part A (the liquid that passed through the filter).

3. Use the dropper to place 20 drops of this liquid into Cup 1 of your SEPUP tray and 20 drops into Cup 2. Return any unused liquid to your teacher.

4. Add one level stir-stick scoop of iron powder to the liquid in Cup 1.

5. Add 3 level stir-stick scoops of iron powder to the liquid in Cup 2.

Procedure

(cont.)

6. Stir the contents of Cups 1 and 2 continuously for 5 minutes. Observe and record any changes that occur in each cup.

7. Label one of your filter-paper circles "Cup 1" and the other "Cup 2." Find and record the mass of each. As you did in Part A, fold and place one of your filter circles into one side of the filter funnel. Fold and place your other filter circle in the other side of the filter funnel. Position your filter funnel so that the filter labeled "Cup 1" drips into Cup A, on the upper row of the SEPUP tray, and the filter labeled "Cup 2" drips into Cup B.

8. Use your dropper to transfer the contents of Cup 1 into the filter labeled "Cup 1" and the contents of Cup 2 into the filter labeled "Cup 2." Allow the liquid to drip through the filters into the large cups of the SEPUP tray.

9. Describe the solids trapped by each of the filters. Place the filters on a paper towel with your group members' names on it and give them to your teacher so that they can dry overnight.

10. Describe the liquid in Cups A and B.

11. Use your dropper to transfer the liquid from Cups A and B to your graduated cup. Pour this liquid into the container provided by your teacher.

12. Write down your observations of the flame test that your teacher performs.

Analysis

?

Group Analysis

1. Crushing rocks costs money. The process of crushing the ore must be advantageous to the refining process if processors are willing to spend money doing it. Why do you think it is advantageous?

2. What do your observations from the flame test tell you about the chemical composition of the two liquids produced in this activity?

3. How does the solid waste produced by the refining process differ from the crushed malachite you began with? Why do you think these differences exist?

4. What do you think happened when the iron powder was added to the blue liquid? What is your evidence?

5. What differences did you observe in the solid and liquid produced in Cups 1 and 2? Why do you think these differences exist?

6. Other types of metal can be used in place of iron powder to extract the copper from the blue solution. What considerations would you need to think about before choosing which metal to use?

Analysis

?

Individual Analysis

7. What evidence indicates that chemical reactions took place during this investigation?

8. What percentage of the original rock sample ended up as solid waste?

9. If the sulfuric acid removed all the copper from your sample of malachite ore, how much copper should your refining process have produced? Explain.

10. Using what you learned in this activity, explain how a representative from a copper refining company might answer the questions listed below.

    a. What type of wastes are produced during the copper refining process used by your company?

    b. What concerns do you have about these wastes?

    c. What do you propose to do with these wastes?

    d. What are the advantages and disadvantages of your proposed method of waste disposal as compared to other possible methods?

Extension
Once the two filter papers from Part B have dried, determine the mass of the solid in each. Use this quantitative evidence to expand your answers to Analysis Questions 4–6.

24.2 The Changing Technology of Materials Science

Purpose ▶ **G**limpse the history of human understanding and utilization of Earth's resources.

Introduction

The richness and diversity of our material world allows us to experience a highly technological lifestyle. All the materials that humans use have their beginnings in Earth's natural resources. It is often easy to forget where common materials come from. For example, pottery is made from clay; glass, transistors, and microchips are made from sand; concrete is made from shells; metals are extracted from rocks; oil, gasoline, and plastics are made from petroleum.

Throughout history there have been numerous scientific and technological developments that have expanded our ability to make use of Earth's materials. This increase in knowledge has led to an overall increase in our quality of life. In particular, human understanding of energy and its uses has played an important role in enabling us to transform natural materials into the products that define today's human society.

All of the materials we use come from natural resources. Calcite, an important component of concrete, comes from shells; silicon, the basis for computer chips, comes from sand.

Obtaining Useful Materials

Wood and rock were among the first resources our early ancestors learned to use. People noticed that wood has a grain, which makes it split easily in one direction. Certain rocks also cleave, or split, easily. These characteristics made it fairly easy to fashion tools and building materials from wood and rock. These characteristics also provided the first clues that there are patterns, or an underlying order, in the nature of matter. Many centuries would pass before humans were able to describe the finer, "hidden" structure within matter—the microscopic and atomic world.

Around 400,000 years ago, early humans began to use fire to keep warm, to drive off predators, to cook food, to dry and harden wood, and to help split stones. As our ancestors learned to control and use fire, they began to transform nature. Our understanding of fire enabled us to enlarge our ecological niche. Humans no longer had to adapt to the environment to survive; we began to be able to alter our environment in ways that make survival easier.

Our use of fire led to many developments in our use of natural materials. In fact, the history of materials science is, in part, a history of our use of fire and the subsequent development of energy technology. Early humans discov-

ered, probably by chance, that certain soft, muddy layers of earth, which we now call **clay**, would harden when heated by fire. There is evidence that around 32,000 years ago, early humans began to mold useful objects, such as bowls and water jugs, from wet clay, then use fire to harden them. It takes relatively little heat energy to make pottery; the heat from an open wood fire can create temperatures up to 700°C, which is enough to fire clay into pottery. About 6,000 years ago, fires were built inside stone ovens, or **kilns**, to create the higher temperatures that are needed to make more durable ceramics such as bricks. The use of stone and other earth materials to make tools, pottery, and bricks are hallmarks of the era of human development known as the Stone Age.

Early humans occasionally found rocks containing copper, gold, silver, iron, or other metals in their pure, elemental state. These rocks were apparently treasured for their rarity and physical properties, such as a shiny finish, ability to be shaped, and durability: they were worked into special objects, such as small figurines of native gods or beads for the ceremonial necklaces of tribal leaders. We know that ancient Native Americans and peoples in the Middle East found chunks of copper on the ground. Because of its relative scarcity, there was very little copper available for general use. Then, around 10,000 years ago, people in the Middle East discovered that heat could be used to extract pure copper from rocks that contain a mixture of copper and other substances. Around 7,000 years ago, in Persia and Afghanistan, crude kilns capable of reaching temperatures of around 1,000°C were used to refine large quantities of copper from malachite.

Copper is a relatively soft metal that is malleable, or easily bent. At first, it was used mainly for artistic and cooking purposes. However, about 5,800 years ago, people found that heating rocks containing both copper and tin produced a much stronger, much less malleable metal that could be sharpened into a durable cutting edge. This **alloy**, or mixture of metals, is called bronze. Its discovery, which transformed human society, ended the Stone Age and signaled the beginning of the Bronze Age.

These desert pillars mark the site where copper was mined during the reign of the biblical King Solomon, in the 10th century B.C.E.

Clay, a naturally occurring substance, can be molded and then fired to produce pottery and other useful items.

Continued improvements in kiln design produced ever-higher temperatures. Approximately 3,500 years ago, the Hittites, who lived around the Black Sea, and inhabitants of West Africa independently learned how to refine iron ore to produce pure iron. This process requires temperatures of 1,500°C. By about 3,000 years ago, iron use became widespread, heralding the beginning of the Iron Age. Again, people discovered that another element, carbon, which is often found in iron-bearing ores, can be mixed with iron to form a much more useful alloy, now known as steel. Evidence shows that steel was first made in India. By around 2,800 years ago, the making of steel had become well established in many parts of the world. For example, steel was used to produce Samurai swords in Japan.

As civilizations developed, people continued to explore the properties of metals and other elements. Experiments conducted by alchemists during the Middle Ages and by scientists in the 18th century paved the way for our current understanding of atoms, elements, molecules, non-molecular substances, and the microscopic structure of materials. It was not until the 20th century that we learned how to synthesize materials from crude oil. By then, scientists understood that atoms are the building blocks of matter, and that the kinds of atoms and their arrangement determine the properties of a substance. We have since learned that, by subjecting materials to different temperatures, pressures, and chemical conditions, we can change the microscopic structures and chemical and physical properties of compounds, thereby creating a whole new array of materials to fit many needs.

In ancient times, raw, unprocessed materials determined what kinds of structures, tools, and common goods were available for use in everyday life. Today, we can design and produce new materials with properties that are tailor-made for certain applications. For instance, consider the differences between clay, metal, glass, and plastic containers. Can you see why containers for most household products are now made of plastic? In the past, the range of possible products was limited by the properties of naturally occurring materials; now, the desire for a particular product can lead to the creation of a new material.

Our ability to design new materials gives us greater control over our environment, our health, and our quality of life. If we carefully consider the implications and trade-offs involved in our decisions about how to use them, our ability to design new materials could help achieve sustainable development.

Because many metals melt at high temperatures, they can be shaped and joined to form almost any shape.

Analysis

?

1. Construct a timeline of materials development, from the beginnings of pottery making to the manufacture of steel. Include the temperatures that are necessary to produce each material on the timeline.

2. In what way has the manufacture of modern materials such as plastics and semiconductors (which form the heart of computer chips) affected the quality of your life?

3. **a.** Imagine that a significant amount of copper ore has been discovered within a mile of your home. A large multinational corporation is interested in mining the ore and refining the copper. What concerns do you have about the removal, transportation, and processing of the ore so close to your home?

   b. Now, imagine that you are the public relations representative for the company described above. How can you justify the removal, transportation, and processing of the ore as being good for the community?

24.3 Material Use Around the World

Purpose ▶ **I**dentify common materials used by human societies around the world and investigate why people living in different countries make use of different materials.

Introduction
All humans have the same basic needs, yet the materials used to meet these needs vary from country to country. Many factors—including personal, social, economic, and geographical considerations—explain why different materials are used in different countries. The photographs in *Material World* will help you explore these differences.

In some parts of the world, wood is a common building material. In other places, bricks are widely used. What factors must be considered when choosing an appropriate building material?

Materials ■■ **For each team of two students**

1 copy of *Material World*

Procedure

Table 1 lists several countries included in *Material World*. From this table, choose two less developed and two more developed countries. Read the information in *Material World* about the family from each of the countries you selected.

Table 1 Countries From *Material World*

Less Developed	More Developed
Brazil	Iceland
China	Israel
Cuba	Italy
Ethiopia	Japan
Guatemala	Kuwait
India	Mexico
Mali	Russia
Thailand	United Kingdom

1. Using the categories listed below, determine the three most common materials used to make the families' home and possessions. Display this information in a data table.
   - earth and stone
   - metal
   - plastic
   - glass
   - ceramic (bricks and pottery)
   - animal products (wool and leather)
   - plant products (wood, reeds, straw, paper, cotton)

2. Estimate the percentage of goods that appear to be made from locally obtained materials.

3. Estimate the percentage of goods that appear to be made from highly processed rather than raw, unprocessed materials.

Analysis

Group Analysis

1. Do the less developed countries you read about tend to use different materials than the more developed countries do? Describe your evidence.

2. List the three factors you think play the biggest role in determining which materials are commonly used in a country.

3. With sustainability in mind, what factors might you consider when faced with choosing between products that serve the same purpose but are made of different materials? Explain why these factors would influence your decision.

Individual Analysis

4. A country's political and business leaders can choose to take actions to help preserve (save) that country's natural resources, allow local refining and manufacturing (use) of the resources, or export (sell) the raw materials for refining and manufacturing in foreign countries.

   a. In what ways could the decision to save, use, or sell a country's raw materials affect the environmental quality of the country? Compare and contrast the environmental impact of each alternative.

   b. Describe other aspects of life that would be affected by the decision to save, use, or sell that country's raw materials.

By-Products of Materials Production   25

25.1   Disposing of Toxic Heavy Metals

Purpose ▶ **D**etect low concentrations of dissolved copper and convert this copper-bearing solution into a solid that can be disposed of safely.

Introduction

As you observed in Activity 24, "Material Resources: Metals," acid solutions can release, or **leach**, metals from their ores. Rain water or groundwater can also leach metals from contaminated soil and landfills. Small amounts of heavy metals, such as copper, lead, and chromium, can be toxic to living organisms. Water moves easily through the environment. If it contains dissolved metals or other toxic chemicals, it can contaminate soil and water supplies and cause health problems for organisms throughout the food web.

In this investigation, you will use the waste solution you produced during Activity 24.1, "Extracting Metal From a Rock." In Part B of this activity you will estimate the amount of copper in your waste solution by performing a serial dilution and then using a test to detect low concentrations of copper in solution. A **serial dilution** is a procedure that creates a series of solutions in which each solution has a concentration that is an exact fraction, such as ½ or ⅟₁₀, of the previous one. In Parts A and C you will investigate a method for treating wastewater containing dissolved copper sulfate so that it will resist leaching when disposed of in a landfill.

Part A "Fixing" the Copper Ions in a Solid Mixture

Materials

■■
■■ **For each group of four students**

1   180-mL bottle of sodium silicate
1   sample of Portland cement
1   120-mL dropper bottle of 50,000 ppm
    copper sulfate solution
    paper towels

■■■ **For each team of two students**

1   clear plastic cup
1   plastic spoon
1   30-mL graduated cup

■ **For each student**

1   pair of safety glasses

Safety Note

Portland cement can irritate your eyes and lungs. Avoid breathing its dust. Wear safety glasses. Do not wash glassware containing Portland cement in the sink, since it will permanently clog the drain. Use wet paper towels to wipe out cups and clean your hands. Then throw the used paper towels in the trash.

Procedure

1. Your teacher will assign you an amount of sodium silicate and Portland cement to add to the copper sulfate solution. Prepare a data table similar to the one below and circle the amounts of Portland cement and sodium silicate your group has been assigned. You will use this data table again in Part C.

Table 1   Solid and Filtrate Observations

Copper Sulfate Solution (mL)	Cement (cm³)	Sodium Silicate (mL)	Appearance of Mixture		Color of Filtrate		Estimated Copper Concentration
			When Wet	After Drying	From Simulated Acid Rain	After Adding Ammonia	
10	20	0					
10	0	20					
10	5	15					
10	10	10					
10	15	5					

Procedure
(cont.)

2. Using your 30-mL graduated cup, measure your assigned amount of Portland cement and pour it into the clear plastic cup. Use a paper towel to wipe off any cement powder that clings to the inside of the graduated cup.

3. Use your graduated cup to add 10 mL of 50,000 ppm copper sulfate solution to the cement in the plastic cup. Stir the mixture well using a plastic spoon. Clean the graduated cup.

4. Use your graduated cup to add your assigned amount of sodium silicate to the cement-copper sulfate mixture. Stir the mixture well until no further change is observed. Describe the wet mixture in your table.

5. Label your sample with the names of your group members and the amounts of Portland cement and sodium silicate you used. Follow your teacher's instructions for storing the mixture while it solidifies.

6. Clean your equipment thoroughly.

Analysis

?

1. Which material or combination of materials do you think will do the best job of preventing the copper from being leached out? Explain.

Part B   Determining the Concentration of Copper Sulfate Solutions

Materials

■■
■■ **For each group of four students**

  1   20-mL sample of copper sulfate solution from Activity 24

  1   120-mL dropper bottle of 50,000 ppm copper sulfate solution

  1   30-mL dropper bottle of 5% ammonia solution

  1   30-mL dropper bottle of water

  paper towels

■■ **For each team of two students**

  1   30-mL graduated cup

  1   dropper

  1   SEPUP tray

■ **For each student**

  1   pair of safety glasses

Procedure

Decide which team of two in your group of four will test the 50,000 ppm copper sulfate solution (Team A) and which team of two will test the "waste" copper sulfate solution produced when you refined malachite in Activity 24.1 (Team B).

1. Prepare a data table similar to the one below. Label it "Data Table 2A" or "Data Table 2B" (depending on which team you are in) and make sure the title indicates which solution you are testing, standard or waste.

Table 2   Determining the Concentration of _____ Copper Sulfate Solution

Cup	Color	Color After Ammonia Is Added	Concentration (ppm)
1			
2			
3			
4			
5			

2. Add 10 drops of your team's copper sulfate solution to Cup 1 of your SEPUP tray.

3. Using the dropper, transfer one drop of your solution from Cup 1 to Cup 2. Add 9 drops of water to Cup 2. Use the dropper to mix the solution.

4. Continue this dilution procedure up through Cup 5 as outlined in Table 3.

5. Add 10 drops of water to Cup 6.

Table 3   Serial Dilution of the Copper Sulfate Solution

Liquid	Cup 1	Cup 2	Cup 3	Cup 4	Cup 5	Cup 6
Copper Sulfate Solution	10 drops	1 drop from Cup 1	1 drop from Cup 2	1 drop from Cup 3	1 drop from Cup 4	0 drops
Distilled Water	0 drops	9 drops	9 drops	9 drops	9 drops	10 drops

6. Record the color of each cup in your data table.

7. Ammonia is an indicator for the presence of dissolved copper. Add 10 drops of 5% ammonia to each of the six cups.

8. Record your observations in your data table.

▶

Procedure
(cont.)

9. Compare your results with the results obtained by the other team in your group and record their data.

10. Using the results from Team A as a guide, estimate the concentration of the undiluted copper sulfate waste from Activity 24.1 tested by Team B. Record your estimate.

11. Dispose of the materials as directed by your teacher.

Analysis
?

Group Analysis

1. How does the addition of ammonia help determine the concentration of the copper in a solution? Is it necessary?

2. What is the lowest concentration of copper that can be detected using the ammonia test?

3. Compare your group's estimate of the concentration of copper in the undiluted waste solution from Activity 24.1 (Part A, Procedure Step 9) with the estimate made by another team. How close are the two estimates? Explain reasons you see for any variations. What do you think you should do if the variations are large?

Individual Analysis

4. Describe in detail how you determined your estimated value for the concentration of copper in the undiluted waste solution from Activity 24.1.

5. An independent laboratory performed the same test as Team A on a 50,000 ppm copper sulfate solution. They report the data shown in the table below. Describe the similarities and differences between their results and those your group reported. Explain why these similarities and differences might have occurred.

Cup	Color	Color After Ammonia Is Added	Concentration (ppm)
1	blue	blue-green	50,000
2	light blue	blue	5,000
3	clear	light blue	500
4	clear	clear	50
5	clear	clear	5

Part C   Testing Your Solidified Mixture Under Acidic Conditions

Materials

■■
■■ **For each group of four students**

| 1 | 120-mL bottle of simulated acid-rain solution (0.5M HCl) |
| 1 | 30-mL dropper bottle of 5% ammonia solution |

■■ **For each team of two students**

1	5-mL sample of solid produced in Part A
1	SEPUP tray
1	30-mL graduated cup
1	plastic spoon
1	dropper
1	filter funnel
1	piece of coarse filter paper
1	mortar and pestle or other crushing device

■ **For each student**

| 1 | pair of safety glasses |

Procedure

1. Use your graduated cup to obtain a 5-mL sample of the solidified mixture you prepared during Part A of this activity.

2. Prepare a filter set-up like the one shown in Figure 1 in Activity 24.1 (page 287).

3. Add 10 mL of simulated acid-rain solution to the solid in your graduated cup. Swirl it gently for 30 seconds.

4. Transfer the contents of your graduated cup to the filter funnel. Clean the graduated cup.

5. Locate the data table that you made for Part A of this activity. Record the color of the filtrate as it drips through the filter and collects in the large cup of your SEPUP tray.

6. Use your dropper to transfer 10 drops of the filtrate to Cup 1 of your SEPUP tray, then add 10 drops of 5% ammonia to it. Record the color of the solution after adding the ammonia.

7. Using the data you collected in Part B, estimate the concentration of copper in the filtrate. Record this in your data table.

8. Obtain the results from the other groups and add this information to your data table.

Analysis **?**

Group Analysis

1. Which cement and sodium silicate mixture did the best job of solidifying and fixing the waste? What is the evidence?

2. How does your answer to Analysis Question 1 compare to your answer to Analysis Question 1 in Part A?

Individual Analysis

3. Portland cement costs about $.04/cm^3; sodium silicate costs about $.02/mL. If your company produced 100 liters per day of liquid waste containing 5,000 ppm dissolved copper, how would you dispose of it? Describe two or more options you would consider, the trade-offs of each, and the additional information you would want to have before making your final choice.

4. Do you think solidification is an appropriate method for disposing of liquid wastes containing dissolved heavy metals? Provide evidence for your decision.

5. As you have learned, leaching is useful in the metal refining process, but needs to be prevented when placing materials containing heavy metals in the environment. The leaching process can also be used to clean up soil contaminated with heavy metals. Briefly describe the process of leaching and how it could be used as a method for cleaning up contaminated soils.

25.2   Who Wants Waste?

Purpose ▶ **E**xamine the pros and cons of a relatively new business, the international trade of hazardous waste.

Introduction ▼

Advances in industrial processing have provided us with many useful and socially beneficial products and materials. However, **by-products** of some of these manufacturing processes are detrimental to the health of the environment. When these by-products need to be disposed of, they may be classified as toxic waste. **Toxic waste** is waste that can cause death or serious injury. The more developed countries produce a large portion of the world's industrial toxic wastes and tend to have stricter environmental regulations for their disposal than less developed countries. These strict regulations have led to an international waste trade, in which more developed countries pay to dispose of their toxic wastes in less developed countries.

Industrial waste is often stored in barrels and tanks until it can be properly disposed of. Many storage containers leak, releasing contaminants to the soil, water, and air.

Where, Oh Where, Should the Toxic Waste Go?

The production of toxic wastes has increased dramatically in recent years. In the United States alone, toxic waste production has risen from 9 million tons to 286 million tons between 1970 and 1986. There is growing concern in some countries about the toxicity of waste products. New regulations in these countries reflect this concern: more waste products are being classified as toxic than ever before. When a waste is classified as toxic, it must go through more expensive treatment and disposal procedures than non-toxic waste, which can be disposed of in a sanitary landfill.

The increased cost of waste disposal has forced industries to change their strategy for dealing with toxic wastes. Commonly, industries in more developed countries implement alternative production methods to reduce their production of toxic waste, find less expensive means of disposing of it, or implement a combination of both. Production methods that create less toxic waste usually involve significant short-term investments,

Common products such as paint, motor oil, and car tires contain toxic chemicals. Because of their potential to damage human health and the environment, many communities do not allow these products to be disposed of in landfills.

making the products more expensive for the consumer. Companies in more developed countries have discovered that many less developed countries are willing to dispose of toxic wastes at bargain rates. For instance, in the late 1980s some African countries were charging $2.50 per ton to dispose of waste that would cost $2,000 per ton in the U.S. Because of these low costs, the exportation of toxic waste, especially to less developed countries, has become the choice of many industries. Less developed countries are willing to accept the toxic waste because it is a good source of much-needed revenue to pay off national debt, to fund industrial development, and to provide social services to residents.

Unfortunately, disposal of the wastes in these countries is often poorly regulated and may cause considerable pollution, environmental dangers, and health problems. When toxic waste is delivered to a country that does not have adequate treatment or disposal facilities, the waste goes untreated.

In 1989, during the Basel Convention, issues surrounding the international waste trade were considered. Alternatives proposed during the convention ranged from a complete ban on waste trade to the lifting of all restrictions. The compromise that was eventually worked out requires companies wishing to export toxic waste to apply to the importing country for a permit. To offer a permit, the importing country must already have adequate disposal facilities. If not, the company applying for the permit must agree to build disposal facilities. Under the compromise agreement, permits cannot be granted by individual land owners, but must be obtained from governmental officials.

Critics of the compromise focus on the fact that there are still many loopholes in the regulatory process. The first loophole exists because countries have different standards for determining the concentration at which a substance is considered toxic. As a result, a substance can be labeled toxic in one country and non-toxic in another! Second, government officials often have little knowledge of chemistry, but they are responsible for setting up trade and disposal agreements. In some cases, small lots of assorted chemicals from different industries—which may require very different methods of disposal—are combined into one larger lot in an attempt to save space and money when transporting the waste. Further, the lifespan and toxicity of some wastes is unknown. As a result, disposal sites that a country expects to reclaim for other uses could remain toxic and unfit for those uses for many years, possibly indefinitely.

Because the practice of buying and selling toxic wastes is so new, relatively few regulations exist. In fact, many countries have not ratified the Basel Convention agreement, which means that even these limited regulations are not being followed in some regions. In the coming decades, it is likely that your vote will have an influence on the future regulation of international trade in toxic wastes.

Analysis
?

Group Analysis

1. Describe the issues directly related to chemistry that are relevant to making decisions about the international toxic-waste trade.

2. Should international trade in toxic waste be banned? Explain both sides of the issue, and justify your answer with evidence and sound reasoning.

Individual Analysis

3. If international trade in toxic waste is not banned, describe a set of regulations that you think might be effective in preventing one country from becoming polluted with toxic waste produced in another country.

25.3 Chemical Production Poster Presentation

Purpose ▶ **I**nvestigate the trade-offs involved in the use and manufacture of chemical products.

Introduction

The colors of fireworks are often produced by chemical compounds containing heavy metal elements.

In our everyday lives we frequently use and dispose of many chemicals that are classified as toxic, such as bleach, drain cleaner, and gasoline. The manufacture of many familiar products, including paper, matches, and leather, also involves the use and disposal of toxic chemicals. Which chemicals are produced in the greatest quantity in the U.S.? What are these chemicals used for? What sort of health or environmental risks are associated with their use and disposal? How and where are these chemicals disposed of?

Procedure

1. Choose a chemical from the list provided by your teacher.

2. Using the resources identified by your teacher, look up the following information about your chosen chemical.

   a. The chemical formula

   b. The amount (mass or volume) of the chemical used nationwide and worldwide each year

   c. At least five products that either contain the chemical or involve its use during the manufacturing process

   d. At least two health or environmental risks associated with the use of the chemical

   e. Recommendations or regulations concerning the disposal of wastes that contain the chemical

> You may wish to gather more information about your chemical on the Internet. Go to the *Science and Sustainability* page on the SEPUP website to find updated sources.

3. Prepare a visual aid, such as a poster, transparency, or multimedia presentation, that illustrates the information you gathered.

4. Use your visual aid when you share information about your chemical with the class.

Analysis

?

Individual Analysis

1. Do you think that continued use of your chemical should be encouraged or discouraged? Discuss the trade-offs involved.

Catalysts, Enzymes, and Reaction Rates

26

26.1   Exploring a Catalyst

Purpose **I**nvestigate the action of catalysts by using copper to catalyze the reaction of ethanol and oxygen to form acetaldehyde and water.

Introduction

You have been studying chemical reactions and their significance in the production of the materials we rely on. Often, useful chemical reactions require energy to get them started—sometimes a lot of energy. **Catalysts** are substances that can decrease the energy needed to get a reaction started, which makes it easier and more economical to carry out the reaction. In this activity, you will observe a catalyst reacting and then returning to its original form. This characteristic is what makes a catalyst different from a reactant: catalysts participate in a reaction, but are not changed by it; reactants are changed by the reaction.

Materials

For each group of four students

1   30-cm copper wire with coiled end
1   small test tube
1   60-mL bottle of denatured ethanol
1   alcohol burner (or Bunsen burner)
1   250-mL beaker
matches or lighter

For each student

1   pair of safety glasses

Safety Note

In this activity, you will be using ethanol as a reactant. You will also have an open flame. Since ethanol is flammable, this is a potential hazard. Keep flammable materials away from open flames. Tie back long hair. Wear safety glasses. Test tubes and wires can be hot even when they do not appear to be. Be careful with hot lab equipment.

Procedure

1. Fill the test tube with ethanol (ethyl alcohol) to a height of 3–4 cm. Keeping your face to one side of the test tube, waft some of the vapor from the test tube towards your nose. Describe the smell of the alcohol. Place the partially filled test tube in the empty beaker. Do not pour the alcohol into the beaker!

2. Bend the non-coiled end of the copper wire over the top of the test tube so that the coil is suspended inside the test tube, about 1 cm above the alcohol. Place the beaker containing the test tube at least 50 cm from the alcohol burner.

3. Light the alcohol burner. Remove the copper wire from the test tube and hold it in the flame until it glows a dull orange color. Remove it from the flame and observe it as it cools. Describe all changes you observed in the wire during the entire process.

4. Reheat the wire until it once again glows a dull orange color. Now quickly place it in the test tube so that it is suspended above the alcohol. Observe and record what happens. Repeat this procedure several times, so that all members of your group can closely observe the reaction. You may want to try placing the heated coil at different heights above the surface of the ethanol.

5. Remove the wire and, by wafting, carefully smell and describe the odor coming from the top of the test tube.

6. Dispose of the alcohol as directed by your teacher.

Catalysts, Enzymes, and Reaction Rates

Analysis

?

Group Analysis

1. What evidence do you have to indicate that a chemical reaction occurred when the hot wire was placed in the test tube with the alcohol? What evidence do you have to indicate that copper wire catalyzed this reaction?

2. The first part of this reaction occurs when the copper (Cu) is heated. The hot copper reacts with oxygen (O_2) in the air to form copper oxide (CuO). Write a balanced chemical equation for this reaction. (**Hint:** Make sure there are equal numbers of atoms of all elements on both sides of the equation.)

3. The second part of this reaction takes place when the copper oxide reacts with ethanol to form acetaldehyde (a flavoring ingredient) and copper metal. The chemical formula for ethanol is C_2H_5OH, and the formula for acetaldehyde is C_2H_4O. Draw structural formulas for these molecules and write a chemical equation for this part of the reaction.

Comparing Catalysts

Purpose

Compare the rate of decomposition of hydrogen peroxide when it is catalyzed by inorganic catalysts and by organic catalysts.

Introduction

Hydrogen peroxide (H_2O_2) spontaneously undergoes a chemical reaction to yield water and oxygen. This is why hydrogen peroxide "goes bad" if it is stored in open bottles too long. The chemical equation for the decomposition of hydrogen peroxide is shown below.

$$2H_2O_2 \text{ (l)} \rightarrow O_2 \text{ (g)} + 2H_2O \text{ (l)}$$

In the presence of a catalyst, this reaction can be made to occur much more rapidly. In this activity, you will add a catalyst to H_2O_2 and observe its decomposition; bubbles will form in the H_2O_2, indicating the production of oxygen gas (O_2).

You will use three different catalysts to perform the same decomposition reaction. Two of the catalysts are from inorganic sources, and one is from a biological source. Catalysts are essential to biological systems because they reduce the amount of energy required to start many of the chemical reactions that take place within the cell. Life would probably not be possible without these biological catalysts, which are called **enzymes**. In this activity, you will compare the effectiveness of using catalase, an enzyme found in potatoes, with the effectiveness of two inorganic catalysts, iron chloride and manganese dioxide. In Part A, you will explore the reactions qualitatively. In Parts B and C, you will measure the concentration of hydrogen peroxide left after decomposition and the amount of oxygen gas produced during the reaction.

Part A   Catalyzing the Decomposition of Hydrogen Peroxide

Materials

 For each group of four students

1	30-mL dropper bottle of 0.1M iron (III) chloride solution
1	30-mL dropper bottle of 0.01M potassium permanganate solution
1	30-mL dropper bottle of distilled water
1	iron cylinder
1	bottle of 3% H_2O_2 solution

For each team of two students

1	SEPUP tray
1	piece of white paper
1	stir stick
1	dropper
1	10-mL graduated cylinder
2	oxy pellets (manganese dioxide)
1	250-mL beaker
1	paper towel
1	30-mL graduated cup
1	piece of potato

For each student

| 1 | pair of safety glasses |

Safety Note

Keep flammable materials away from open flames. Wear safety glasses. Hydrogen peroxide can bleach hair and clothes. Rinse off any spilled hydrogen peroxide solution with plenty of water.

Procedure

1. Put approximately 10 mL of 3% H_2O_2 solution in your 30-mL graduated cup.

2. Place the white paper under your SEPUP tray. Using your dropper, add 20 drops of 3% H_2O_2 to each of Cups 1–4 of your SEPUP tray. Rinse out your dropper with water.

3. Make a data table to record your observations of each cup for 5 minutes.

4. Place your potato piece in Cup A, on the upper row of your SEPUP tray, and then add about 20 drops of distilled water. Use the iron cylinder to mash the potato and the water together so that liquid comes out of the potato pieces. Push the potato pieces to one side of the cup to separate the solids from the liquids.

5. Add 2 drops of 0.1M iron (III) chloride solution to Cup 2, 2 drops of potato liquid (which contains the organic enzyme catalase) to Cup 3, and 2 pellets of manganese dioxide to Cup 4. Make and record observations for 5 minutes.

Analysis

?

Group Analysis

1. Based on your observations in this activity, which catalyst is the most effective in catalyzing the decomposition of hydrogen peroxide? Explain.

Part B   Determining the Concentration of H₂O₂ in a Solution

One way to measure how much H_2O_2 is contained in a solution is to use the chemical potassium permanganate ($KMnO_4$). When a solution of $KMnO_4$ reacts with H_2O_2, the color of the solution changes from purple to colorless.

Safety Note

 Potassium permanganate can stain skin and clothes. Be careful!

Procedure

1. Add one drop of 3% H_2O_2 to Cup 6.

2. Carefully add, one drop at a time, 0.01M potassium permanganate ($KMnO_4$) solution to Cup 6. Stir after each drop. Continue adding drops until the solution remains colored after stirring.

3. Record the total number of drops of 0.01M $KMnO_4$ you added.

4. Repeat Steps 1–3 for Cup 7 and then calculate the average number of drops of 0.01M $KMnO_4$ needed to turn the 3% H_2O_2 solution purple.

5. Add one drop of water and one drop of 3% H_2O_2 to Cup 8. You now have 2 drops of 1.5% H_2O_2 in Cup 8.

6. Use your dropper to transfer one drop of the solution in Cup 8 into Cup 9, so that both Cup 8 and Cup 9 contain one drop of 1.5% H_2O_2.

7. Repeat Steps 2–3 for the liquid in Cups 8 and 9. Calculate the average number of drops of 0.01 M $KMnO_4$ solution necessary to turn the 1.5% H_2O_2 solution purple.

8. Carefully clean out the cups in your tray and dry with a paper towel. Do not throw out the manganese dioxide pellets. Rinse them with water and return them to your teacher.

Analysis

Group Analysis

1. Based on your results from Part B, how many drops of 0.01M $KMnO_4$ do you predict would be needed to react with one drop of 1% H_2O_2? Explain how you made your prediction.

2. How could potassium permanganate be used to help you answer the Analysis Question in Part A?

Part C   Measuring the Effectiveness of Two Catalysts Quantitatively

Materials

 For each group of four students

1	30-mL dropper bottle of 0.1M iron (III) chloride solution
1	30-mL dropper bottle of 0.01M potassium permanganate solution
1	bottle of 3% H_2O_2 solution
1	plastic dishpan
1	30-mL dropper bottle of distilled water
1	piece of potato
1	iron cylinder

■■ **For each team of two students**

1	SEPUP tray
1	pipet
1	dropper
1	10-mL graduated cylinder
1	large test tube
1	rubber stopper with attached plastic tubing
1	250-mL beaker
1	50-mL graduated cylinder
1	paper towel
1	30-mL graduated cup
1	wooden splint (optional)

■ **For each student**

1	pair of safety glasses

Safety Note

Keep flammable materials away from open flames. Wear safety glasses. Hydrogen peroxide can bleach hair and clothes. Rinse off any spilled hydrogen peroxide solution with plenty of water.

Procedure

1. Set up a water reservoir, to be shared by both teams in your group, by filling the dishpan with water to a height of at least 10 cm.

2. Decide which team will use potato liquid and which team will use iron (III) chloride as the catalyst.

3. Read the rest of the Procedure and prepare a data table to record your data and observations.

4. Fill your 50-mL graduated cylinder with water and place it sideways in the dishpan, submerging it in the water. You will use this set-up to measure how much oxygen gas is produced during the decomposition of hydrogen peroxide.

5. Add 4 mL of 3% H_2O_2 solution to your large test tube, place it in the beaker, and then cap it with the rubber stopper with attached plastic tubing. Thread the other end of the plastic tubing into your submerged 50-mL graduated cylinder, as shown in Figure 1.

Procedure

(cont.)

Figure 1   Experimental Set-up for Measuring Oxygen Production

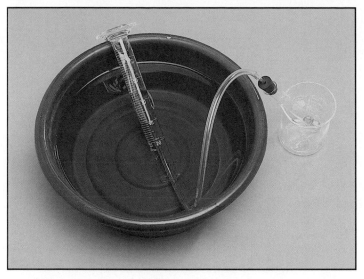

6.   One partner should now carefully turn the 50-mL graduated cylinder upside down in the pan, making sure that no, or very little, air gets trapped inside, as shown in Figure 1. If you cannot get all of the air out, record the volume of air trapped inside. Someone will need to hold the cylinder in this inverted position until the experiment is completed.

7.   If your team is using potato liquid, prepare it as described in Part A, Procedure Step 4.

8.   Add 10 drops of your team's catalyst to the test tube and re-stopper it immediately. Record your observations and the volume of oxygen gas (O_2) trapped in the graduated cylinder every minute for 10 minutes.

> **Note:** At the end of this experiment you may need to perform a test on the collected gas. Carefully pull the plastic tube out of the graduated cylinder and leave the graduated cylinder upside down in the water so that no gas escapes or gets in.

9.   Remove the stopper from the test tube and use the pipet to place one drop of the liquid into each of Cups 1, 2, and 3 of your SEPUP tray.

10.   Perform the potassium permanganate test on the solution in Cups 1, 2, and 3 as described in Part B, Procedure Steps 2–4.

11.   Report the total amount of O_2 gas produced and the results of the potassium permanganate test to your teacher. Make a data table to record the results for the whole class.

12.   Carefully clean all the equipment used during this activity.

Analysis

Group Analysis

1. Determine the average rate of production of O_2 gas for each of the liquid catalysts tested. Express these data in mL/minute.

2. For each liquid catalyst tested, calculate the rate of O_2 gas production during the first 5 minutes and the rate during the second 5 minutes. Compare the calculated rates for the two catalysts.

3. Using the class data, estimate the percent H_2O_2 remaining in solution after 10 minutes of decomposition with each liquid catalyst.

4. Which liquid catalyst was most effective in decomposing H_2O_2? Use all possible evidence to support your decision.

Individual Analysis

5. Which method do you think is better for determining how completely the hydrogen peroxide has decomposed—measuring the O_2 gas produced or measuring the percent H_2O_2 remaining in solution? Explain.

6. If you worked for a company that needed to decompose hydrogen peroxide on an industrial scale, which one of these catalysts would you use? Give reasons to support your choice.

7. Design an experiment that would allow you to test quantitatively the effectiveness of enzymes found in grass to catalyze the decomposition of H_2O_2. Describe your experiment so that students taking this class next year could carry it out.

Extension   Refer to a biology textbook or another library resource to determine why many cells contain the enzyme catalase.

26.3 Catalysts and Food Production

Purpose ▶ **E**xplore the relationship between the use of catalysts and the production of ammonia, which is often used in the manufacture of fertilizers.

Introduction ▼

Food is basic to survival. The amount and quality of food available to people in different regions of the world is one of the most important indicators of the quality of life. Much of the food produced today couldn't be grown without the addition of nitrogen-rich **fertilizers** to the soil. Ammonia gas (NH_3), which can be used to make fertilizer, is produced by a chemical process that requires nitrogen and hydrogen gases, as shown in the equation below.

$$N_2 \quad + \quad 3H_2 \quad \rightarrow \quad 2NH_3$$
$$\text{nitrogen} \quad + \quad \text{hydrogen} \quad \rightarrow \quad \text{ammonia}$$

Nitrogen and hydrogen react to form ammonia only under certain conditions. Until the early 1900s, this reaction took place only in the cells of certain bacteria or in the laboratory at extremely high temperatures and pressures. Achieving the necessary temperature and pressure conditions was very dangerous and very costly. In 1913, the German chemist Fritz Haber discovered a method of producing ammonia at much lower temperatures and pressures, thus allowing it to be manufactured commercially. Haber's discovery of an appropriate catalyst was key to the success of his process.

The Haber Process

Before the invention of the Haber process, converting a mixture of nitrogen and hydrogen gas into ammonia was a very expensive procedure that required highly specialized equipment. Fritz Haber performed over 6,500 experiments before he discovered an iron-based catalyst that significantly decreased the energy necessary to initiate the ammonia synthesis reaction. Even using Haber's catalytic process, obtaining a reasonable quantity of ammonia gas requires heating the reactants to around 500°C and putting them under a pressure of 200 atmospheres (~3000 pounds per square inch).

Haber was motivated to find a cheaper way to produce ammonia because it was in great demand for use in fertilizers and explosives and in short supply. At the time, much of the world's ammonia was produced from guano, the nitrogen-rich waste product of birds. Local guano deposits were becoming depleted and international trade was difficult, due to tensions that eventually led to World War I.

Ammonia is the component of fertilizers that adds nitrogen to the soil. Although nitrogen is abundant in the air, this gaseous form cannot be used by plants. Plants must get the nitrogen they need from the soil. Some types of

The world's farmers add millions of tons of fertilizer to the soil each year. Fertilizer can be applied to fields in many ways, including by airplane.

bacteria living near plant roots can convert gaseous nitrogen from the air to usable nitrogen-based compounds. Some nitrogen also comes from the decomposition of dead organisms. Decomposers produce gaseous nitrogen and other nitrogen-based compounds. Without these sources of nitrogen, plants would not be able to produce proteins or reproduce.

Nitrogen, like all material resources on Earth, is not created or destroyed. Nitrogen cycles through soil, plants, animals, and the atmosphere over and over again. The Haber process uses gaseous nitrogen obtained from the atmosphere and returns it to the soil in the form of nitrogen-based compounds in fertilizers.

Nitrogen compounds are also important in the manufacture of explosives. Gunpowder is composed of about 75% potassium nitrate (KNO_3) and 25% sulfur and charcoal. As this mixture burns, it produces vast amounts of hot gases that expand, creating extremely high pressures that can lead to an explosion. You will learn more about the way gases behave under pressure in Activity 29.3, "Cramped and Hot."

The Haber process supplies the vast majority of the world's manufactured nitrogen-based compounds. The ammonia it produces can be applied directly to the soil. It can be converted to ammonium sulfate, ammonium nitrate, or ammonium phosphate, all of which are common fertilizers used in the production of major food crops around the world. To supply the food for an ever-increasing world population, the amount of fertilizer produced by the Haber process grew from 14 million tons in 1950 to over 130 million tons in 2000. This increase is directly responsible for recent increases in the world's food supply and played a crucial role in the Green Revolution. As with many great inventions, the chemical process Haber discovered is not without trade-offs. Despite the huge energy reduction accomplished by the Haber process, producing ammonia still requires vast amounts of energy. Fertilizer use contributes to water pollution problems, and the increased availability of ammonia has led to an increase in the production of military explosives.

The Haber process can produce fertilizers cheaply because most of the energy used to fuel the process comes from fossil fuels, which are plentiful and fairly inexpensive in some parts of the world. Countries such as the United States, Saudi Arabia, Iraq, Iran, Mexico, the United Kingdom, and Indonesia have large petroleum resources. They can manufacture fertilizers cheaply for export to Japan, China, India, and countries in South America and Africa that do not have such abundant fossil-fuel resources. Fertilizers have greatly increased the productivity of poor soil and helped many countries develop an agricultural base and enrich their

quality of life. However, fertilizer use has also contributed to water pollution. Water that runs off or is leached from agricultural fields often transports dissolved chemicals from fertilizers into rivers, lakes, and groundwater.

The Haber process is considered by some to be the most important chemical reaction humans use to synthesize and control a material that has very beneficial as well as harmful effects on society. It demonstrates how discovery and scientific innovation present us with opportunities, decisions, trade-offs, and challenges.

Analysis

Group Analysis

1. Why is nitrogen so important to life processes?

Individual Analysis

2. Describe one issue that involves the trade-offs associated with use of the Haber process to produce ammonia for fertilizer. Explain the evidence on both sides of the issue, and state your recommendation for how the issue might be resolved.

Catalysts, Enzymes, and Reaction Rates

26.4 Catalysis Paralysis

Purpose ▶ **I**nvestigate the effect of temperature on the enzymatic breakdown of a sugar.

Introduction

One condition for survival of humans and other organisms is an appropriate temperature range. Your observations in Activity 3.1, "Bubble-Blowing Fungi," hold true for most organisms: there is an optimal temperature range for survival, and at extremes of temperature most organisms cannot survive. For example, high temperatures change the structure of proteins, which prevents normal cell function. One important group of proteins is the **enzymes** that catalyze reactions in cells. In this activity you will study the effect of temperature on the enzyme **lactase**.

Many people around the world avoid eating milk products. Do you know why?

Lactase is produced in the intestine, where it breaks down **lactose**, a sugar found in milk products. Lactose is composed of two smaller sugars—glucose and galactose—that are linked together. Lactase splits lactose into these two smaller sugars, which are more easily absorbed by the intestinal walls and transported to the blood. Note that glucose and galactose have the same chemical formulas, but slightly different structures, as shown in Figure 2.

Introduction
(cont.)

People with lactose intolerance have difficulty digesting milk products. This common condition, which can be inherited genetically or develop as people grow older, occurs when the body does not produce enough lactase. Lactase tablets, available in supermarkets and drug stores, can be taken to provide the lactase enzyme and aid digestion. Another alternative is to decrease consumption of lactose by purchasing lactose-reduced milk products or avoiding milk products altogether. You will use lactase tablets as a source of the lactase enzyme for your investigation.

Prediction

What do you think will be the optimal temperature for the breakdown of lactose by lactase?

Figure 2   Catalysis of Lactose Breakdown by the Enzyme Lactase

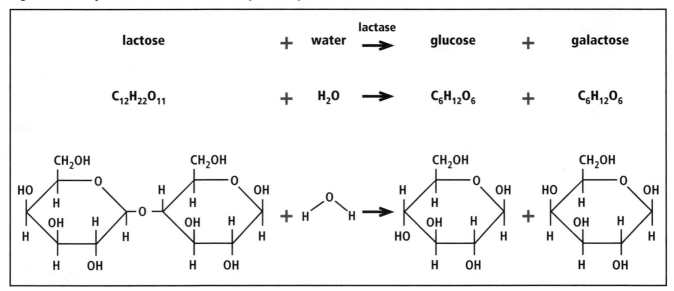

Materials

For each group of four students

1   hot plate
1   graduated cup containing
     lactose solution
2   150-mL beakers

For each team of two students

1   SEPUP tray
3   glucose test strips
2   small test tubes, each containing
     1-mL of lactase enzyme
2   droppers
1   immersion thermometer
1   stir stick

For each student

1   pair of safety glasses

Safety Note

Wear safety glasses. Glassware can be hot even when it does not appear to be. Be careful with hot lab equipment.

Procedure

1. Using the materials listed above, design an experiment to determine the optimal temperature for the lactase reaction.

    Hint: Each group of students in the class can be assigned to collect data at a different temperature.

2. With the help of your teacher and classmates, decide on a procedure that will result in the collection of enough data to determine the optimal temperature for the lactase reaction.

3. Carry out your procedure as discussed in class. Make sure to clean up all the lab equipment you used when you have finished.

4. Report your data so that they can be compiled to create a class data table.

5. Prepare your own data table to use in recording the data collected by all the other teams in the class.

Analysis
?

Group Analysis

1. What is the purpose of the glucose test strips? What information do they provide about the solution before and after the reaction with lactase?

2. Are the results of your experiments with lactase qualitative or quantitative? Explain.

3. Compare the results of this investigation with the results from the investigation of yeast you performed in Activity 3.1, "Bubble-Blowing Fungi." What evidence do these experiments provide concerning the effect of temperature on living organisms?

Individual Analysis

4. Use the class data to formulate and then write a conclusion about the effect of temperature on the enzyme lactase. How does your conclusion compare with your prediction?

27.1 Degradability: Solution or False Promise?

Purpose ▶ **R**elate the use of materials to their disposal and degradability.

Introduction ▼

Decomposing materials can generate large quantities of heat and other products. Conditions within landfills and compost heaps are often monitored to ensure safety and to maintain optimum conditions for degradation.

In past activities you have studied how materials are used to make products. After a product is made and used, it must be disposed of. In some cases, disposal involves reuse or recycling. Aluminum cans and glass bottles are frequently recycled. Natural organic waste materials—food and yard trimmings—can be composted. However, many used products are disposed of in landfills. Disposing of wastes in landfills poses many potential problems. In Activity 25, "By-Products of Materials Production," you learned about toxic-waste landfills and researched a method for treating heavy metal wastes to prevent them from contaminating surrounding soil and water. Some landfills are designed to prevent

Introduction
(cont.)

degradation of the waste material so that it does not cause environmental damage; other landfills are designed to promote degradation so that the elements in the waste materials can eventually be reused. In Activity 25, you investigated chemical reactions that were meant to prevent degradation reactions. Here, you will look at the degradation reactions themselves.

A material that is "degradable" is one that decomposes or can otherwise be broken down. Many people believe that the degradable materials disposed of in a landfill will eventually decompose or break down into reusable substances. Do they?

In most cases, degradation involves chemical reactions that require either the presence of microorganisms (biodegradation) or sunlight (photodegradation). For example, food and yard wastes are considered biodegradable because they can be broken down by microorganisms living in soil. The products of degradation reactions are smaller, simpler substances, such as water, carbon dioxide, and other compounds that are reused via natural cycles. This reading describes what may—or may not—happen to the wastes that are disposed of in sanitary landfills, which are used for the disposal of non-toxic household wastes.

The Degradation Dilemma

Experts agree that we must reduce our use of landfills as the primary method of disposing of our solid waste. Landfill space is limited, and existing landfills are rapidly being filled to capacity. In the future, waste management will most likely combine waste reduction strategies with a variety of waste disposal methods.

Waste reduction strategies include the reduction or elimination of unnecessary materials use, such as packaging that serves no purpose other than advertising, combined with an increase in the reuse and recycling of materials like glass, aluminum, and plastic. Increasing the life of reusable products, such as appliances and automobiles, by making them from sturdy, long-lasting materials can also reduce the volume of solid waste that enters our landfills.

Innovative waste disposal approaches include high-temperature incineration, which not only dramatically reduces the volume of solid waste but also produces potentially useful energy. Making items such as disposable plates, diapers, and packaging materials from easily degradable materials is also considered a good way to

help reduce the amount of space required for municipal solid-waste disposal. How promising are degradable products as a solution to the landfill problem?

Most natural organic materials, such as food and yard waste, degrade relatively quickly. However, many synthetic organic materials, such as plastics and other petroleum products, degrade very slowly. Conventional plastics typically require 100–400 years to degrade naturally. New specially formulated plastics are designed to be quickly biodegradable or photodegradable. Unfortunately, the fact that a material can degrade does not mean that it will. Biodegradation requires an environment favorable to the growth of microorganisms; photodegradation requires exposure to strong ultraviolet light.

Degradation within a landfill is not without its trade-offs. It can produce environmental problems, such as the growth of harmful bacteria or the production of toxic by-products. Some degradable products undergo partial breakdown in landfills, releasing substances that can then leach out and pose a hazard to surrounding soil and water. For example, if rain or garbage moisture seeps through the landfill, it may cause heavy metals (such as lead and mercury) and hazardous plastic polymers to leach out and contaminate surrounding soil and water. Consequently, many landfills are lined and covered to minimize the flow of materials into and out of the landfill, thus reducing the possibility of environmental contamination, especially through the process of leaching. The interior of a lined and covered landfill is dark, dry, and low in oxygen. These conditions do not favor degradation. In fact, they inhibit it.

Dr. William Rathje, a renowned expert on U.S. landfills and garbage, speaks of the "myth of biodegradation." In the U.S., he cautions, "we cherish the faith that this process (of degradation) flourishes in every landfill. It's a false promise." His excavations show that most of the easily biodegradable organic material deposited in landfills does not decompose; instead, it is actually preserved. The evidence he has gathered includes 40-year-old newspapers that are still readable, 15-year-old steaks with the fat intact, 25-year-old tree leaves, 35-year-old lumber, and even 20-year-old hot dogs. "Their preservatives really work!" exclaims Rathje. Given our current waste disposal options, Rathje's findings mean that the public's perception of degradability as an advantage for consumer products may be a false hope.

Decisions about when and how to use degradable products are not simple. Composting has the potential to substantially reduce the volume of food and yard waste in landfills. However, composting is not a viable method of handling other forms of waste, such as plastics and other manufactured products. Since we can't be sure that all degradable products will decompose after disposal, we can't necessarily assume that degradable products are preferable to non-degradable products. When purchasing a product, consumers need to consider factors other than cost and the material the product is made from. We should take into account how much waste is produced and, when disposed of, whether or not it will degrade or be reused or recycled.

Analysis

?

1. Think of a product that might last longer, and therefore create less waste, if it were made from a different material. Describe the properties of the ideal material to use in making this product.

2. Describe a disposable item that you think is made of the most appropriate material currently available. Explain why you think this is the best material to use for this item.

3. Describe a disposable item that you think is not made of the most appropriate material currently available. Propose an alternative material, and explain why that material would be better.

4. Describe one or more situations in which biodegradation or photodegradation of substances might affect the environment, either positively or negatively, and relate this to sustainable development.

27.2 A Closer Look at Chemical Degradation

Purpose ▶ **D**etermine how temperature affects the chemical degradation of acetylsalicylic acid.

Introduction ▼

Aspirin is a common household medication. Its chemical name is acetylsalicylic acid. Most medications, including aspirin, come in a package with a clearly marked expiration date. You may have been told that it's a good idea to dispose of unused medications once the expiration date has passed. Why? Over time, medications degrade, or break down, and lose their effectiveness. In some cases, the breakdown products can even be harmful. The process of degradation occurs at different rates for different materials. The chemical equation that describes the degradation of acetylsalicylic acid is shown below.

$$C_9H_8O_4 \quad + \quad H_2O \quad \rightarrow \quad C_7H_6O_3 \quad + \quad C_2H_4O_2$$

acetylsalicylic acid + water → salicylic acid + acetic acid

In Part A of this activity, you will perform a serial dilution to establish a standard set of salicylic acid solutions with known concentrations. In Part B, you will use the standard set of solutions to estimate unknown concentrations of salicylic acid produced by the breakdown of aspirin.

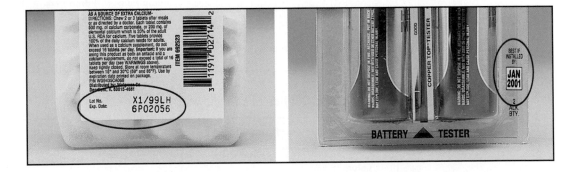

Many consumer items contain chemicals that degrade over time. The expiration date printed on the package indicates the point at which the amount of degradation that has occurred will reduce the product's effectiveness.

Part A Developing a Standard for Comparison—Performing a Serial Dilution

Materials

⊞ For each group of four students

2 aspirin tablets, one each of two different brands

1 30-mL dropper bottle of 0.5% salicylic acid solution

1 30-mL dropper bottle of iron (III) chloride solution

1 30-mL dropper bottle of distilled water

▦ For each team of two students

1 SEPUP tray

1 stir stick

1 clean plastic cup

1 dropper

■ For each student

1 pair of safety glasses

Procedure

1. Decide which team is going to test which brand of aspirin, then use a plastic cup to dissolve your team's aspirin tablet in 100 mL of water. This may take a few minutes, so set the solution aside until you need it for Part B.

2. Use the dropper bottle to add 0.5% salicylic acid to Cups 1 and 2 of your SEPUP tray as indicated in Table 1.

3. Use the dropper bottle to add distilled water to Cups 2–6 as indicated in Table 1.

4. Make the solution for Cup 3 by using the dropper to remove some of the solution from Cup 2 and adding exactly one drop to Cup 3. Return any solution remaining in the dropper to Cup 2. Stir Cup 3.

5. Repeat Step 4 for Cups 4 and 5, always taking one drop from the previous cup.

Table 1   Standard Serial Dilution of Salicylic Acid

Liquid	Cup 1	Cup 2	Cup 3	Cup 4	Cup 5	Cup 6
Salicylic Acid	10 drops 0.5%	1 drop 0.5%	1 drop from Cup 2	1 drop from Cup 3	1 drop from Cup 4	0 drops
Distilled Water	0 drops	9 drops	9 drops	9 drops	9 drops	10 drops

Table 2   Serial Dilution of Salicylic Acid

Liquid	Cup 1	Cup 2	Cup 3	Cup 4	Cup 5	Cup 6
Color						
Salicylic Acid Concentration						

Procedure
(cont.)

6. Add 1 drop of iron (III) chloride solution to each of the six cups.

7. Prepare a data table similar to Table 2 and record in it the color of the solution in each cup. (You will fill in the concentration data when you answer Analysis Question 1.)

8. Have one team in your group save the SEPUP tray for later. Do not clean it out, because both teams will need to compare the solutions in Part B with this standard set of solutions.

Analysis
?

Group Analysis

1. Calculate the concentration of salicylic acid in each cup of your serial dilution. Record your answers in your data table.

2. Which cup contains the control?

Part B   Determining the Effect of Temperature on Aspirin Degradation

Materials

■■
■■ **For each group of four students**

1	30-mL dropper bottle of 0.5% salicylic acid solution
1	30-mL dropper bottle of iron (III) chloride solution
1	100-mL graduated cylinder
1	hot plate
1	150-mL glass beaker (for hot water bath)
1	immersion thermometer
2	10-mL graduated cylinder
1	SEPUP tray with standard solutions from Part A

■■ **For each team of two students**

1	fresh SEPUP tray
1	plastic pipet
1	small test tube
1	stir stick
1	plastic cup with aspirin solution from Part A

■ **For each student**

1	pair of safety glasses
1	sheet of graph paper

Safety Note Wear safety glasses. Hot plates and glassware can be hot even when they do not appear to be. Be careful with hot lab equipment.

Procedure

1. Prepare a data table you can use to record one measurement of color and concentration at room temperature and six measurements of color and concentration at your assigned temperature.

2. Fill a 150-mL beaker with about 100 mL of tap water, place a thermometer in the beaker, and place the beaker on the hot plate. Turn the hot plate on at a setting appropriate to your assigned temperature.

3. Stir the aspirin solution in the plastic cup. Using a pipet, place 10 drops of your dissolved aspirin sample in Cup 1 of your team's fresh SEPUP tray. Test the solution by adding one drop of iron (III) chloride solution to Cup 1.

4. Move your SEPUP tray close to the tray from Part A (containing your standard solutions) so that you can estimate the concentration of salicylic acid. Record these room-temperature results in your data table.

5. Place 5 mL of your aspirin solution in a labeled test tube. When your water bath reaches your assigned temperature, place the test tube in the water bath and record the exact time as well as the temperature. Proceed immediately to the next step.

6. Use the pipet to remove some aspirin solution from the test tube in the water bath and place 10 drops in Cup 2 of your SEPUP tray, as shown in Table 3. Return the unused solution to the test tube.

Table 3   Distribution of Solutions in the SEPUP Tray

Cup	1	2	3	4	5	6	7
Temperature	Room temperature	Your assigned temperature					
Time	——	Start	After 3 min.	After 6 min.	After 9 min.	After 12 min.	After 15 min.
Aspirin Solution	10 drops	10 drops	10 drops	10 drops	10 drops	10 drops	10 drops
Iron (III) Chloride Solution	1 drop	1 drop	1 drop	1 drop	1 drop	1 drop	1 drop

Procedure
(cont.)

7. Add one drop of iron (III) chloride to the new sample in your SEPUP tray. Stir. Determine the approximate concentration by comparing the color of this solution to your standard solutions. Record the results in your data table.

8. Repeat Steps 6 and 7 five more times, once every 3 minutes, until a total of 15 minutes have elapsed. Each time place the 10 drops of solution into a different small cup of your SEPUP tray.

9. Report to your teacher the highest concentration of salicylic acid and the time at which the solution first reached this level.

Analysis
?

Group Analysis

1. Prepare a graph of your salicylic acid concentration (in ppm) over time.

2. Based on your experimental results,

   a. what is the effect of time on the degradation of acetylsalicylic acid?

   b. what is the effect of temperature on the degradation of acetylsalicylic acid?

3. How do chemical degradation reactions relate to material resource use issues?

Individual Analysis

4. Based on the results from the whole class, predict what would happen if your water bath had been 100°C. Explain the evidence on which you based your prediction.

5. Based on what you have learned in this activity, what recommendations would you make for the storage of aspirin in the home? Explain why you would make these recommendations.

6. How could the results of this degradation experiment be useful to someone concerned with global sustainability? Describe specific example(s) in your response. (**Hint:** Efforts toward sustainable development involve wise use of resources.)

Extension   Perform Part B of this activity with an aspirin tablet that is already past its expiration date, and compare the salicylic acid concentration with your data on tablets that have not yet expired.

27.3 Give It to Them, They'll Eat It

Purpose ▶ **E**xplore the use of bioremediation in industrial clean-up of organic pollutants.

Introduction

As discussed in previous activities, toxic substances can, and occasionally do, contaminate groundwater and soil at toxic-waste disposal sites, around underground storage tanks, or in other areas where toxic materials are used, produced, and stored. Over time, these substances can be transported long distances and spread over large areas by water flowing along the surface or through the ground. Several alternatives exist for the clean-up of these materials. Traditional methods include chemical and physical techniques that remove a contaminant from soil or water. For example, a large percentage of spilled oil can be cleaned using a special vacuum cleaner. One chemical clean-up method involves dissolving the contaminant in a liquid, then removing the liquid from the site for further treatment or disposal. Recently, environmental engineers have been making increased use of biological treatment methods that use microorganisms to break down the contaminant into non-toxic products, primarily minerals, that do not have to be removed from the site. The choice of which techniques to use differs from one contaminated site to another, depending on the type of contaminant, the intended future use of the site, the extent and size of the contaminated area, and state and local regulations.

These circular and rectangular structures are major components of a large urban water treatment facility. Contaminants are removed from the wastewater using physical, chemical, and biological treatments.

Cleaning Up Contaminants With Microorganisms

Bioremediation is the process of using microorganisms to treat organic contaminants by breaking them down into non-toxic compounds. The final products of biological treatment are usually carbon dioxide and water, although a number of intermediate products may be generated during the degradation process. Often, the use of microorganisms can significantly reduce the cost of a clean-up project. Bioremediation has long been used in wastewater treatment to break down large organic molecules. It is now being used in a number of other applications as well, including the clean-up of contaminated groundwater and soil. Some naturally occurring organisms can be used for bioremediation. Recently, scientists have engineered new strains of microorganisms designed to degrade specific contaminants.

Bioremediation is suitable for the clean-up of organic contaminants, such as oil and sewage waste, but it cannot usually be used to clean up metals, because they are already very simple elemental substances. Some organic pesticides and industrial solvents contain heavily chlorinated compounds or complex cyclic structures that are not suitable for bioremediation. The very properties that make them useful to humans make them toxic to the microorganisms. In some cases, as toxic materials are being degraded by microorganisms, intermediate products can be produced that are even more dangerous than the original substances. Careful monitoring must be performed during such procedures. Eventually the microorganisms significantly reduce the toxicity of both the original and the intermediate substances.

Organisms suitable for bioremediation are often found in large numbers at contaminated sites. This situation occurs naturally because their food supply (the contaminant) is abundant, and their predators have decreased, due to the toxicity of the contaminant. However, to complete the clean-up effort more quickly, additional microorganisms, oxygen, and nutrient supplies are often introduced to the contaminated site. Special considerations must be taken into account to enable these organisms to thrive in their toxic surroundings. Oxygen must be provided for those that undergo aerobic respiration, waste products must be removed, and an additional food supply may also be necessary. Some of the species used in bioremediation thrive on methane or methanol; they may require additional supplies of these compounds. The organisms that are added may be the same species that were already there or species that have been genetically engineered to more efficiently and effectively degrade the contaminant.

Generally, bioremediation is used in combination with other clean-up techniques. A related technique is phytoremediation, the use of plants in toxic-waste clean-up efforts. The soil region near plant roots provides an ideal environment for microorganisms, so the presence of plants can increase the efficiency of bioremediation efforts. Plants can also remove some non-degradable substances from soil or water simply by absorbing them through their roots. More traditional chemical and physical techniques can be used in conjunction with biological action to increase the speed and depth of treatment.

Clean-up of contaminated areas involves the work of teams of scientists and engineers from many fields, including chemistry, physics, geology, biology, and ecology. These projects also involve specialists in law, environmental policy, and commercial development, who work together to solve a problem that affects the larger community.

Analysis

Group Analysis

1. If you had to decide whether bioremediation would be appropriate for the clean-up of a toxic waste spill in your neighborhood, what additional information would you want to have?

2. Describe some of the trade-offs involved when deciding whether to use bioremediation or traditional chemical and physical strategies for waste removal.

Food Preservation

28.1   Return of the Bubble-Blowing Fungi

Purpose ▶ **E**xplore, analyze, and compare different methods for slowing down the growth of microorganisms that can cause food spoilage.

Introduction ▼

The earliest humans had to hunt and gather food continuously because they had no means of preserving it. If they didn't eat it within a short time, it spoiled. The technology of food preservation has evolved along with human civilization. Some of the earliest methods involved storing food in cool places, such as caves, cool streams, or underground cellars, often called "root cellars." Humans who traveled and needed to carry food with them used food preservation methods such as drying and salting. Long-term food preservation was not available until the late 18th century, when canning, which involves boiling food and storing it in a sealed container, was developed.

Currently, the world produces enough food to adequately feed the entire human population. However, even today, a large percentage of the world's food spoils before it can be eaten. Much of the spoilage occurs because of the action of microscopic organisms, such as mold or bacteria. In this activity, you will use yeast as a model for any microorganism that spoils food. You will investigate how well yeast grows in a variety of different physical and chemical environments.

A cook's journal, 1779

> **May 10, 1779**
>
> One of the hardest things for me to do is to plan well enough for each day so that I need make only one trip to the root cellar. The cellar was constructed two years ago, and to make it plenty large and cool enough to hold the fresh foodstuffs and jars of preserves we need, it was dug one hundred yards from the kitchen door. Most days I end up taking two or three trips to the cellar, and this is especially hard when it is cold and rainy. However, I am thankful for the job I have and the ample resources with which I currently work. It was not always that way for me.

Introduction
(cont.)

As you may remember from previous activities, yeast produce carbon dioxide (CO_2). When CO_2 dissolves in water, it produces carbonic acid (H_2CO_3). In this activity you will estimate the rate of yeast activity by measuring changes in acidity associated with the production of H_2CO_3. The indicator bromthymol blue turns green and then yellow as the solution becomes increasingly acidic. The careful addition of drops of sodium hydroxide (NaOH), a base, makes the solution neutral. The number of drops of base needed to neutralize the acid is a measure of the amount of acid that was in solution. This method of determining the amount of acid (or base) in a solution is called a **titration**.

Safety
Note

All students should wear safety glasses in both parts of this activity.

Part A Temperature as a Food Preserver

Materials

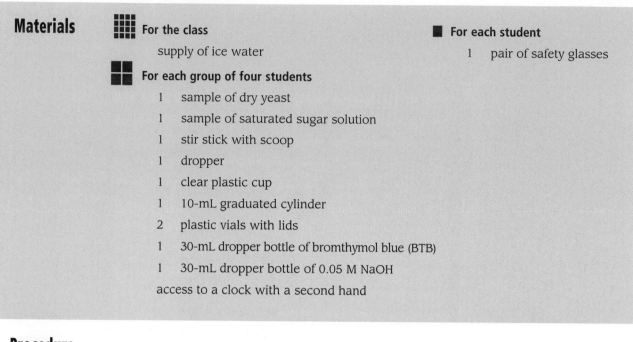

For the class

supply of ice water

For each group of four students

1 sample of dry yeast

1 sample of saturated sugar solution

1 stir stick with scoop

1 dropper

1 clear plastic cup

1 10-mL graduated cylinder

2 plastic vials with lids

1 30-mL dropper bottle of bromthymol blue (BTB)

1 30-mL dropper bottle of 0.05 M NaOH

access to a clock with a second hand

For each student

1 pair of safety glasses

Procedure

1. Read Steps 2–5 carefully and prepare a data table to record your data.

2. Fill your cup halfway with ice water. Make sure you have one or two ice cubes in your cup.

3. To each of the two vials, add 5 mL of tap water, 2 drops of saturated sugar solution, and 5 drops of BTB. Gently shake each vial and record the appearance of each.

4. Place one stir-stick scoop of yeast in each vial. Cap and gently shake each vial. Place one vial in the cold water bath (you will have to apply pressure to the lid to keep the vial submerged) and allow the other to remain at room temperature. Record the exact time.

5. Every minute for 10 minutes, swirl each vial and record its color. If the color is greenish or yellowish, add 0.05 M NaOH one drop at a time until the color turns blue again. Record the number of drops added.

Analysis

Group Analysis

1. What is the purpose of performing one experiment at room temperature?

2. What is the purpose of adding the saturated sugar solution?

3. What is the purpose of adding the bromthymol blue?

4. Explain the purpose of Procedure Step 5.

Part B   Chemicals as Food Preservers

Prediction

How do you think yeast growth in a very salt-rich or sugar-rich environment at room temperature will compare with your results from Part A? Explain your reasoning.

Materials

■ **For each group of four students**

1   sample of dry yeast

1   sample of saturated sugar solution

1   sample of saturated salt solution

1   stir stick with scoop

1   dropper

1   clear plastic cup

1   10-mL graduated cylinder

2   plastic vials with lids

1   30-mL dropper bottle of bromthymol blue (BTB)

1   30-mL dropper bottle of 0.05 M NaOH

access to a clock with a second hand

■ **For each student**

1   pair of safety glasses

Procedure

1. Design an experiment to test your prediction. Write out the procedural steps for your experiment.

2. Perform your experiment and record all your collected data.

Analysis

▼
?

Group Analysis

1. Which of the three environments—cold, salty, or sugary—best reduced yeast growth? What is your evidence?

2. Use your knowledge of cells and chemistry to explain the effects that an environment with a high concentration of salt or sugar has on yeast growth.

3. What are the advantages and disadvantages of using each of the methods—cooling, salting, or sugaring—for preserving foods?

4. Before 1900 many homes had a root cellar, yet there is a good chance you had never heard of one before this activity. Have root cellars disappeared because they are no longer useful, no longer appropriate, or no longer popular? Explain your reasoning.

Individual Analysis

5. Which method—cooling, salting, or sugaring—would you prefer to use for the preservation of your food? Explain your reasoning.

6. Would eliminating food spoilage guarantee that no one would starve? Explain.

28.2 Chilling Choices

Purpose ▶ **I**nvestigate some of the trade-offs associated with various food preservation technologies.

A cook's journal, 1840

October 20, 1840

The recent years have been very difficult for all of us here in the kitchen. We have been expected to keep everything going smoothly, but nothing is as it was! I hope I will become more accustomed to all the new things before long.

The one new thing I am truly happy about is having an icebox here in the house—I can store fresh food for the whole week! At first I feared that it would be unsafe to keep food inside because it would spoil too fast, not to mention that melting ice would be untidy. It is true, I do have to keep a close eye on the ice, but surely it is better than the daily trip down to the dank old root cellar!

Introduction ▼

The historical readings on the following pages introduce some of the trade-offs associated with choosing a method of food preservation for use in the home. In the U.S., root cellars were popular until the mid-1800s, when iceboxes became widespread. Iceboxes are insulated containers built to contain ice and fresh food. For nearly 100 years, the ice business flourished and most homes relied upon a regular delivery of ice. Near the end of the 19th century, electric refrigerators became available to the general public. As these two readings illustrate, public concern can be aroused when new technologies are introduced.

Think Twice—Depend Upon Ice!

THERE was an old woman who lived in a shoe
Decided her old icebox never would do.
Installed a high polished and fancy affair
That cooled by the draft of a marvelous air
No more would her ice man come around every day;
No more would the bill be a bother to pay.

HER monthly account was decidedly small
But the new device claimed it cost nothing at all.
Said nothing of noise, mentioned nothing of waste,
Would dry up the meat but improve on its taste,
Was colder than ice by a dozen degrees
And plastered all over with gold guarantees.

HER troubles were over, she surely was glad,
She knew it was true for so said the ad.
The thing was installed and it ran for a while,
Calm was her soul and broad was her smile.
It looked awfully pretty, it certainly was nice,
But the cute little pans wouldn't make enough ice!

WHENEVER she wanted a cool lemonade
She waited ten hours till more ice was made;
The thing wouldn't run 'cause the parts were not oiled,
Then the current went off and her vegetables spoiled.
The coils sprung a leak and the smell was a fright,
Couldn't get in the kitchen till late in the night.

AWOKE in the morning, the milk was all sour,
Phoned for some ice; it was there in an hour.
Greeted her ice man, a friend staunch and true,
"I'll never again try to be without you."
So listen to this tale, tho' the ad stuff looks nice,
For reliable service "DEPEND UPON ICE."

circa 1920, author unknown

Iceless Electric Refrigerators Make Ice for You

Everybody knows at least one thrifty housewife who "economizes on ice." She buys only fifty or seventy-five pounds when she should buy one hundred. She saves ice—but spoils food! Not badly enough to make it inedible, perhaps. The butter is only tainted. The meat requires the addition of a pinch of baking soda "to make it sweet" again. These she accepts as the inevitable spoilage of hot weather. She does not recognize the connection between the various summer ailments suffered by her family and this partly spoiled food which she has served to them.

To preserve food it must be kept cold and dry—cold and dry enough to stop the growth of poison-making organisms. One can't get rid of them entirely. Cold keeps them quiet, but they need only warmth and moisture to get busy. At 44°F (7°C) they are dead to mischief. At 46°F (8°C) they begin to stir. At 60°F (16°C) they are multiplying by thousands. At temperatures above 60°F (16°C) they make short work of any sort of animal food or delicate fruit, gorging and reproducing themselves by millions.

Excess moisture can cause molds even at low temperatures. Have you ever cleaned the drain pipe of an icebox refrigerator? If you have you know something of the unwholesomeness for which ice is responsible. To safeguard your food both the temperature and moisture inside your refrigerator should be held within certain defined limits.

Testing 110 icebox refrigerators, only twenty-seven were below 45°F (7.5°C) and forty-nine were above the positive danger line of 50°F (10°C). Each time a door is opened the inrush of warm air must be cooled. No wonder, then, that the icebox refrigerator which marks 45°F (7.5°C) in its coldest food chamber three hours after the iceman has called may be up in the dangerous sixties before his next visit.

Now comes a home-size electric refrigerating unit by which both heat and moisture are extracted from your box. It is automatic, self-regulating, self-contained. You can set the thermostat—a simple control device—so that the cold-maker will recognize when your refrigerator has warmed up and will at once begin to reduce the temperature. In a short time the food chambers below will be as cold as they ever need to be and they will remain at that temperature, keeping food fresh, pure, and wholesome for a week if necessary.

From the *Ladies' Home Journal*, 1919

Analysis

?

1. List the advantages and disadvantages of iceboxes and mechanical refrigerators.

2. Choose one of the passages. Do you think this passage is a fair representation of the facts? In your response, comment on the facts presented that support the author's point, any information about the drawbacks of the author's preferred technology that has been left out of the passage, and the details included that are intended to provoke an emotional response from the reader.

3. What other information would you like to have to better understand the trade-offs involved in choosing between these two technologies?

4. Name a new technology that has recently been introduced and discuss the trade-offs involved with its use.

28.3 Frigidly Steamy

Purpose ▶ **M**easure the amount of thermal energy required to change ice to water and water to steam.

Introduction

When a solid melts, it absorbs heat from its surroundings, cooling them down. The same thing happens as a liquid evaporates. The low-temperature environments of iceboxes and refrigerators are created when a refrigerant goes through a change in state. Before any substance can change its state from a solid to a liquid (melt) or from a liquid to a gas (evaporate, or boil), energy must be added to that substance. At temperatures not equal to the substance's melting point or boiling point,

In this photograph, water is present in each physical state—solid, liquid, and gas. Can you identify them?

almost all added energy results in an increased speed of molecular motion and a rise in temperature. At the melting or boiling point, however, added energy does not increase the speed of molecules; instead, it is used to break bonds between molecules. This allows the molecules to move more freely and independently and changes the physical state of the substance. Thus, adding energy at a substance's melting or boiling point changes its state, but not its temperature.

▶

Introduction
(cont.)

The energy needed to change one gram of a substance from a solid to a liquid (and vice versa) is called the **heat of fusion** for that substance. The energy needed to change one gram of a substance from a liquid to a gas (and vice versa) is called the **heat of vaporization** for that substance. In an icebox or a refrigerator, the energy required to change the state of the refrigerant comes from the food placed inside. This transfer of energy out of the food reduces the temperature of the food.

Part A   Investigating Changes of State

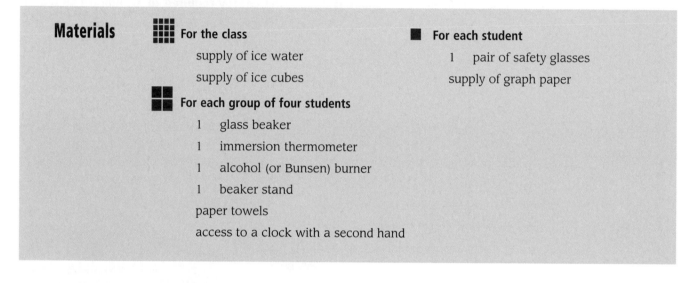

Materials

 For the class

    supply of ice water

    supply of ice cubes

 For each group of four students

  1   glass beaker

  1   immersion thermometer

  1   alcohol (or Bunsen) burner

  1   beaker stand

  paper towels

  access to a clock with a second hand

■ **For each student**

  1   pair of safety glasses

  supply of graph paper

Safety Note

All students should wear safety glasses.

Procedure

1. Read each step of the Procedure carefully and make a data table that will allow you to record the data you will collect.

2. Light your burner. Adjust the beaker stand so that when you place a beaker on the stand (Step 5), the bottom of the beaker will be in the hottest part of the flame. Extinguish the flame. Figure 1 illustrates the correct experimental set-up.

Procedure
(cont.)

3. Find the mass of your empty beaker.

4. Obtain two ice cubes. Pat them dry with a paper towel and place them in the beaker, then find the combined mass of the beaker and ice.

5. Add 50 mL of ice water to your beaker. Take the temperature of the ice water, note the time, and record this as time "0." Re-light your burner and put the beaker over the heat source. Stir gently but constantly, keeping the thermometer bulb in the center of the liquid.

6. Record the temperature every 30 seconds until you have data that includes 2–3 minutes of boiling, then remove the beaker from the heat and extinguish the flame.

7. Once your beaker has cooled enough for you to handle it safely, find the combined mass of the beaker and hot water.

Figure 1   Correct Positioning of Beaker

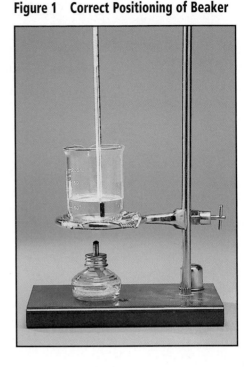

Analysis

Group Analysis

1. Calculate the mass of the ice from your results in Procedure Steps 3 and 4 and the mass of the water from your results in Procedure Steps 3 and 7.

2. Compare the combined mass of the ice and water at the beginning with the mass of the water at the end. What do you notice and how can you explain this?

3. Use your data to make a graph of temperature (y-axis) vs. time (x-axis).

Part B   Calculating Energy Transferred During Phase Changes

Procedure

1. Use the value of the slope of the central segment of the graph from Part A (in °C/min) to calculate the average thermal energy transferred from your burner to the beaker, in calories per minute (cal/min). Use the following equation:

$$\text{thermal energy transfer} \quad = \quad \text{slope of central segment} \quad \bullet \quad C_p \text{ of } H_2O \quad \bullet \quad \text{mass of } H_2O$$

$$\left(\frac{cal}{min}\right) \qquad\qquad \left(\frac{°C}{min}\right) \qquad\qquad \left(\frac{cal}{g \bullet °C}\right) \qquad \left(g\right)$$

$$\text{where} \quad C_p \text{ of } H_2O \quad = \quad 1.0 \ \frac{cal}{g \bullet °C}$$

2. Using the following equations, calculate your experimental value for the heat of fusion of ice.

$$\text{heat of fusion} \quad = \quad \frac{\text{energy absorbed by ice (cal)}}{\text{mass of ice (g)}} \quad = \quad \frac{cal}{g}$$

$$\text{energy absorbed by ice} \quad = \quad \text{thermal transfer energy} \quad \bullet \quad \text{total time of melting}$$

$$\left(cal\right) \qquad\qquad\qquad \left(\frac{cal}{min}\right) \qquad\qquad\qquad \left(min\right)$$

3. Using the following equations, calculate your experimental value for the heat of vaporization of water.

$$\text{heat of vaporization} \quad = \quad \frac{\text{energy absorbed by vapor (cal)}}{\text{mass of vapor (g)}} \quad = \quad \frac{cal}{g}$$

$$\text{energy absorbed by vapor} \quad = \quad \text{thermal transfer energy} \quad \bullet \quad \text{total time of boiling}$$

$$\left(cal\right) \qquad\qquad\qquad \left(\frac{cal}{min}\right) \qquad\qquad\qquad \left(min\right)$$

Analysis

?

Group Analysis

1. Compare your calculated values for the heat of fusion of ice and the heat of vaporization of water. Why do you think the heat of fusion is so different from the heat of vaporization?

2. Ice has been used to cool food for thousands of years, yet iceboxes did not become a common home appliance until the 1800s. Why do you think it took so long for iceboxes to become widely used?

Individual Analysis

3. Describe the slope of each of the three segments of your graph and explain what was happening to the contents of the beaker during each of the segments.

4. What information does the slope of each segment provide?

5. How does the energy transferred from the burner affect the molecules of ice and water during the time span of each of the three segments of your graph?

6. Many technologies take advantage of a change in state to produce a cool environment. Explain how a change in state can be used to cool things down.

28.4 Keeping Cool

Purpose ▶ **E**xplore some of the major changes in food preservation technology, especially those that occurred in the first half of the 20th century.

A cook's journal, 1925

April 5, 1925

I just spoke with my ice delivery man and he assures me that there will be enough ice this winter to make it all through the summer. Last summer was such a problem. The people in charge here say that I should look into getting one of those new "ice-less" electric refrigerators. The way those things work frightens me. First off I don't know if I trust that electricity and then there's the machine itself. They're big and noisy, they break down, and sometimes they even blow up from the ammonia inside the pipes. I have to say that if they were smaller and safer, I'd be interested in trying one out. It sure would be nice not to have to worry about ice this coming summer.

Introduction

Drying, salting, and canning have the side effect of significantly altering the taste and nutritive value of food. Using ice to cool or freeze food has less impact on taste and nutrition. As a result, when iceboxes were introduced in the 1800s they quickly proved their usefulness and became quite popular. The ice used to keep iceboxes cool was supplied from icehouses—large, insulated buildings in which ice that was collected from frozen lakes and ponds during the winter could be stored, even throughout the warmer months in some regions. Once electric refrigerators became safe, reliable, and compact enough for household use, they quickly replaced iceboxes as the most common household refrigeration device. New technologies now under development may eventually replace the refrigerators of today.

19th- and 20th-Century Advances in Food Preservation

The importance of ice to meatpackers, mill companies, brewers, and grocers everywhere made the ice business prominent among early industries in the United States. For example, in 1850 New York City alone required 300,000 tons of ice. The winter ice harvest provided enough ice for year-round use in climates cold enough to freeze lakes. Unfortunately, there was no way to provide a full year's supply of ice to warm climates, where the need for ice was even greater. However, in the late 1800s a number of discoveries and inventions initially developed for the beer industry led to the development of refrigeration machines that could keep a room cold or turn water into "artificial" ice in almost any climate.

The first manufactured ice was made by lowering 150–200 kg containers of pure water into a tank filled with brine (saltwater). The tank of brine also contained continuous coils of pipe, through which ammonia circulated. Within the pipes, as liquid ammonia vaporized and turned into a gas, heat was transferred from the brine to the ammonia, thus lowering the temperature of the brine. The brine usually had sufficient salinity to permit its temperature to fall to at least –18°C (0°F). But early mechanical refrigerators were large and noisy; they also used steam power, required skilled operators to keep them running, and contained refrigerants, such as ammonia, that were not safe for home use.

Between 1920 and 1930, four developments made the home refrigerator popular: a self-contained refrigerator with all parts in one machine, automatic controls, a safe refrigerant, and an effective advertising campaign. The following passage, published in *Fortune* magazine in 1937, describes the breakthrough discovery of the "perfect" refrigerant.

One morning in the mid-1930s Thomas Midgley Jr. phoned his scientific and personal friend, Charles Kettering. They chatted for a moment about minor matters and then Mr. Kettering said, "Midge, I was talking to Lester Keilholtz last night and we came to the conclusion that the refrigeration industry needs a new refrigerant if it ever expects to get anywhere."

Mr. Midgley listened and was dubious. The two best bets for use in domestic refrigerators were sulfur dioxide and methyl chloride. Both substances were toxic; one was flammable. He doubted that any one substance could be found that would be both non-toxic and non-flammable, and still be obliging enough to boil between 0° and -40°C—the essential range. What volatile but stable organic compounds might they discover in the International Critical Tables—tabulations of melting and boiling points and other properties—that they could hope to synthesize without too much trouble?

Discarding the Critical Tables as too incomplete for his purposes, Midgley turned now to a table of elements arranged according to Nobel Prize–winner Irving Langmuir's theories of atomic structure. "It was," said Midgley himself, "an exciting deduction. Seemingly, no one had considered it possible that fluorine might be non-toxic in some of its compounds. If the problem before us were solvable by the use of a single compound, then that compound would certainly contain fluorine." Still working in the library, the Midgley team decided to synthesize dichlorodifluoromethane (CCl_2F_2), a water-clear liquid which was said to boil at –20°C, but whose other properties were entirely unknown. Once they successfully synthesized a few grams of dichlorodifluoromethane, they placed a guinea pig and a tea-spoonful of the chemical, furiously boiling at room temperature, under a bell jar. The guinea pig did not collapse but instead breathed the dichlorodifluoromethane vapors with equanimity and uncon-cern. After only three days the predictions of the Midgley team were triumphantly fulfilled!

Because of this discovery, Midgley was awarded the Perkin Medal—next to a Nobel Prize, the most resounding acclaim a U.S. chemist can receive. While making his acceptance speech, Midgley, who had an instinct for the dramatic, demonstrated two important qualities of his new substance. He casu-ally drew in a lungful of the warmed gas, held his breath for a moment, and then blew it softly through a rubber hose into a beaker containing a lighted candle. The candle flame wavered and went out.

Dichlorodifluoromethane was soon manufactured in large quantities and sold under the trade name Freon.

Adapted from Fortune, 1937

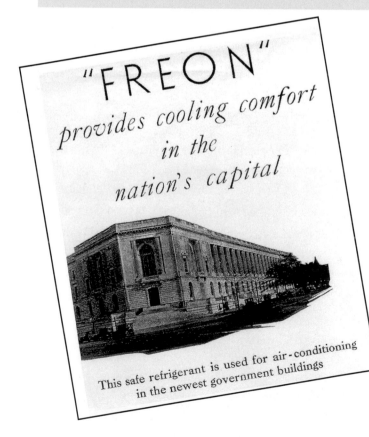

With the discovery of Freon and other similar chlorofluorocarbon gases (CFCs), refrigerator manufacturers began aggressively promoting their new technology. In 1925, one of the first self-contained refrigerators was introduced by the Kelvinator Company. It took its name from British scientist Lord Kelvin, who did some of the early research on refrigeration. The "Kelvinator" had a thermostat that regulated its temperature. The Kelvinator Company had been making iceboxes for years, and its first refrigerators could be operated either with ice or with electricity. Sales of the new home refrigerators soared. In 1929, an equal number of refrigerators and iceboxes could be found in American homes; by 1935, there were far more refrigerators. The year 1953 was a special one in the history of refrigeration: the last home icebox in the United States was manufactured that year, 150 years after the icebox was invented. Continued advances in chemical, mechanical, and electrical technology allowed refrigerators to become safer, more efficient, and easier to operate. Nonetheless, today's refrigerators operate using the same scientific principles underlying the original mechanical ice-makers, which used ammonia.

Today, nearly every household in the United States has at least one refrigerator. Unfortunately, refrigerators aren't the perfect answer to food preservation. They are costly to manufacture, they require the use of potentially hazardous substances, they require a continuous supply of energy, and, even with widespread refrigeration, it is estimated that around 33 million cases of food-borne illness occur each year in the United States. Worldwide, one-quarter to one-third of the food supply is lost to microbial spoilage, insect infestation, and rodents. Less developed tropical countries experience the highest levels of loss. This is due in part to the climate and the lack of availability both of refrigerators and of the consistent energy supply needed to operate them.

Analysis

1. As described in the cook's journal entry from 1925, when refrigerators were first invented many people were reluctant to use them. What were their concerns? Do you think these concerns were justified?

2. Why was Freon considered the "perfect" refrigerant when it was discovered?

3. How are Mendeleev's and Midgley's work similar?

4. Which method of food preservation—chemical preservatives, iceboxes, or refrigerators—do you think is best suited for use in less developed countries? In more developed countries? Describe some of the trade-offs involved and explain your reasoning.

Refrigeration Technology

29.1 — Wet and Cold

Purpose ▶ **D**iscover what happens to the temperature of the surroundings as a variety of liquids evaporate, and apply this information to food preservation.

Introduction

As you demonstrated in Activity 28.1, "Return of the Bubble-Blowing Fungi," keeping food cool extends its useful life because the lowered temperature inhibits spoilage. Humans discovered long ago that wrapping a container with a damp cloth would chill its contents, especially during warm weather. If you've ever gotten your clothes wet on a hot day, you have probably noticed this cooling effect. Wet cloth feels cool, even if it is the same temperature as your body, because when liquid evaporates it removes heat from its surroundings—in this case, your skin.

In this activity you will investigate the cooling effects of four different liquids. At any given temperature, different liquids will evaporate at different rates. The evaporation rate depends on the liquid's physical properties. Two physical properties—the boiling point and the heat of vaporization—determine the overall effect of evaporation on the temperature of the surroundings.

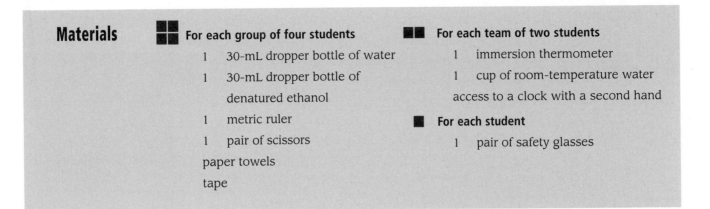

Materials

For each group of four students

1 30-mL dropper bottle of water

1 30-mL dropper bottle of denatured ethanol

1 metric ruler

1 pair of scissors

paper towels

tape

For each team of two students

1 immersion thermometer

1 cup of room-temperature water

access to a clock with a second hand

For each student

1 pair of safety glasses

Safety Note

Safety glasses are mandatory for both parts of this activity. You will be using ethyl alcohol and acetone, which are both flammable. Always keep alcohol and acetone away from heat sources, especially open flames.

Part A The Effects of Water and Alcohol

Prediction

During this activity, you are going to place drops of water or ethyl alcohol on a thermometer bulb to see if either of these liquids has an effect on temperature. What do you predict will happen? Explain your reasoning.

Procedure

1. Decide which team of two from your group will investigate water and which will investigate ethanol. Carefully read Steps 3 and 4, then set up a data table to record your results. Proceed with the experiment.

2. Cut out a 0.5 cm x 2 cm rectangle of paper towel and carefully place a small piece of tape on one end. Tape the paper towel to the thermometer so that the untaped portion of the paper towel covers one side of the bulb of the thermometer.

Procedure
(cont.)

3. Place the thermometer on the table so that about 5 cm of the bulb end hangs over the edge, as shown in Figure 1. Determine the temperature of the air in the room and record this as the starting temperature.

4. Carefully put 2 drops of the liquid your team is testing onto the paper towel covering the thermometer bulb. Observe the paper towel and record the temperature every 30 seconds for 10 minutes.

5. Rinse the thermometer in the cup of water and dry it with a paper towel.

Figure 1
Position of Thermometer

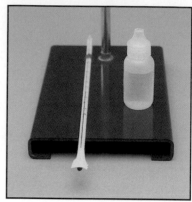

Part B   The Effects of Acetone and Vegetable Oil

Prediction

Examine Table 1. Predict how you think the evaporation of acetone and vegetable oil will affect the temperature. Explain your reasoning.

Table 1   Some Physical Properties of Liquids

Substance	Boiling Point	Heat of Vaporization
ethanol	78.5°C	210 cal/gram
water	100°C	540 cal/gram
acetone	56°C	132 cal/gram
vegetable oil	150°C	_____

Procedure

1. Ask your teacher for dropper bottles of acetone and vegetable oil. Test these liquids in the same manner as you did alcohol or water in Part A.

2. Record all your observations and measurements in a fully labeled data table.

Analysis

?

Group Analysis

1. Using a single set of axes, make a graph of temperature vs. time for each of the four liquids your group tested.

2. Which liquid had the greatest cooling effect? Explain.

3. What conclusions can be drawn about the relationship between the data you collected in this activity and the physical properties of the liquids listed in Table 1?

Individual Analysis

4. Predict the heat of vaporization for vegetable oil and provide scientific rationale for your prediction. (**Hint:** You may want to make a graph.)

5. How might an understanding of the properties of liquids be useful to someone interested in developing a sustainable means of keeping food cool? Be as specific and detailed as possible.

Extension Test different ways to increase the cooling effect of the evaporating liquids. Describe your experiments and record your results.

29.2 __In Search of the "Perfect" Refrigerant__

Purpose ▶ **A**nalyze the properties of various liquids to evaluate their potential for use as refrigerants.

Introduction

As you determined in Activity 29.1, "Wet and Cold," liquids absorb thermal energy as they evaporate. It is the process of evaporation that removes thermal energy from the interior of a refrigerator, making the refrigerator's contents cold. The refrigerant is contained in a system of coiled tubing. As the refrigerant liquid flows through the tubing inside the refrigerator, it absorbs heat and evaporates. Once it has evaporated, the gas flows out of the refrigerator into a compressor, where it is put under pressure. The increased pressure condenses the gas back into a liquid, releasing the heat of vaporization of the refrigerant. This process further increases the temperature of the refrigerant and propels it through the coils of tubing located on the exterior of the refrigerator. As the refrigerant travels through the coils, it transfers thermal energy to the air outside the refrigerator as it cools. (If you hold your hand beneath or behind the coils of a refrigerator while the compressor is running, you can feel this heat.) After flowing through the exterior coils, the cooled refrigerant is recirculated through the refrigerator to absorb more thermal energy from the interior.

During the early stages of refrigerator design and development, a number of liquids were tested to determine which one was best suited for use as a refrigerant. Each of the tested liquids had benefits and drawbacks. Refrigeration engineers had to weigh the trade-offs involved when choosing which liquid to use.

Procedure

1. Use the data in the table below to select the two substances most appropriate for use as refrigerants in home refrigerators.

Table 2   Physical Properties of Potential Refrigerants

Substance	Boiling Point	Pressure Required	Heat of Vaporization	Ignition Temperature	Toxicity
A	−78°C	960 p.s.i.	88 cal/g	non-flammable	low
B	−10°C	45 p.s.i.	97 cal/g	non-flammable	moderate
C	−30°C	120 p.s.i.	40 cal/g	non-flammable	non-toxic
D	−33°C	140 p.s.i.	339 cal/g	651°C	high
E	1°C	25 p.s.i.	92 cal/g	430°C	low
F	−43°C	130 p.s.i.	100 cal/g	466°C	low
G	−24°C	73 p.s.i.	103 cal/g	632°C	moderate

Analysis

?

Individual Analysis

1. Explain the reasons behind the choice you made in Procedure Step 1.

2. Describe the properties of a "perfect" refrigerant for each of the five categories listed in Table 2. Explain why each of the properties you described would contribute to the characteristics of a "perfect" refrigerant.

3. With global sustainability in mind, describe two other characteristics, not necessarily chemical or physical properties, of a "perfect" refrigerant.

29.3 Cramped and Hot

Purpose ▶ **D**iscover how a change in the pressure of a gas affects the temperature and volume of the gas.

A cook's journal, 1931

> July 25, 1931
>
> We've been having quite a hot spell this last week. And on account of that explosion, I've got no ice in the kitchen and I'm spending all my free time in the blazing sun looking for my cat! I guess I should feel lucky that I was at work when that ice factory blew up a block from the house. I heard the blast from ten blocks away and there's still a nasty ammonia smell in the air. It's no wonder the cat won't come back!

Introduction ▼

Modern refrigerators include machinery that is used to alter the pressure of the circulating refrigerant gas. The temperature and volume of gases, unlike liquids and solids, change significantly in response to relatively small changes in pressure. There are mathematical equations that describe the relationship between changes in temperature, volume, and pressure of a gas. These equations can be used to accurately predict how a change in one of the conditions—pressure, temperature, or volume—will affect the other two.

Part A   The Effects of Pressure on Temperature

Prediction

Do you think pumping air into a bicycle tire affects the temperature inside the tire? What evidence, if any, do you have to support your hypothesis? Explain any physical principles that would support your hypothesis.

Materials

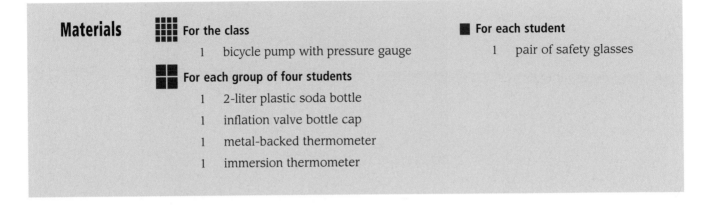

▦ **For the class**

   1   bicycle pump with pressure gauge

▦ **For each group of four students**

   1   2-liter plastic soda bottle

   1   inflation valve bottle cap

   1   metal-backed thermometer

   1   immersion thermometer

■ **For each student**

   1   pair of safety glasses

Safety Note

All students should wear safety glasses.

Procedure

1. Prepare a data table with the following headings:

Trial	Unpressurized		Pressurized to ___ psi		Depressurization	
	Air Temperature Inside	Air Temperature Outside	Air Temperature Inside	Air Temperature Outside	Air Temperature Inside	Air Temperature Outside

2. Carefully place the metal-backed thermometer inside your 2-liter bottle and tightly screw on the inflation valve cap.

3. Use the metal-backed thermometer to measure the air temperature inside the bottle. Use the immersion thermometer to measure the air temperature outside the bottle. Record both measurements in the "Unpressurized" column of your data table.

4. With your teacher's assistance, pressurize the 2-liter bottle to your assigned level. Read and record the air temperature inside the pressurized bottle.

Procedure
(cont.)

5. Carefully use the bulb of your immersion thermometer to depress the valve and release the air in a steady stream from the bottle, as shown in Figure 2. As the stream of released air is passing over the bulb of the immersion thermometer, read and record the temperature of each thermometer.

Figure 2 Measuring Depressurization Temperatures

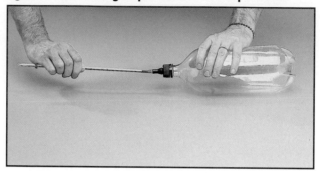

6. Repeat Steps 3–5 for two more trials.

7. Calculate the average of each temperature reading you took and report these data to your teacher to add to the class data table.

8. Make a copy of the completed class data table.

Part B The Effects of Pressure on Volume

Prediction

Do you think the volume of air in a bicycle tire changes when you get on or off the bike? What evidence, if any, do you have to support your hypothesis? Explain any physical principles that would support your hypothesis.

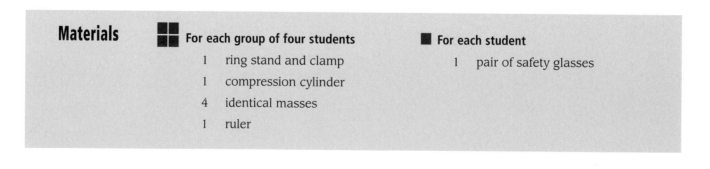

Materials

■ For each group of four students

1	ring stand and clamp
1	compression cylinder
4	identical masses
1	ruler

■ For each student

1	pair of safety glasses

**Safety
Note** All students should wear safety glasses.

Procedure

1. Remove the plug from the tip of the compression cylinder and insert the piston until the black piston plug is near the 50 cc-mark. Replace the plug securely on the cylinder tip.

2. Attach the compression cylinder to the ring stand as shown in Figure 3. Make sure the cylinder is vertical and resting solidly on its plug.

3. Push down on the piston, then let it rise back up to a resting position. Record the level of the piston ring.

4. Carefully balance one of your masses on the top of the piston plunger.

Figure 3   Positioning of Compression Cylinder and Support Block

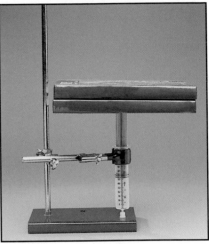

5. Gently press down on the mass so that the piston depresses slightly. Let the piston, with the mass still balanced on it, rise back up to a resting position. Record the level of the piston ring.

6. Place another mass so that it balances on top of the first mass, then repeat Step 5.

7. Repeat Step 6 until all four masses have been used.

8. Remove all of the masses from the top of the plunger.

9. Calculate the change in volume that occurred as each mass was added.

10. Prepare a data table to organize your measurements and then report your data to the teacher to add to the class data table.

11. Make a copy of the completed class data table.

Analysis

Group Analysis

1. Using evidence from your experiments, describe as precisely as you can the effects of a pressure change

    a. on the temperature of a fixed volume of gas.

    b. on the volume of a fixed mass of gas.

Individual Analysis

2. Explain how the laws of thermodynamics and the kinetic theory support your conclusions from Analysis Questions 1a and 1b.

3. Use averages from both class data tables to determine the mathematical relationship between

    a. the pressure and temperature of a gas.

    b. the pressure and volume of a gas.

    (**Hint:** Graphs can help determine mathematical relationships.)

4. Describe, in as much detail as you can, how iceboxes and refrigerators reduce the temperature of stored food. Include a discussion of how the cooling effect of each appliance is related to a change of state.

Unforeseen Consequences

Purpose ▶ **L**earn about the pros and cons of some modern food preservation technologies.

A cook's e-mail, 1999

Subject: Fridge problems
Date: 26 Jan 1999 20:00:49

The main refrigerator broke and I don't know what to do. I know it cost me quite a bit just to throw out my old air conditioner and they'll probably charge me ten times that for hauling away that big old fridge. I've heard that the Freon that's inside eats a hole in the atmosphere but I'm not sure why that is. As far as I know, that fridge hasn't leaked since I bought this place 20 years ago. Why should I have to pay for a mistake that someone else made and why didn't they test it carefully in the first place? I think I need to know more about my options before I buy a new one.

Introduction ▼ In the early days of mechanical refrigeration, one of the major drawbacks to the new technology was the lack of a suitable, safe refrigerant. The discovery of Freon in the 1930s and the subsequent use of other chlorofluorocarbon gases (CFCs) seemed to have solved the problem. CFCs were widely used as refrigerants and in other applications for over 50 years. In the 1980s it was discovered that CFCs contribute to the destruction of part of Earth's atmosphere, and public policies were adopted to ban or limit their use. As a result, today scientists are developing new refrigerants and other methods of food preservation. As with the introduction of many new technologies, scientists, engineers, and the public must carefully examine the trade-offs involved in using a new technology before committing to its widespread use.

Half a Century Later—What's Wrong With Freon?

Freon is one member of the chlorofluorocarbon (CFC) group of chemicals. In addition to their use as refrigerants, CFCs have been widely used in other industrial and consumer products, such as aerosol propellants, solvents, and cleaners.

The ozone layer is a thin part of the stratosphere, a layer of Earth's upper atmosphere. Ozone (O_3) in the stratosphere reflects much of the sun's ultraviolet (UV) radiation back out into space. While the sun's radiation is responsible for heating the planet to a livable temperature, some of its high-energy components, including UV radiation, can damage living cells. Because the ozone layer prevents much of the sun's UV radiation from reaching Earth's surface, it protects living organisms from disease and death.

There is evidence that the protective ozone layer is being damaged by CFCs. This evidence comes from careful measurements of ozone and chlorine levels in the stratosphere, as well as UV radiation levels on Earth. According to these measurements, the level of chlorine in the stratosphere is rising, the concentration of ozone in some parts of the ozone layer is decreasing, and UV radiation at Earth's surface is increasing. The mechanism for CFCs' attack on ozone involves chlorine atoms, which arrive in the stratosphere as part of CFC molecules. In this part of the atmosphere, radiation from the sun is powerful enough to break apart the CFC molecules and produce a very reactive type of chlorine. This chlorine can react with atmospheric oxygen (O_2) to form chlorine monoxide, which is also very reactive. It can break down O_3 molecules, causing them to degrade to form O_2. This mechanism for ozone depletion was first proposed by F. Sherwood Rowland and Mario J. Molina in an article published in 1974 in the journal *Nature*.

By the late 1970s, the nature and extent of this environmental threat was widely recognized. However, because no alternatives to CFC use were readily available and numerous industrial processes and consumer products relied upon the use of CFCs, they continued to be produced and used. Although some individual countries and industries enacted policies to reduce CFC use, worldwide production continued to increase until it peaked in 1988. The use of CFCs has decreased steeply in recent years due to international agreements, signed by the leaders of many countries, that require reductions in their use.

New Refrigerator Technologies

As an alternative to CFCs, many "ozone-friendly" refrigerants have been developed. Currently, the most widely used are hydrochlorofluorocarbons (HCFCs). HCFCs are about 95% less damaging to the ozone layer than CFCs. However, because they do contribute to ozone depletion, HCFCs provide only a short-term solution; their use is to be phased out by 2030. Refrigerants made from hydrofluorocarbons (HFCs) and hydrocarbons show more promise as long-term alternatives because they have no known damaging effects on the ozone layer. It is possible, though, that we will discover unforeseen negative impacts caused by the widespread production and use of new refrigerant chemicals. In addition, some alternative refrigerants have known problems—they are flammable or interact unfavorably with other components of the refrigeration system, or both.

In addition to alternative refrigerants, new refrigerator designs are also being developed. These new refrigerators employ a variety of innovations, including computerized controls, "super insulation," and the use of sound waves in place of a chemical refrigerant as the means of cooling.

Food Irradiation

Another technology for reducing food spoilage, food irradiation, does not involve creating a cold environment. For many years, high-energy radiation produced by radioactive materials or electronic devices (powerful relatives of X-ray machines) has been used to sterilize medical supplies. If food products are briefly exposed to the same radiation, they spoil at much slower rates.

When high-energy radiation interacts with matter, the energy transferred to the matter is enough to break molecular bonds and strip electrons from atoms. DNA and RNA, carriers of the "blueprints" for cell function and reproduction, are very large molecules that are easily damaged by radiation. Irradiation can be used to kill microorganisms, or at least destroy their ability to reproduce. In either case, the population of the organism is exterminated or severely reduced.

Unlike chemical treatments, irradiation leaves no harmful residue or unpleasant taste. But the lack of residue does have a negative side effect: irradiation acts only on organisms living on the food at the time of treatment. As a result, irradiated food is not protected from future contamination.

However, irradiation does have some long-term effects. It causes chemical changes in some food molecules that result in the formation of new substances called radiolytic products. Studies using a dose of 10 kilogray (kGy), the maximum allowed for food irradiation, found production of radiolytic products to be 3–30 parts per billion or 3–30 grams per 1,000 metric tonnes. Most studies indicate that these radiolytic products pose no health risk. Current technology detects no nutritional difference between non-irradiated foods and foods irradiated at less than 1 kGy. Even at high dose levels, nutrient loss is very small. There is substantially less nutrient loss in irradiated foods than in foods preserved with methods that involve high temperatures, such as pasteurization or canning.

Analysis

1. Describe the unanticipated side-effect of CFC use.

2. The reading in Activity 28.4, "Keeping Cool," contains a quote from the 1930s: "The refrigeration industry needs a new refrigerant." Why is this statement still pertinent today, more than 60 years later?

3. Who, if anyone, should be held responsible for the damage done by CFCs? Explain your reasoning.

4. Briefly describe how food irradiation works.

5. In the 1930s, people were concerned about using new refrigeration technologies. Today, people are concerned about using new irradiation technologies. What are their concerns? Do you think these concerns are justified?

6. Consider the four food preservation technologies discussed so far—chemical preservatives, iceboxes, refrigerators, and irradiation. Which do you think is best suited, technically, for use worldwide, in both less developed and more developed countries? In your response, make sure to discuss the trade-offs involved in the use of each technology.

Economy of Material Use

30

30.1 Material Resource Use and Sustainability

Purpose ▶ **C**ompare data on metal production and consumption in the United States and China and investigate the implications for global sustainability.

Introduction ▼

In this activity, you will evaluate the effects of current resource use on future development possibilities in the United States and China. The U.S. is a more developed country; China is a less developed country. Both are included in *Material World*. Both countries are very large in terms of population and land area. Both have growing economies, although China is growing more rapidly than the U.S. Each country has its own strategy for controlling population growth and providing for its citizens.

Extracting Earth's resources can have a significant impact on the environment.

Introduction
(cont.)

In your analysis, you will use estimated values for the world's reserves of various material resources. The values given for reserves do not represent all deposits of that material; they represent the amount of the material that is known to be recoverable in the foreseeable future. It is unlikely that humans will ever extract all of any material from the environment. Throughout history, for technological and economic reasons, human reliance on certain materials has changed as more desirable materials are used in place of less desirable ones. These changes in patterns of material use can cause, or be caused by, the fact that once the more highly concentrated and more easily accessible deposits of a given material have been depleted, the cost of extracting and refining the remaining material will be greater than its worth. Also, if individuals and community leaders (either by choice or by law) consider long-term environmental costs, they may decide to use materials that are more expensive but less damaging to the environment.

Procedure

Examine the data in the table below. Use this information to answer the Group Analysis Questions.

Table 1 **Production and Consumption of Three Metals in the U.S. and China**
(in thousands of metric tonnes)

		United States		China		World Total		Known World Reserves
		1985	1994	1985	1994	1985	1994	1994
Copper	Production	1,105	1,795	185	432	8,088	9,523	310,000
	Consumption	1,958	2,674	420	745	9,700	11,084	
Lead	Production	424	374	200	376	3,431	2,765	63,000
	Consumption	1,142	1,375	220	214	5,237	5,342	
Iron Ore	Production	49,533	55,651	80,000	234,660	860,640	988,797	150,000,000
	Consumption	64,679	63,039	140,354	222,771	860,640	970,422	

Analysis

?

Group Analysis

1. Look at the production and consumption of lead in the U.S. over the 10-year period from 1985 through 1994. How is this trend different from that of copper over the same time period? Is the trend the same in China as in the U.S.?

2. Is the trend in world use of these three metals reflected in the production and consumption figures for the U.S. and China?

3. If the world continued to produce copper metal from ore at the 1994 rate, and if no further investments were made to search for undiscovered deposits of copper ore or to develop known ones, when would the world's copper reserves be used up?

4. What kinds of factors could lead to changes in the projections you made in Analysis Question 3 based on the 1994 data? Explain your answer.

5. What factors, other than increases in reserves, might lead to a longer economic lifetime for extraction of these ores?

Individual Analysis

6. Iron is the only metal produced and consumed by China in larger quantities than by the U.S. Why do you think this might be so? What additional information would be helpful in answering this question?

7. When mining of a metal stops, does the world have no more of that metal? Explain.

30.2 Additional Information on the Production and Use of Metals

Purpose ▶ **U**se additional information to analyze the sustainability of current metal production and consumption practices.

Introduction ▼ The statistics you used in Activity 30.1, "Material Resource Use," were incomplete. To help you understand more about how the use of metals might affect global issues of sustainability, this reading describes the techniques used to process copper, lead, and iron. It also discusses some of the environmental impacts of these techniques and outlines the most common uses of each metal.

Heavy Metal Rock

Copper

Copper use is increasing worldwide, in more developed and in less developed countries alike. A multitude of applications for copper exist in the electrical, communications, construction, transportation, and coinage industries. For example, most of the wiring for cars and trucks is made of copper. Newly minted silver U.S. coins are mostly copper with a thin silver plating.

In Activity 24, "Material Resources: Metals," and Activity 25, "By-Products of Materials Production," you explored the process of refining copper from malachite and methods for disposing of the resulting wastes. You extracted the copper by treating copper sulfate solution with iron. Industrial copper refining uses a different refining process, known as **electrolytic refining**. This process uses large quantities of electricity to extract copper from ores and produces both solid

and liquid wastes. A third method of refining copper is **smelting,** in which copper ore is heated repeatedly to very high temperatures to remove impurities. Smelting produces both toxic gaseous wastes and **slag**, a solid waste containing a variety of toxic metals. The toxicity of the gases can be reduced in a catalytic converter, which requires the consumption of additional energy and resources. As toxic gases pass through a catalytic converter, they come in contact with a variety of chemical catalysts. These catalysts enable toxic compounds to react and form non-toxic compounds.

Lead

About 70% of global lead use is in the production of lead-acid batteries. Research on battery-powered vehicles in the western hemisphere has contributed significantly to a recent increase in lead consumption. Researchers are

▶

currently working to develop small, lightweight batteries that can store enough energy to power a vehicle. As these batteries become available, lead consumption can be expected to increase. However, new battery technologies that do not involve lead are also being developed. If these batteries become dominant, lead consumption may decrease.

Lead is also used to protect underground and underwater cables, including transoceanic cable systems; to prevent exposure to gamma radiation and X-rays, especially in healthcare fields; and as a component of the solder used to make electronic devices.

Lead is almost always obtained from lead sulfide ores, which are crushed and smelted to remove impurities; it can be further refined electrolytically. As with copper, lead smelting produces slag and toxic gases, which must be treated to remove sulfur and metal contaminants, and electrolytic refining uses large amounts of electricity.

Pyrometallurgy, another lead refining process, involves additional heating of the ore and also produces toxic gases and slag. Lead production is highly regulated, both in the U.S. and globally, because any lead that is allowed to enter the environment can have extremely toxic effects. Increasingly, lead is being recycled—in fact, it is the most recycled metal. Using lead from recycled sources reduces the need for direct mining of new lead deposits.

Iron

Iron is used primarily in the production of steel for construction projects. It is also used as ballast to weigh down underwater oil pipelines and in many applications that involve welding. In the refining process, iron ore is crushed, then purified using a blast furnace or an electric arc furnace. As in the processing of copper and lead, waste products from iron refineries include solid slag and toxic gases.

Analysis

Group Analysis

1. Organize the information from the reading into a table summarizing the processing technologies, uses, and overall environmental impact of each of the three metals.

2. Suggest changes that you think would lead to more sustainable practices for metals processing and use.

Individual Analysis

3. Do you think that pollutants associated with the extraction and processing of ores should be taxed to encourage companies to make choices that will lead to sustainable metal processing and use? What consequences would your decision have for the company and for the consumer?

4. Using evidence from this activity and from other activities in this course to support your answer, explain how resource use affects sustainable development. As a part of your answer, compare one of the more developed countries in *Material World* with one of the less developed countries. Discuss decisions made by citizens and by governments that might affect the sustainability of a country's approach to resource use.

30.3   <u>You Can Bank On It</u>

Purpose ▶ **S**imulate the decision-making process the World Bank uses for allocating funds to assist nations with sustainable development projects.

Introduction

One of the missions of the World Bank is to reduce poverty and promote sustainable development around the world. Although it is supported by many nations, seven countries, known as the Group of Seven, contribute the most money to the World Bank's operation. The Group of Seven ("G7") countries include Canada, France, Germany, Italy, Japan, the United Kingdom, and the United States.

To fulfill its mission, the World Bank issues loans to developing countries. These loans must be repaid by the country receiving the loan. Although the World Bank does not issue grants, there are two classes of World Bank loans. One type of loan is made to countries able to repay the loans at a competitive interest rate. The other type of loan is made to less affluent countries and is repaid at a very low interest rate. In most cases, a loan from the World Bank does not cover the entire cost of a project. The country's government, with help from commercial industries in the country, usually contributes some of the needed funds. This is reflected in the project descriptions you will read in this activity—in each case, the amount of the loan request is less than the total project cost.

In this activity, you have two roles. As a member of your group, you will research one of the many loan requests made to the World Bank by less developed countries and make a presentation about it to the class. The presentation should provide information about the country you are representing, its need for the loan, and its ability to repay the loan at either a competitive or low interest rate. As an individual, you will make decisions about how the World Bank should spend a budget of $100 million. You may choose to fund one or more of the proposals described, as long as you do not exceed the $100 million budget. For each

▶

Introduction

(cont.)

loan you decide to issue, you must indicate whether it will be at a competitive interest rate or a very low interest rate. You may also decide not to spend all the money. Instead, you may want to look for a different type of proposal, not represented here, that you think would make a significant difference toward global sustainable development. All of your decisions should take into consideration the information you've learned so far in this course.

Figure 1 Locator Map

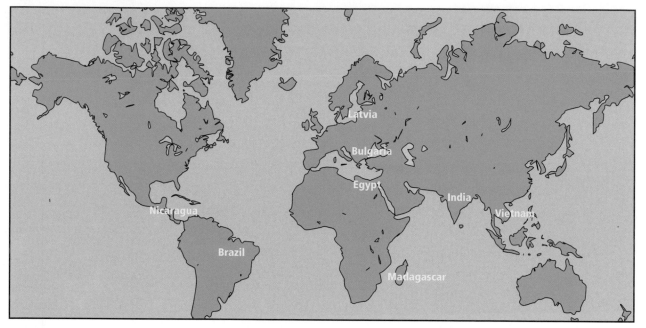

Procedure

1. Based on your teacher's instructions, determine which country your group will research for this project.

2. Collect pieces of evidence to support your case that the World Bank should provide the loan your country has requested. Carefully consider the science principles and ethical issues involved in your group's proposal.

3. Collect evidence that may indicate that the funding you are requesting is unnecessary or undeserved. As a group, discuss the trade-offs involved in making a decision about whether to fund your country's request. Be prepared for questions about these issues during your presentation.

4. Prepare a presentation that will explain your evidence, scientific and ethical considerations, and decisions to the class.

> You may wish to gather information about these countries and their proposals on the Internet. Go to the *Science and Sustainability* page on the SEPUP website to find updated sources.

Project Descriptions

1. Nicaragua

Project: Road improvement

This project proposes to improve Nicaragua's system of roads. It has the following goals: to make travel routes between commercial centers more accessible; to build the necessary roads and bridges to provide essential goods and services to people living in isolated rural areas; and to enhance the country's ability to export products.

Loan request: $25 million   **Total project cost:** $47.4 million

2. India

Project: Healthcare for women and small children

This project proposes to establish an effective and sustainable system for providing healthcare to women and small children in rural areas. This will be accomplished through the creation of a series of small local clinics, which will provide basic healthcare and immunizations, distribute food supplements, and monitor malnutrition.

Loan request: $76.4 million   **Total project cost:** $90.7 million

3. Brazil

Project: Gas pipeline

This project proposes to construct a gas pipeline from Rio Grande in Bolivia to São Paulo in Brazil. It will help develop a market for gas in Southern Brazil and create an export opportunity for Bolivia.

Loan request: $50 million   **Total project cost:** $2 billion

4. Madagascar

Project: Irrigation and water management

This project proposes to support the use of improved irrigation and agricultural water-management practices. These practices will enable long-term, sustainable use of the country's water resources. By working directly with individual farmers, the program will contribute to increased local agricultural production, improving rural incomes and reducing hunger.

Loan request: $24 million   **Total project cost:** $27.1 million

5. Bulgaria

Project: Smelting plant clean-up

This pilot project proposes to clean up a major environmental hazard—the waste surrounding a copper smelting plant. In addition to addressing environmental problems at one of the most polluted sites in Bulgaria, this project will provide a model for determining how to clean up and repair existing environmental damage that has resulted from past industrial practices.

Loan request: $16 million   **Total project cost:** $25 million

6. Egypt

Project: Pollution reduction

This project proposes to reduce industrial air and water pollution, which causes adverse health effects and ecological degradation. This will be accomplished by financing pollution reduction plans for the 30–40 worst industrial polluters, strengthening existing methods for monitoring and enforcing pollution control measures, and promoting awareness and education.

Loan request: $35 million   **Total project cost:** $48.7 million

7. Latvia

Project: Landfill improvement

This project proposes to enhance the country's largest solid waste disposal site so that improved treatment options are available for degradable wastes. Improved lining technologies will be used to protect groundwater.

Loan request: $7.9 million   **Total project cost:** $15 million

8.   **Vietnam**

Project: Forest protection and rural development

This project proposes to protect and manage natural forests with high biodiversity by creating and enforcing strict guidelines for development in and around them. These funds will also be used to encourage sustainable land use and to support rural infrastructure improvements, including expanding irrigation systems, upgrading roads, and improving road access.

Loan request: $21.5 million   **Total project cost:** $32.3 million

Analysis

Group Analysis

1. What aspects of the presentations in your class were most effective? How will you improve your presentation for the next project?

Individual Analysis

2. If you were responsible for deciding how the World Bank should spend its $100 million among these proposals, how would you spend it? Explain your reasoning, using evidence.

Moving the World

Energy drives most aspects of society. Without a plentiful and affordable source of energy, the systems we use to meet our basic survival needs—such as agriculture, healthcare, and construction—would not be possible. Transportation, communication, commerce, and many other systems that we "just couldn't do without" require energy, as well.

The energy sources currently used by most societies—primarily fossil fuels and nuclear energy—are non-renewable, and their use is associated with the production of potentially harmful waste products. If increases in global population and per capita energy use continue into the future as projected, energy reserves will diminish more rapidly and the volume of waste products will increase.

Technologies are being developed that can produce more energy and less waste, while at the same time depleting fewer resources. However, these new technologies are often too expensive for widespread use in less developed countries. If a clean, renewable energy source is to be developed, promoted, and used, it must be affordable. Although many of the more developed countries can afford to develop and use new energy-production technologies, current public policies and economic pressures do not promote their adoption. As a result, these technologies are not widely used.

In the long term, new technologies will most likely replace our current ones, and companies that invest in the development of more efficient, cleaner technologies stand to gain. If this occurs, the economy in communities in which these companies are based would benefit and, more importantly, so would all living things on Earth.

The U.S. is by far the world's largest per capita consumer of energy, accounting for nearly 25% of total global consumption by less than 5% of the world's population. Even with advances in energy-production technologies, each community and each individual—particularly in more developed countries—may have to change their habits if we are to achieve sustainable energy production and use.

Part 4

Moving the World

Throughout this course, you have been exploring scientific and social issues surrounding global sustainability. Many of your investigations have involved the production, use, and transfer of energy. The activities in Part 4, "Moving the World," focus directly on this theme, introducing scientific concepts that are fundamental to the study of energy and its use. Energy is responsible for all physical motion, including that which occurs during chemical interactions. Throughout history humans have used energy to make survival easier, more comfortable, and more convenient. In this part of *Science and Sustainability*, you will consider a wide range of questions related to the use of energy in its various forms. All raise the critical interplay between energy issues and sustainable development:

- How do we obtain the energy that we use to fuel our society?

- Have changes in technology affected our patterns of energy use?

- What trade-offs are associated with the use of various energy sources?

- How do our efforts to obtain and utilize energy affect our health, safety, and environment?

- Are there ways that we can change our energy use patterns to make them more sustainable?

- What information do we need to make informed decisions regarding future energy use?

- How can we best use our understanding of energy to maintain or improve our quality of life, the quality of life for people around the world, and the quality of life for future generations?

Fueling Trade-offs

31.1 How Much Energy Is There?

Purpose ▶ **D**etermine which of two fuels—kerosene or ethanol—releases more energy as it combusts. Decide which fuel is better for use in automobiles and identify the trade-offs in attempting to create the perfect fuel.

Introduction

All human activity requires energy—from sleeping to constructing roads and buildings, from watching television to generating electricity. In this activity, you will explore combustion reactions. **Combustion**, commonly called burning, is the source of much of the energy we use for transportation, cooking, heating, generating electricity, and other activities. Combustion also produces chemical products, some of which are pollutants. Two commonly combusted fuels are kerosene, which is a petroleum product, and ethanol, which is an alcohol often produced from crops. You will burn each of these fuels to compare the amount of energy released and some of the chemical wastes produced.

Prediction

Which fuel—kerosene or ethanol—do you think will release the most energy when it combusts? Explain your reasoning.

Materials

 For each group of four students

1	glass fuel burner containing kerosene
1	glass fuel burner containing ethanol
1	book of matches
1	balance
1	100-mL graduated cylinder
1	metric ruler

access to a clock with a second hand

access to water

For each team of two students

1	aluminum soda can
1	can holder (for soda can)
1	immersion thermometer
1	calculator (optional)

For each student

| 1 | pair of safety glasses |

Safety Note

Exercise extreme caution when using an open flame in the classroom. Be sure you know how to extinguish the burner and where to find a fire blanket or fire extinguisher to put out any accidental fires. Always wear safety glasses during this activity, and wash your hands before leaving the lab.

Procedure

1. Within your group of four, discuss and decide upon an experimental procedure that uses only the equipment listed in the Materials section and will allow you to collect evidence that indicates whether or not your prediction is correct.

   Hint: Recall Activity 1.3, "Burn A Nut," and Activity 4.1, "Are You In Hot Water?"

2. Write out the step-by-step procedure that your group will follow. Make sure that the procedure is stated clearly enough so that any other group in your class could perform your investigation.

3. Prepare a data table to record your observations and measurements.

4. Carry out your investigation.

5. Analyze your results, making sure to record any calculations made. Prepare a written report that includes your prediction, your experimental procedure, your collected data, your analysis, and your conclusions.

Analysis

Group Analysis

1. The results of any experiment may be affected by a variety of errors. Errors can often account for variations in data. Identify some potential sources of error in this experiment that may have affected your results.

2. The energy content of a fuel can be measured as the amount of energy per mass (cal/g or J/g) or the amount of energy per volume (cal/mL or J/mL). Which of these measures of energy content do you think is more useful when comparing fuels? Explain your reasoning.

Individual Analysis

3. Gasoline is chemically very similar to kerosene. How could the results of this experiment affect your decision to buy fuel for your car that contains ethanol rather than pure gasoline?

4. If you were considering the use of one of these fuels for your car, what other information would you like to have before deciding which to use?

31.2 | Fuels for the Future

Purpose ▶ **E**xplore some of the more promising sources of energy that are not fossil fuels.

Introduction Vast amounts of energy are used every day to heat and cool buildings, to provide light, to produce the goods and services people rely on, and to transport people and goods. Most of this energy is supplied by the combustion of fossil fuels. Currently, these fuels are well accepted by consumers, in part because they are abundant, relatively low in cost, and perceived as safe to use. Petroleum is a versatile fuel, and convenient for many different purposes: it is easily transported, it can produce electricity, it can heat homes, and it can supply power for transportation and industrial processes. In many ways, the physical properties of petroleum have shaped the ways we use and produce energy. In fact, we have come to rely upon petroleum for many of our energy needs, and even take it for granted. However, petroleum fuels are not perfect. Their production and use contributes to environmental degradation, and they will not always be available in such abundance or at such low cost.

Wind power can be used to propel boats, pump water, grind grain into flour, and generate electricity.

For sustainable development to occur, energy must be supplied in a fashion that is socially acceptable, dependable, and economical. To be sustainable, energy use must result in minimal

Introduction
(cont.)

negative environmental impact—now and in the future. There are a number of **alternative energy sources** that can replace or supplement fossil fuels as human society's main source of energy. Several of these alternatives are briefly described here.

In general, it can be said that all alternatives to fossil fuels are more expensive, at least in the short term, and less convenient, less versatile, and less adaptable to existing patterns of use. However, the full "cost" of providing and using energy includes the cost associated with impacts on global environmental and human health. If the full cost of today's energy were in fact paid by the users—rather than by people in other countries or by future generations—prices would be substantially higher. As a result, people would use less energy, and preferences for technology would shift toward options that create less risk of environmental damage.

Alternative Energy Sources

Solar

Solar energy is available wherever sunlight strikes. Because sunlight does not shine constantly on any one part of our planet, solar energy cannot provide a continuous supply of useful energy unless it is collected and stored for later use. Most living organisms would not survive without Earth's many natural systems for the collection and storage of solar energy—the atmosphere, surface water and rocks, plant and animal tissues. The functioning of present-day human society requires more intense, more continuous, or different forms of energy than nature provides. While the solar energy supply is "unlimited and free," the structures required for collecting, storing, and converting the sun's rays into useful forms of energy can be expensive. Environmental impacts and risks associated with the construction and operation of solar technologies are low, but not negligible.

Sunlight can be used to directly heat collectors filled with water or other materials that are then used to heat other items. For example, a solar collector installed on the roof of a house can heat water that is then used to heat the air inside the house. Attaining temperatures that are high enough to be useful typically requires several hours of direct sunlight and a solar collector with a surface area of at least 10 m². In many warm-weather countries, on-site solar collectors can be used to meet a significant percentage of household heating needs. It is not economically practical to move the heated materials any significant distance for use in other, "off-site" locations. Also, without equipment that can significantly intensify natural solar radiation, the sun's rays cannot provide all of the energy required to heat most large buildings, much less provide the amounts of energy required for industry or transportation purposes.

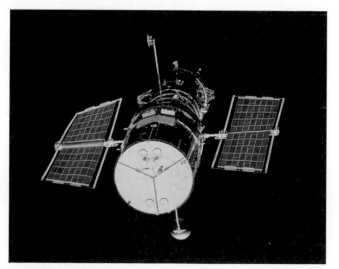

Electricity needed to operate this satellite is generated when sunlight strikes the photovoltaic cells located on its "wings."

Another type of solar technology can generate electricity directly from sunlight. This technology takes advantage of a phenomenon known as the **photovoltaic effect**. Photovoltaic cells, which are usually manufactured from silicon, produce electric current when exposed to light. Small arrays of photovoltaic cells, some no larger than 2 cm², can provide enough electricity to operate devices such as solar calculators and emergency roadside phones. Much larger photovoltaic arrays, of at least 10 m², are needed to provide all the electricity required by a satellite, space station, or typical home. Presently, high cost is a key barrier to the use of photovoltaics as a major source of society's energy. Photovoltaic electricity will not be competitive until the cost of manufacturing photovoltaic cells decreases or the cost of fossil-fuel derived electricity increases, or both.

Nuclear

Nuclear energy can be produced in two ways: fusion and fission. **Nuclear fusion** involves the combining, or fusing, of atomic nuclei. Fusion reactions produce the energy emitted by the sun and other stars. **Nuclear fission** involves the breaking apart of atomic nuclei. Fission reactions are currently used in many countries to supply significant amounts of energy, primarily for generating electricity, for heating purposes, and for military uses, such as nuclear-powered ships.

All nuclear reactions convert a small amount of mass into a large quantity of energy without producing any of the chemical emissions, such as soot or CO_2, that are associated with the combustion of fossil fuels. Albert Einstein quantified the relationship between mass and energy in the equation $E = mc^2$, where E is energy, m is mass, and c is the speed of light. Because the speed of light is such a large number (3×10^8 meters per second), nuclear reactions produce huge amounts of energy—in the form of heat and intense radiation—from tiny amounts of fuel. Earth contains enough nuclear fuel to meet all our present and future energy needs.

The safe use of nuclear energy for peaceful purposes requires careful management. Fission reactions produce large amounts of "waste" heat and radiation that can cause damage if released directly into the environment. These reactions must be enclosed in centralized structures that contain expensive equipment and systems designed to control the release of heat and radiation. Nuclear power has not become as widespread as many people once predicted it would. There are a number of reasons for this, including high cost and fear of exposure to radiation or nuclear explosions caused by accidents or sabotage. In addition, some people associate the use of nuclear power with the spread of nuclear weapons technology.

Fission reactions generate energy by splitting apart the nuclei of heavy, unstable, radioactive elements. The elements most commonly used as nuclear fuels are isotopes of uranium or plutonium. Safety concerns include the potential for accidental release of radioactive compounds during the production and storage of radioactive fuel, during the routine operation of nuclear power plants, or from disastrous accidents that could lead to the overheating, or "meltdown," of the reactor. Radioactive wastes produced during nuclear power generation can remain dangerous for thousands of years, leading to concerns about ensuring the safe, long-term storage of these materials. Plutonium is especially toxic; small quantities can poison an entire city's water or air supply. There is also the potential for sabotage, and for the authorized or unauthorized use of radioactive materials for the construction of atomic bombs. Safe operation of the entire nuclear power generation process, from mining and processing of fuels to waste disposal, requires stable, highly centralized political and governmental control. Despite the promise of newer reactor designs and other technological innovations that will reduce risks to public health and safety, overcoming public concern about the use of nuclear reactors will not be easy.

Fusion reactions generate energy when the nuclei of light elements, most commonly hydrogen, combine or fuse together. Humans have been able to create explosions derived from uncontrolled fusion reactions, but have not yet been able to initiate a controlled, long-lasting fusion reaction suitable for producing heat and electricity. Using the hydrogen atoms in one liter of seawater as fuel, fusion could theoretically generate the energy equivalent of 300 liters of gasoline. The elements used as fuel for fusion reactions do not pose any of the radiation risks associated with fissionable fuels. However, fusion reactions produce intense radiation that bombards all the materials in the reactor, causing them to become intensely radioactive. Depending on the design of the reactor, these irradiated materials can be very dangerous and potentially as expensive to handle as the fuels and by-products of fission reactions. Although fusion reactors hold some promise for the future, they are prohibitively expensive. With current technology, more energy is consumed in the initiation and control of the reaction than is produced. Practical fusion reactors are not expected to be developed for at least several decades.

Wind and Water

Throughout history, technologies as varied as water wheels, wind mills, rafts, and sailboats have harnessed the energy of wind and flowing water for use in manufacturing or transportation. Today's turbines and generators can efficiently convert the energy of wind and water into electricity, which has thousands of uses.

Wind and water power have many of the same drawbacks as solar power. These energy sources vary with season and geographic location, which means that a continuous energy supply would require storage capabilities. The large-scale generation of electricity by "wind farms" and hydroelectric facilities requires large initial

Although hydroelectric dams are an important source of electricity, there are significant environmental concerns associated with their construction and use.

investments, covers large geographic areas, and contributes to habitat destruction and other environmental damage.

Hydroelectric dams currently produce a significant fraction of the electricity used in the U.S. and other countries, in part because dams can provide year-round water storage. Unfortunately, the lakes created by these dams gradually fill with silt, slowly but surely reducing both the amount of water stored behind the dam and the capacity to generate power. Dams severely disrupt local ecosystems. They also have the potential to break and cause catastrophic floods.

"Wind farms" currently exist and continue to be built in many regions of the world that have consistently strong winds. However, because wind cannot be stored, all energy generation stops when the wind is not blowing. As a result, wind is used mainly to supplement other energy sources. Ecological impacts of wind farms include habitat destruction, noise pollution, and unintended killing of birds that may be caught in the rotating propellers.

Earth's oceans offer several possibilities for energy production. One promising source of energy is the daily rise and fall of the tides. Technology capable of converting this motion into electricity already exists. Another possibility is to use the difference in temperature between surface

In regions with consistent winds it is not uncommon to use arrays of windmills called "wind farms" to generate large quantities of electricity.

waters, which are heated by the sun, and much colder, deeper waters. A temperature difference of at least 15°C can be harnessed for the generation of electricity. Serious problems with this technology include the need for stabilization and mooring of large, submerged structures and the impact such structures would likely have on ocean-dwelling organisms.

Biomass

Biomass, or organic material, can be burned as a fuel or used to create other fuels. Biomass comes from many different sources, including municipal wastes, crop residues, manure, and by-products of lumber and paper industries. These organic materials can be burned directly, fermented to create alcohol, anaerobically digested to produce **biogas** (a mixture of gases containing methane and CO_2), or gasified to produce methane and hydrogen. Methane and alcohol release thermal energy when combusted; methane and hydrogen can be used in fuel cells that generate heat and electricity. A power plant in southern California that burns 40 tons of cow manure each hour generates enough electricity for 20,000 homes. Each year, Chinese biogas generators produce the energy equivalent of 22 million tons of coal, which is enough energy to meet the needs of 35 million people. If all the waste from cattle in U.S. feedlots were converted to biogas, it would produce enough energy to heat a million homes.

Although there are many reasons to convert biomass to energy, there are also trade-offs. Large-scale use of farm and forest residue to produce energy would interrupt the natural recycling of nutrients and could contribute to loss of soil fertility. Currently, most farm and forest residues are returned to the soil, where they decompose and release nutrients back into the soil. Another drawback to the conversion of biomass to energy is the production of large quantities of pollutants. The production of biogas, for example, can generate large quantities of air pollutants. Biogas reactors can be constructed with air pollution control devices, but these are often very expensive and require consistent maintenance. Direct combustion of biomass also produces large quantities of air pollutants. To prevent the release of these pollutants into the atmosphere, biomass combusters must be fitted with effective air pollution control devices.

Alcohol is produced by the fermentation of biomass materials that have a fairly high sugar content. Biogas can be produced from almost any organic matter, regardless of its sugar content, that can be digested by certain types of anaerobic bacteria. The CO_2 and other gaseous components of biogas can be removed with fairly simple technology to yield nearly pure methane. Nitrogen, phosphorus, and other nutrients contained in the organic matter used to produce alcohol or biogas remain in the residue but need not be wasted; these nutrient-rich residues can be used as fertilizer.

Geothermal

Earth contains an immense amount of heat, most of which lies buried deep beneath the planet's crust. In regions with hot springs or volcanic activity, this heat comes close to the surface in the form of hot water and gases. These forms of **geothermal energy** can be used to heat homes and water and to generate electricity for local communities. New technology now being developed could be capable of drilling wells up to six miles deep to tap Earth's thermal energy in areas that are not volcanically active; these technologies are still in an experimental stage. Geographic restrictions, cost, and environmental impact are the major factors that limit current use of geothermal energy. Environmental impacts include disturbance of habitats and the need to dispose of large quantities of noxious gases and very saline (salty) water.

Hydrogen

Hydrogen gas reacts with oxygen to produce only heat and water. This reaction produces none of the SO_x, NO_x, or CO_2 pollutants associated with hydrocarbon combustion. However, unlike wood, coal, and other carbon-based fuels, hydrogen fuel must be manufactured. At present, hydrogen gas is most commonly made from methane gas (CH_4), in a process called steam reforming, or from the decomposition of water by electricity. New technologies may allow us to extract hydrogen from water or methanol more efficiently and at less cost.

Because hydrogen is a gas, its energy density (ratio of energy to volume) is much lower than that of most liquid fuels. A low energy density means that hydrogen-fueled vehicles, particularly heavy ones, must either be equipped with very large fuel tanks or be refueled frequently. Hydrogen gas can be compressed or cooled to a liquid state, which reduces the volume of the fuel considerably. The decrease in volume increases energy density, but the technologies required for this process are expensive.

Fuel cells, a technology now under development, makes use of the energy released when hydrogen and oxygen react chemically to form water. Fuel cells require a constant supply of hydrogen gas as a fuel, but water is the only waste they produce. Small groups of fuel cells could eventually be used to produce all of a building's heat and elec-tricity needs. Fuel cells also have potential for use in hybrid fuel cell–electric vehicles that produce no air pollutants. Fuel cells are discussed in more detail in Activity 33.3, "Energy as You Like It."

Major drawbacks to the widespread use of hydrogen as a fuel are its high cost of production and its extreme flammability, which requires the use of costly safety equipment. Careful precautions must be taken during its storage, transportation, and use. Uncontrolled combustion of hydrogen, however, is much less destructive than uncontrolled hydrocarbon combustion. For example, the Hindenburg, a hydrogen-filled air ship similar to a blimp, burst into flames while landing in New Jersey in the 1930s. People survived the crash despite being very near the fiery wreckage.

Analysis ?

Individual Analysis

1. List the characteristics of an ideal energy source for use in vehicles. What energy source has characteristics most similar to those you listed?

2. What characteristics would you look for in the ideal source of energy for generating electricity? Which of the energy sources discussed in the reading has characteristics most similar to those you described?

3. Fossil-fuel combustion is currently used to produce a large percentage of our electricity. Compare the characteristics of the energy source you chose in Analysis Question 2 to those of fossil-fuel combustion. Explain the trade-offs that should be considered when deciding whether to build an electricity-generation facility that consumes fossil fuels or one that uses your chosen energy source.

4. Which of the alternative sources of energy presented in this reading do you think provides energy in the most sustainable way? Why is this energy source not more widely used? Explain your reasoning.

31.3 Combustion

Purpose ▶ **M**odel the chemical and structural changes that occur when a hydrocarbon molecule reacts with oxygen, and relate these changes to the energy released during combustion.

Introduction

Combustion is a chemical reaction that can occur between a fuel and oxygen. **Complete (100%) combustion** of a pure hydrocarbon fuel results in the production of nothing other than water, carbon dioxide, and energy. Energy is released during a combustion reaction because bonds between atoms in the reactants store more energy than bonds in the products. Part of the reason that hydrocarbon molecules make such good fuels is that the amount of energy stored in a carbon-carbon bond or carbon-hydrogen bond is much larger than the amount of energy stored in the bonds of the waste products—oxygen-hydrogen or carbon-oxygen bonds.

Combustion of commonly used carbon-based fuels—gasoline, diesel fuel, natural gas, propane, fuel oil, coal, wood, and charcoal—produces not only water, carbon dioxide, and energy, but also small amounts of other products, such as solid ash particles and gases containing sulfur and nitrogen. These products are created because elements other than hydrogen and carbon are found in the fuel, and air is not pure oxygen gas. Additional waste products are

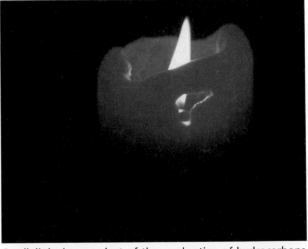

Candlelight is a product of the combustion of hydrocarbons found in beeswax or paraffin, a petroleum product.

▶

Introduction
(cont.)

formed if complete combustion does not take place. **Incomplete combustion** results from a lack of sufficient oxygen. This lack of oxygen causes the production of **carbon monoxide** (CO). According to the *Journal of the American Medical Association*, inhalation of CO is the leading cause of accidental poisoning deaths in America. When air containing a large amount of CO is inhaled into the lungs, it enters the bloodstream and takes the place of oxygen (O_2). When blood transports CO rather than O_2, the supply of oxygen to the body's cells is reduced. Depending upon the concentration and duration of CO inhalation, CO poisoning can cause tissue damage, loss of consciousness, and death.

Materials

■■ **For each team of two students**

1 molecular model set

Procedure

1. Make models of one methane (CH_4) molecule and two oxygen (O_2) molecules. Break the bonds, rearrange the atoms, and form new bonds to construct water and carbon dioxide molecules.

   Count how many of each "new" molecule is produced, then fill in the blanks in front of the products in the chemical equation that describes this reaction:

   $$CH_4 + 2O_2 \rightarrow \underline{\quad}CO_2 + \underline{\quad} H_2O + energy$$

   Note: You have just written a balanced chemical equation. Each atom of each reactant appears in the products, and the products do not contain any atoms that do not appear in the reactants.

2. Make models of one ethanol (C_2H_5OH) molecule and eight O_2 molecules.

3. Use as many oxygen molecules as needed to completely combust the ethanol molecule into CO_2 and H_2O molecules. Count the number of O_2 molecules needed and the number of CO_2 and H_2O molecules produced.

4. Write a balanced chemical equation to describe the combustion of a single ethanol molecule.

5. Repeat Steps 2–4 for

   a. benzene (C_6H_6).

   b. hexane (C_6H_{14}).

6. Take apart your molecules and put all the pieces back in their container.

Analysis

Group Analysis

1. The incomplete combustion of methane (CH_4) produces carbon monoxide (CO), carbon dioxide (CO_2), and water (H_2O). Write a balanced chemical equation for this reaction. (**Hint:** Try starting with more than one methane molecule.)

Individual Analysis

2. Draw structural formulas for the reactant and product molecules you constructed in Procedure Step 1.

3. **a.** Find the molecular mass for each of the reactant molecule(s) you drew for Analysis Question 2.

   b. Find the molecular mass for each of the product molecule(s) you drew for Analysis Question 2.

   c. Calculate the total mass of all the reactant molecules in your balanced equation from Procedure Step 1.

   d. Calculate the total mass of all the product molecules in your balanced equation from Procedure Step 1.

   e. Write a sentence comparing the total mass of the products to that of the reactants.

4. What is the source of the energy released during combustion?

5. Use your knowledge of chemistry to rank four hydrocarbon molecules—methane, ethanol, benzene, and hexane—according to which releases the most energy during combustion. Explain your reasoning.

Fuel From Food

32.1 Biofuels

Purpose ▶ **P**roduce a fuel through the fermentation and distillation of plant products.

Introduction

Biofuels—renewable fuels obtained from plant materials—are an alternative to non-renewable fossil fuels such as petroleum and coal. One method of producing biofuels is by fermentation. **Fermentation** is the process by which some organisms, such as yeast, obtain energy from sugars and starches. In addition to energy, fermentation produces ethanol and carbon dioxide, as shown in Equations 1 and 2. Fermentation proceeds most efficiently in the absence of oxygen. It can produce solutions containing up to about 12% ethanol, which can be purified by fractional distillation to produce a mixture composed of 95% ethanol and 5% water. This hydrated ethanol mixture is suitable for use as a fuel in automobiles, for heating and cooking, and for generating electricity.

Equation 1   **Fermentation of Glucose**

$$C_6H_{12}O_6 \quad \rightarrow \quad 2C_2H_5OH \quad + \quad 2CO_2 \quad + \quad energy$$

glucose   (yeast)   ethanol   carbon dioxide

Equation 2   **Fermentation of Sucrose**

$$C_{12}H_{22}O_{11} \quad + \quad H_2O \quad \rightarrow \quad 4C_2H_5OH \quad + \quad 4CO_2 \quad + \quad energy$$

sucrose   water   (yeast)   ethanol   carbon dioxide

Prediction

Which of four food products—corn, table sugar, molasses, or corn syrup—do you think will produce the most ethanol? Explain your reasoning.

Part A Fermentation

Materials

▦ **For the class**

supply of pureed corn

supply of table sugar (sucrose)

supply of molasses

supply of corn syrup

supply of yeast suspension

access to water

▤ **For each group of four students**

1   30-mL graduated cup

1   50-mL graduated cylinder

1   250-mL flask

1   air trap apparatus:

2-holed rubber stopper

5-cm length of glass tubing

immersion thermometer

plastic tubing

1   beaker

1   30-mL dropper bottle of bromthymol blue (BTB)

Procedure

1. Record which plant product your group will be investigating. Depending on your group's product, complete Step 1a or 1b.

    a. If your group is assigned table sugar, molasses, or corn syrup, place 30 mL of your plant product in the flask and add 150 mL of water and 15 mL of yeast suspension.

    b. If your group is assigned corn, place 150 mL of the pureed corn in the flask and add 15 mL of yeast suspension.

Figure 1
Experimental Set-up for Fermentation

2. Swirl the flask contents to mix well, then use an air trap apparatus to stopper the flask.

3. Fill your beaker half-way with water. Submerge the free end of the tubing on the air trap apparatus in the water to prevent air from entering your container during fermentation. Figure 1 shows the correct set-up for this experiment.

▶

Procedure
(cont.)

4. Add 20 drops of bromthymol blue to the water in the beaker.

5. Label your flask and beaker to identify the members of your group, your class period, and today's date. Carefully move the flask and beaker to a warm (>20°C), safe place designated by your teacher.

6. Make and record daily observations for at least 5 class sessions. For each observation, describe

   a. what you see happening in the flask.

   b. the rate that bubbles are emitted from the tubing submerged in water.

   c. any other changes that occur.

Part B   Distillation (at least 5 days later)

Materials

 For the class

   supply of crushed ice

 For each group of four students

   1   distillation apparatus:

      condenser

      2-holed rubber stopper for flask

      1-holed rubber stopper for condenser

      immersion thermometer

      5-cm length of glass tubing

      plastic tubing

   1   ring stand and clamp

   1   hot plate

   1   balance

   1   50-mL beaker

   1   10-mL or 25-mL graduated cylinder

■ **For each student**

   1   pair of safety glasses

**Safety
Note**

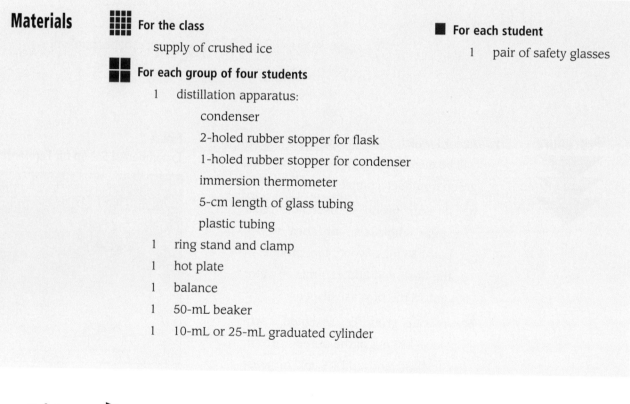 Do not use open flames as a heat source for this activity. Only hot plates should be used. Never touch the hot plate or hot flask! Use safety glasses at all times!

Procedure

1. Describe the contents of your flask. Remove the stopper and use your hand to waft some of the fumes toward your nose. Describe the odor. Record the color of the solution in your beaker. Compare your observations with those from the other groups.

2. Set up your distillation apparatus as shown in Figure 2. Find and record the mass of a clean, dry 50-mL beaker. Place the beaker so that it will collect the distillate as it drips from the condenser.

Figure 2 Experimental Set-up for Distillation

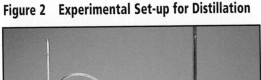
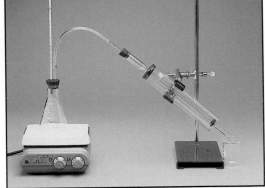

3. Turn on your hot plate to medium heat, and make observations of what happens in the flask and throughout the distillation apparatus. If the liquid begins to boil too vigorously, adjust the temperature of the hot plate so that it boils slowly.

4. Collect distillate until the temperature of the thermometer reaches 95°C, then turn off the hot plate. Describe the stillage (material remaining in the flask) and the distillate.

5. Accurately measure the volume and mass of the distillate.

6. Have your teacher test your distillate for flammability.

7. Prepare a data table and record the plant product used and the volume and mass of the distillates obtained by the other groups.

8. Clean up as directed.

Analysis

Group Analysis

1. How did your plant product's rate of ethanol production compare with that of the products tested by other groups in the class? Describe the evidence that supports your answer.

2. How does the amount of distillate you obtained compare with the amount obtained by the other groups?

3. What is the density of the distillate you produced?

4. What substance(s) do you think are in the distillate? Look back at your results from Activity 21, "Identifying and Separating Hydrocarbons," and give at least three pieces of evidence that support your conclusion.

5. Account for any similarities or differences in the amount of distillate produced by the other groups.

Individual Analysis

6. On the basis of your experimental results, which plant product (corn, table sugar, molasses, or corn syrup) would you recommend for use as the raw material for producing ethanol?

7. If you were interested in starting an ethanol distillation factory, what other information would you want to know before deciding which plant product you would ferment?

32.2 The Ethanol Alternative

Purpose ▶ **B**ecome an "expert" on one aspect of the issue of whether or not the production and use of ethanol fuels should be promoted. With the other members of your group, formulate an energy policy that does or does not include ethanol as an alternative fuel.

Introduction ▼ Crude oil, or petroleum, is the source of many fuels, including gasoline, diesel, and jet fuel (kerosene). Because crude oil is a non-renewable resource, at some time in the future it will become expensive and in short supply. A limited supply of costly fossil fuels is likely to make the use of other fuels more economically competitive. Ethanol, a renewable bio-fuel, can be used in place of some petroleum-derived fuels. During World War II, when petroleum was in short supply, some countries relied on ethanol as a fuel. As you saw in the last activity, ethanol can be produced from the fermentation of products derived from corn or sugar cane. There are economic, social, political, and environmental trade-offs involved in the decision to use food products to produce ethanol fuel.

Supporters of the development of ethanol fuels point out that ethanol is a renewable source of energy and that it might potentially

- decrease military and economic tensions over fossil-fuel supplies.
- improve air quality.
- save petroleum for uses other than fuels.
- increase rural agricultural job opportunities.
- provide new exports for less developed countries.

Detractors of its development have raised the following questions:

- Is ethanol really a viable alternative to petroleum fuels?
- Will using farmland to grow crops to make ethanol cause food shortages and push up food prices?
- Is large-scale ethanol production an economically sound practice?
- Is ethanol really an environmentally sound energy source?

Procedure

1. Select one member from your group of four to become the group "expert" for each of the following issues:

   Issue 1: Using ethanol as a fuel

   Issue 2: Environmental impact of ethanol use

   Issue 3: Using crops for food vs. using crops for fuel

   Issue 4: Using ethanol as a pollution-reducing fuel additive

2. Read about your issue, then discuss it with the members of the other groups who are researching the same issue. Take notes so that you can report back to your group.

   You may wish to gather more information about ethanol and your issue on the Internet. Go to the *Science and Sustainability* page on the SEPUP website to find updated sources.

3. Rejoin your original group to report on your issue. Review and discuss each of the points outlined in the Introduction to this activity.

Analysis

Individual Analysis

1. Rank the four issues from most important to least important. Explain your reasoning.

2. Prepare a report outlining your views on the production of ethanol for fuel. Be sure to include any evidence that supports your position and any areas where additional evidence would be helpful.

Issue 1   Using Ethanol as a Fuel

"Pure" ethanol, or ethanol blended with petroleum-based fuels, can be used as a liquid fuel for various types of gasoline and diesel engines. In these internal combustion engines, burning fuel produces hot gases that expand and exert enough pressure to cause the appropriate parts of the engine to move. In a gasoline engine, an electric spark from the spark plug ignites the fuel. In a diesel engine there are no spark plugs. Instead, the fuel is ignited by heat created by the high compression pressures of the pistons.

The Use of "Pure" Ethanol

Most automobile engines built to use gasoline can, with a few minor modifications, be fueled with hydrated ethanol (95% ethanol, 5% water). The use of ethanol as a replacement for diesel fuel in industry and agriculture is also possible, but this substitution requires more extensive and expensive equipment modifications. To achieve satisfactory performance from cars using hydrated ethanol, the carburetor or fuel injection system has to be modified to allow more fuel into the engine. In Brazil in March of 1980, General Motors, Volkswagen, Ford, and Fiat began selling ethanol-fueled cars; by the end of 1980, over 25% of the cars sold in Brazil were ethanol-fueled.

Researchers in Brazil compared the performance of cars fueled by hydrated ethanol with those fueled by gasoline. They found that the mileage of a "standard" 1300-cc engine was 10.3 km per liter (29.3 mpg) when using ethanol, compared with 12.8 km per liter (36.4 mpg) for gasoline. The use of ethanol cuts carbon monoxide emissions by 66%; it also reduces hydrocarbon emissions significantly. Sulfur dioxide (SO_x), an emission from gasoline-fueled vehicles, is not present in the exhaust gases of ethanol-powered engines, but nitrogen oxides (NO_x) are present. Air is 78% nitrogen gas. At high temperatures, nitrogen reacts with oxygen to produce NO_x. Nitrogen oxides are produced during all high-temperature combustion reactions that use air as the source of oxygen, including those occurring in car engines, regardless of the type of fuel used. At lower temperatures, N_2 and O_2 remain stable and do not react to form NO_x compounds.

A significant problem with hydrated ethanol-fueled engines is that they are difficult to start in cold weather. This problem occurs because ethanol does not vaporize as easily as gasoline. Engineers have tried to overcome this problem in a number of ways, including

- adding small, auxiliary fuel tanks for gasoline, to use only when starting the engine.

- adding 10% ether or 5% butane to the ethanol to increase its volatility.

- using pre-heating systems.

The Use of Ethanol Blends

Gasohol, also called E10, is a blend of 90% gasoline and 10% ethanol. Tests show that most unmodified engines run as well or better when fueled with gasohol as they do with gasoline. The State of Nebraska ran its fleet vehicles on gasohol for 2 million miles at different altitudes and under varying conditions. Their published results on performance state that drivers reported no problems with starting or driving the vehicles. Examinations of the engines during and after the program revealed no unusual engine wear or carbon build-up. Fuel consumption with gasohol was at least as good as with pure gasoline; figures produced by Professor William Scheller of the University of Nebraska showed a 5.3% improvement over gasoline. **Flexible fuel vehicles**, which can use gasoline, ethanol, or an ethanol-blend fuel, have 5% greater fuel economy when using E85 (85% ethanol, 15% gasoline) than when using gasoline.

Technological Innovations

The U.S. Department of Agriculture in California has tested a continuous fermentation process that can ferment 92%–96% of the sugars in a plant product to yield a 10% ethanol solution in four hours. This continuous fermentation process is faster than the traditional batch process and cuts costs. Creating a vacuum inside a distillation column, which reduces the boiling temperature of ethanol, can further cut production costs by decreasing the energy required for distillation. It also allows the use of less expensive plastic columns, rather than metal. The Solargas Company markets this idea in the United States. In addition, Misui Shipbuilding Co. of Japan is

developing heat- and ethanol-tolerant strains of yeast. Yeast strains that can tolerate heat could allow continuous and simultaneous fermentation and distillation of ethanol. Most yeast strains die when the ethanol concentration gets above about 15%; yeast strains that can tolerate higher ethanol concentrations could further increase the yield of ethanol.

Issue 2 Environmental Impact of Ethanol Use

Air

Research has shown that the burning of ethanol fuels produces fewer atmospheric pollutants than gasoline or diesel. However, ethanol fuels are not perfect. Table 1 shows emissions from engines using gasoline, gasohol (also called E10, made up of 10% ethanol and 90% gasoline), and "pure" hydrated ethanol.

Table 1   Emissions From 1.4-Liter Engine (g/km)

Emission	Gasoline	Gasohol	Ethanol
Hydrocarbons	2.05	1.51	1.55
Carbon monoxide	30. 61	12.78	10.54
Nitrogen oxides	1.15	1.77	1.03
Aldehydes: methanal	0.03	0.04	0.02
ethanal	0.03	0.10	0.32

Note that, according to Table 1, carbon monoxide and hydrocarbon emissions are lower for both ethanol fuels. The emissions of nitrogen oxides (NO_x), however, are increased by the use of gasohol fuels. NO_x compounds contribute to both acid rain and photochemical smog. Ethanol fuels also produce increased levels of other exhaust emissions. Ethanol's low boiling point leads to an increase in emissions of unburned fuel because of evaporation from carburetors and fuel tanks. NO_x compounds and unburned ethanol are reactive in sunlight and contribute to the formation of hazy clouds of photochemical smog. Exhaust systems of many vehicles include **catalytic converters** that convert some of these pollutant emissions to less harmful products before they are released into the atmosphere.

Water

The distillation process used to make ethanol produces 10–16 liters of impure water, or stillage water, for every liter of ethanol produced. The Argonne National Laboratory in Illinois investigated the stillage from an ethanol distillery that used corn as the raw material. The distillery produced 76 million liters of stillage water per year. It would cost several million dollars to equip this distillery to meet current U.S. pollution standards. Research suggests that the use of stillage for fertilizer may be possible.

Soil

A report from the U.S. Office of Technology Assessment concluded that "new intensive crop production for ethanol is likely to lead to a transformation of unmanaged or lightly managed ecosystems, such as forests, into intensively managed systems." A problem with highly productive modern farming techniques is that they can cause the soil to deteriorate by both "mining" its nutrients and increasing soil erosion. Soil erosion has many effects. For example, essential nutrients are washed away. Water quality is affected because of increased sediment load and the introduction of fertilizers and pesticides. Increased sediment deposits also limit the spawning of fish, affect the ability of boats to navigate waterways, and complicate the operation of hydroelectric generators.

Issue 3 Using Crops for Food vs. Using Crops for Fuel

Food Supplies

Two major crops used as feedstocks for ethanol production are corn and sugar cane, both of which are important food crops. In 1998, about 5.5 billion liters of corn ethanol were produced in the United States, which is equivalent to slightly more than 1% of total U.S. gasoline consumption. The production of this ethanol consumed about 6% of all the corn grown in this country. Tripling the annual U.S. corn ethanol production to 15 billion liters would require about 20% of the U.S. corn crop. Grain from the United States does not feed Americans only; Asia and Africa are major importers. The U.S. Office of Technology Assessment predicts that a production rate of 15 billion liters of ethanol per year will have a "substantial impact" on crop availability, potentially pushing prices up by one-third. This increase could have a considerable effect on the diet and survival of several hundred million people around the world who already have limited food supplies and spend a large percent of their income on food.

Lester Brown, director of the Worldwatch Institute, calculated the annual grain and cropland requirements of people and cars, as shown in Table 2. The figures can be used to estimate how many people could be fed using the grain that, if turned into ethanol, would be used by a typical car in Europe or in the United States.

Table 2 Annual Grain and Cropland Requirements for Food and Fuel

Consumer	Grain (kg)	Cropland (acres)
Subsistence diet	180	0.2
Affluent diet	725	0.9
Typical European car (7,000 miles @ 37.5 mpg)	2,800	3.3
Typical U.S. car (10,000 @ 15 mpg)	6,620	7.8

Corn is an important food resource for humans. We eat it directly and also feed it to the livestock that provide many other food products.

Economic Pros and Cons

Large-scale ethanol production would provide local economies with increased job opportunities and other economic benefits, but there are economic drawbacks. Producing the crops needed to supply an ethanol plant would require increasing the amount of cultivated land. This change could force poor landowners off their land, especially in less developed countries.

Running an ethanol production plant also uses a considerable amount of energy. One of the controversies surrounding ethanol technology is over the net energy balance. There is some concern that more energy must be used to make ethanol than is justified by the amount of energy provided by ethanol fuel. Although the amount of energy supplied by the sun is not included in energy balance equations for ethanol, there are considerable inputs of support energy. In addition to the energy used during fermentation and distillation, support energy includes the energy consumed in the manufacture and use of the fertilizers, pesticides, tractors, and trucks used to grow and transport the crops. According to a 1999 study conducted by the Center for Transportation Research at the

Argonne National Laboratory, the total amount of support energy required to produce corn ethanol is about 75% of the energy content of ethanol fuel. Although this is a net energy gain of 25%, ethanol must be burned to provide usable energy and much of the energy released during combustion, often over 70%, is lost to the environment.

However, energy balance concerns often do not have much practical significance. For example, the energy content of electricity is about 60% less than the energy content of the fuels that are combusted during electricity generation. Ethanol, like electricity, is a high-quality form of energy that is more versatile and useful than the resources from which it is made. An energy loss is the price that is often paid for the higher quality and greater utility of energy sources such as electricity and ethanol.

The prices of sugar cane, corn, and crude oil tend to vary significantly, which makes the economics of ethanol production uncertain. The World Bank has estimated that the consumer cost of ethanol made from sugar cane would be competitive economically with petroleum fuels if the cost of sugar cane were less than $15 per ton and that of crude oil over $30 per barrel. Even though oil prices have risen lately—above $50 per barrel in 2004—sugar cane production continues to fluctuate unreliably. However, even though ethanol costs more than gasoline or diesel fuel, some experts believe that the costs of its aggregate environmental impacts and risks over the entire production process are less than those of petroleum-derived fuels. Use of ethanol fuels in place of petroleum fuels would also make more crude oil available as a raw material for plastics and other petroleum products. Political instability in some countries is also a potential threat to the reliability of the international system for distributing crude oil. In an emergency, a more expensive, alternative fuel is better than no fuel at all. The U.S. Office of Technology Assessment considers corn to be the cheapest source of ethanol. They estimated that a corn ethanol plant with an output of 190 million liters per year would cost $70.4 million dollars; a sugar cane plant of comparable size would cost $132 million dollars.

Technological Innovations

The University of Pennsylvania has been studying the use of enzymes to convert wood to glucose syrup that could be used for ethanol production. If the development of the enzyme process is successful, trees that grow rapidly on dry or marginal land, such as poplar, aspen, and eucalyptus, could be used in place of food crops to produce ethanol. Scientists at the University of Pennsylvania estimate that it would require a poplar plantation slightly larger than the size of Pennsylvania (about 117,000 km^2 or 45,000 mi^2) to produce about 75 billion liters of ethanol, which equals approximately 15% of the volume of gasoline consumed in the U.S. in 2004. In related research, scientists at New York University have developed a process for treating sawdust or shredded paper with sulfuric acid to make sugars that could be used for ethanol production. The 1999 Center for Transportation Research study projected that in the near future, due to changes in production methods and through the use of biomass other than corn, the support energy needed to produce ethanol could fall to less than 10% of the energy contained in ethanol.

Issue 4 Using Ethanol as a Pollution-Reducing Fuel Additive

As an automobile fuel, ethanol burns much cleaner than gasoline. Many state and local regulations require the addition of a small percentage of ethanol or other chemical additives to gasoline to make it burn more completely and produce fewer pollutants. The decision to use one chemical over another involves assessing the trade-offs associated with each.

In the 1970s and '80s, as leaded gasoline was being phased out and advances in automobile technology were taking place, the pollutant levels in automobile exhaust were lowered significantly. Nonetheless, automobiles still remained a major source of air pollution and the government decided that changes in the chemistry of the fuel were needed.

In 1990, Congress passed an amendment to the Clean Air Act requiring that air-cleaning chemicals be added to gasoline. The chemical additives in this reformulated gasoline, called oxygenates, add oxygen to the fuel so that more complete combustion occurs, thus decreasing the production of carbon monoxide and other pollutants.

Two organic chemicals—ethanol, C_2H_5OH, and **methyl tertiary butyl ether (MTBE)**, $C_5H_{12}O$—were the leading candidates for use as gasoline oxygenators. Both ethanol and MTBE are colorless, flammable liquids. The oil industry overwhelmingly chose to use MTBE mainly because MTBE, unlike ethanol, is synthesized from a waste product of the oil refining process and can be pumped through existing gasoline pipelines. The production of MTBE can reduce wastes generated by the petroleum industry and at the same time reduce air pollution.

According to an automobile and oil industry study begun in 1989, compared with the combustion of an equal volume of 100% gasoline, the combustion of a fuel mixture of 89% gasoline and 11% MTBE produces about ⅓ less ozone and significantly lower levels of carbon monoxide and other toxic exhaust products, such as benzene.

Recent studies have shown that MTBE can cause cancerous tumors in laboratory animals, and scientists now fear that MTBE could jeopardize human health. Since the introduction of MTBE as a fuel additive, drivers, as well as employees of refineries and gas stations, have reported ailments such as breathing difficulties, eye irritation, headaches, dizziness, nausea, rashes, and nosebleeds. For years, tons of MTBE have entered California's air every day from exhaust fumes; leaky underground storage tanks can potentially release thousands of gallons of MTBE into groundwater. MTBE has some properties that make underground leaks a serious threat to water supplies—it dissolves easily in water, allowing it to travel quickly through the soil, and it does not biodegrade, which makes it difficult and expensive to clean up. Due to the potential harmful effects of MTBE, some towns have been forced to stop using local water sources that have been contaminated with MTBE. By 2004, about half the states in the U.S. had adopted or were about to implement bans of MTBE. In other states, local officials continue to have to choose between using MTBE, to improve air quality, and not using MTBE, to safeguard water quality.

Exothermic and Endothermic Interactions

33.1   Interaction Energy

Purpose ▶ **E**xplore the energy transfers involved when two substances dissolve in water.

Introduction

In this course you have been introduced to many different forms of energy; you have also explored the conversion of energy from one form to another. In Activity 3, "Survival Needs: Temperature," you learned about the two laws of thermodynamics. The first law, the Law of Conservation of Energy, states that energy is never created or destroyed. In Activity 4, "Energy Transfer," you explored how mechanical (kinetic) energy can be transformed into thermal (heat) energy and learned that any change in temperature is a measure of change in the motion of atoms or molecules. Activity 16, "Photosynthesis," focused on the conversion of radiant energy (light) into chemical energy. This chemical energy can be transformed into thermal energy by the combustion reactions you studied in Activity 31, "Fueling Trade-offs."

Many chemical processes involve a transfer of energy that takes place as original bonds are broken and new bonds are formed. Chemical reactions, changes of state, and the dissolving of a solid in a liquid all involve bonding changes. In this activity, you will examine the chemical interactions that take place when two different solids dissolve in water. Recall from Activity 21, "Identifying and Separating Hydrocarbons," that water is capable of dissolving a greater variety of substances than methanol or glycerin. Water dissolves

Introduction
(cont.)

many types of compounds because of its molecular structure. As shown in Figure 1, the hydrogen atoms in a water molecule tend to make an obtuse (about 102°) angle with the oxygen atom. This non-linear arrangement results in a molecule with a positive side (near the hydrogen atoms) and a negative side (near the oxygen atom). Molecules that have negative and positive sides are called **polar molecules**.

Figure 1   Non-Linear Structure of a Water Molecule

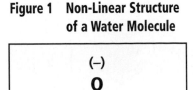

A substance that dissolves in a liquid is called a **solute**; the liquid in which the solute dissolves is called the **solvent**. Polar solvents can more easily form bonds with many common solutes. As a result, it is easier for many solutes to dissolve in polar solvents than in non-polar solvents. Water makes an excellent solvent because its molecules are polar; in fact, water is sometimes referred to as the universal solvent. Liquids like methanol and glycerin consist of non-polar molecules; they cannot dissolve as many different solutes as water can.

As some solutes dissolve, they release energy to the surrounding solvent; other solutes absorb energy from the solvent. Chemical processes that release energy are called **exothermic**. Those that absorb energy are called **endothermic**.

Materials

▦ **For the class**

   2–4 balances
   2–4 spatulas
   80 g sodium bicarbonate ($NaHCO_3$)
   80 g calcium chloride ($CaCl_2$)
   supply of room-temperature water
   weighing papers

◫ **For each group of four students**

   2   calorimeters
   2   styrofoam inserts for calorimeter
   2   immersion thermometers
   1   50-mL graduated cylinder

▪ **For each student**

   1   pair of safety glasses

Safety Note

 All students should wear safety glasses.

Procedure

1. Decide which team of two in your group will investigate $NaHCO_3$ and which team will investigate $CaCl_2$.

2. Measure 50 mL of water and put it in your calorimeter. Measure the temperature of the water and record this as the initial temperature.

3. Obtain 10 g of your team's solid ($NaHCO_3$ or $CaCl_2$).

4. Add your solid to the water in the calorimeter and put the lid on, making sure the thermometer bulb is submerged in the water. Gently swirl the cup until the solid is completely dissolved and the temperature stops changing. Record this as the final temperature.

5. Calculate and record the change in temperature and report this to your teacher.

6. Record the change in temperature obtained by the other team in your group.

Analysis

Group Analysis

1. What do you think would be the effect of adding 20 g of $NaHCO_3$ to 50 mL of water? Explain.

2. What do you think would be the effect of adding 10 g of $CaCl_2$ to 100 mL of water? Explain.

3. What do you think causes the change in temperature when $NaHCO_3$ dissolves in water?

4. What do you think causes the change in temperature when $CaCl_2$ dissolves in water?

Individual Analysis

5. How much $CaCl_2$ combined with how much $NaHCO_3$ would produce no temperature change when added to 50 mL of water? Explain your reasoning.

33.2 Quantitative Investigation of an Exothermic Interaction

Purpose

Discover whether there is a predictable relationship between the amount of solvent and solute chemicals used and the amount of thermal energy transferred during the dissolving process.

Introduction

You have explored how chemical reactions are used to make clothing, paper, fuel, and other everyday products. You have used models to illustrate a chemical reaction: bonds joining the atoms of the reactant chemicals are broken, the atoms are rearranged, and new bonds are formed to make the product chemicals. You have even noticed that the mass of the reactants is exactly the same as the mass of the products. Based on studies similar to yours, scientists have come to understand that although new substances are formed during a chemical reaction, neither mass nor matter is created or destroyed. This observation is called the **Law of Conservation of Matter**.

Chemicals that release energy as they interact with water, such as CaCl$_2$, are used to melt ice on roads in the wintertime.

In the previous activity, you also learned that a product without mass—heat (thermal energy)—can be released or absorbed during chemical processes. These energy transfers occur because different amounts of energy are stored in different chemical bonds. If the bonds in the product chemicals store more energy than the bonds in the reactant chemicals, the process is endothermic (energy is absorbed). If the bonds in the product chemicals store

Introduction
(cont.)

less energy than the bonds in the reactant chemicals, the process is exothermic (energy is released). Can the amount of energy transferred during a chemical process be predicted? In this activity, you will measure the thermal energy transferred during a chemical process as you continue to investigate the relationships between matter and energy and the concept of the conservation of mass and energy.

Materials

For the class

2–4 balances

2–4 spatulas

200 g calcium chloride ($CaCl_2$)

weighing papers

supply of room-temperature water

For each team of two students

1   calorimeter

1   immersion thermometer

1   50-mL graduated cylinder

For each student

1   pair of safety glasses

Procedure

Part A   Prediction

1.   Prepare data tables similar to Tables 1 and 2. In each table, record in the appropriate box the class average for the temperature change that occurs when 10 g of $CaCl_2$ dissolve in 50 mL of water. You determined this temperature change in Activity 33.1, "Interaction Energy."

2.   Since you know what happens when 10 g of $CaCl_2$ dissolve in 50 mL of water, fill in the rest of the first data table with your predictions.

Table 1   Predicting Temperature Changes

$CaCl_2$ (g)	H_2O (mL)	Predicted Temp. Change (°C)
10	50	
5	50	
15	50	
20	50	
25	50	
10	25	
10	100	
5	25	
5	100	

Part B   Experimentation

Safety Note

  Wear safety glasses during this activity. Wash your hands before you leave class.

Procedure (cont.)

1. Measure out the amount of water assigned to your team, pour it into your calorimeter, and measure and record the temperature.

2. Weigh out your assigned amount of $CaCl_2$ and add it to the calorimeter. Put the lid on, making sure the thermometer is immersed in the water. Gently swirl the cup until the solid is completely dissolved and the temperature stops rising. Record the highest temperature reached.

3. Determine the temperature change for your experiment.

4. Report your experiment's temperature change to the teacher for the class data table.

5. Fill in your second data table using data from the class data table.

Table 2   Class Data for the $CaCl_2$–H_2O Interaction

$CaCl_2$ (g)	H_2O (mL)	Team 1 Temp. Change (°C)	Team 2 Temp. Change (°C)	Average Temp. Change (°C)
10	50			
5	50			
15	50			
20	50			
25	50			
10	25			
10	100			
5	25			
5	100			

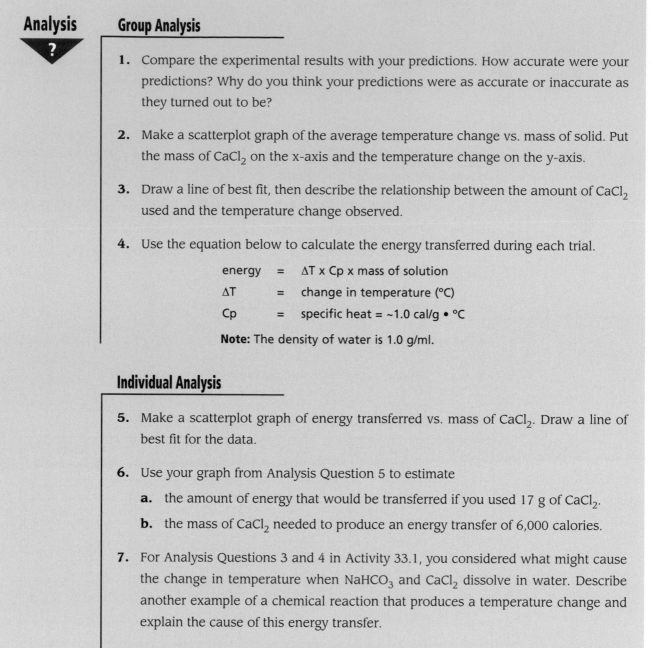

Analysis

?

Group Analysis

1. Compare the experimental results with your predictions. How accurate were your predictions? Why do you think your predictions were as accurate or inaccurate as they turned out to be?

2. Make a scatterplot graph of the average temperature change vs. mass of solid. Put the mass of $CaCl_2$ on the x-axis and the temperature change on the y-axis.

3. Draw a line of best fit, then describe the relationship between the amount of $CaCl_2$ used and the temperature change observed.

4. Use the equation below to calculate the energy transferred during each trial.

energy	=	ΔT x Cp x mass of solution
ΔT	=	change in temperature (°C)
Cp	=	specific heat = ~1.0 cal/g • °C

Note: The density of water is 1.0 g/ml.

Individual Analysis

5. Make a scatterplot graph of energy transferred vs. mass of $CaCl_2$. Draw a line of best fit for the data.

6. Use your graph from Analysis Question 5 to estimate
   a. the amount of energy that would be transferred if you used 17 g of $CaCl_2$.
   b. the mass of $CaCl_2$ needed to produce an energy transfer of 6,000 calories.

7. For Analysis Questions 3 and 4 in Activity 33.1, you considered what might cause the change in temperature when $NaHCO_3$ and $CaCl_2$ dissolve in water. Describe another example of a chemical reaction that produces a temperature change and explain the cause of this energy transfer.

8. Do you think that exothermic interactions could be used to help achieve sustainable development? Explain.

Extension   Determine the mathematical equation that relates the mass of $CaCl_2$ used to the amount of energy transferred.

33.3 Energy As You Like It

Purpose ▶ **I**nvestigate the possibility of using energy sources other than combustible, carbon-based fuels.

Introduction ▼

There are many potential sources of energy on Earth. Over the course of any year, energy derived from the sun—including wind and flowing water as well as sunlight—is capable of supplying more energy than we can use. With technologies available today, we can capture this energy fairly easily for immediate use, but we cannot store much of it for future use. For example, electricity is a very convenient, versatile form of energy that is often generated using solar-derived energy. But electricity must

The sun's rays provide Earth with a vast amount of energy. Technologies that allow us to use solar energy effectively and conveniently could help reduce our reliance on fossil fuels.

be used soon after it is generated. Consequently, most of the world's electricity is generated "on demand" through the combustion of fossil fuels.

Naturally occurring fuels—such as food, petroleum, and uranium—are excellent store-houses of energy. Many common fuels contain stored chemical energy trapped in the bonds among their atoms and molecules. This stored energy can be released for later use. Humans have developed numerous technologies that allow us to release stored energy as

▶

Introduction
(cont.)

needed. We have also developed many technologies for storing energy and continue to look for new energy storage technologies.

Technology that would allow us to take advantage of some of Earth's "unused" energy could greatly increase the world's supply of usable energy. Much of the energy in sunlight goes unused. Abundant chemicals, such as hydrogen, also contain a vast amount of stored energy. A variety of technologies designed to make this "unused" energy available have been, or are being, developed. They include batteries, fuel cells, and a variety of methods for storing solar energy.

Using Chemicals to Store and Release Energy

Solar Energy

Direct solar radiation—not including solar-derived wind and flowing water—is capable of supplying more energy than we can use over the course of any given year. However, direct solar energy is available to us at only certain times of the day or year. If solar energy is to be put to practical use, we must be able to store it for use whenever we need it. Solar energy is a vast, renewable resource that can be stored in a number of different ways. Some examples include storing it in heated materials with relatively high heat capacity; as the latent heat of fusion in solid-liquid phase changes; or as the heat of endothermic reactions.

Most applications of solar energy involve direct heating. The energy carried by the sun's rays is used to heat water, thus storing the energy in the motion of the water molecules. This heated water is then either used as hot water for purposes such as cleaning or piped through buildings to heat the indoor air. A technology that uses underground tanks for seasonal storage of solar heat has been used successfully, even in cold climates like Sweden. In summer, hot water from solar heat collectors is used to heat rocks stored in large underground tanks. In winter, cold water pumped through the tanks absorbs heat from the rocks. The heated water is then circulated through buildings to provide heat.

Solar heat can also be stored as phase changes—that is, as heat of fusion or vaporization. A number of materials have melting points near 40°C, a temperature easily exceeded by solar radiation. Solar collectors containing these materials

can be designed to attain even higher temperatures. When the sun shines, the material melts, absorbing the heat of fusion. When the temperature of the surroundings fall below the melting point, the material "freezes," releasing the heat of fusion back to the surroundings.

Another technology under development is called a solar chemical heat pipe. Heat collected from a solar panel provides energy that is absorbed during an endothermic reaction. The chemical products of this reaction, which contain trapped solar energy in the form of chemical bonds, can be stored until the heat is needed. At that time, the reverse—exothermic—chemical reaction is carried out, releasing the stored heat. In Figure 2, the reactant chemicals of the endothermic reaction are shown as "A" and the product chemicals are shown as "B."

In the exothermic reaction, the "B" chemicals are the reactants and the "A" chemicals are the products. One of the major problems with this technology is that the "B"

Figure 2   Solar Chemical Heat Pipe

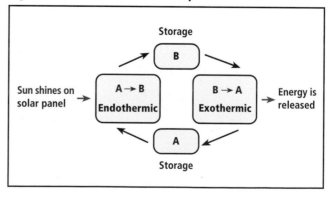

Figure 3 Lithium-Sulfur Battery

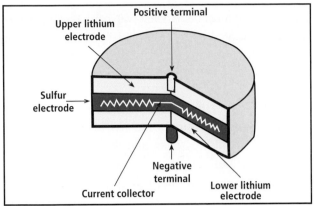

Upper lithium electrode
Positive terminal
Sulfur electrode
Negative terminal
Current collector
Lower lithium electrode

chemicals, which store the energy, cannot easily be transported from one location to another, especially if long distances are involved. Because the energy density (cal/g or J/g) of these reactions is low, they can be used only for on-site purposes, such as supplying heat. Also, finding the most efficient chemical reactions to use for the endothermic and exothermic reactions has proven to be challenging.

Batteries

Electricity produced from any energy source, including photovoltaic or solar thermal processes, can be stored as chemical energy in batteries. Batteries contain chemicals. When these chemicals react, they cause the transfer of energetic electrons. This transfer of electrons creates a flow of electricity. A battery "dies" when most of its reactant chemicals have reacted to form product chemicals and electricity. When there are too few reactant chemicals, the reaction rate and resulting electron flow become too low to be useful.

Batteries can often be "recharged" by reversing this chemical reaction: a battery charger forces electrons to flow in the opposite direction, causing the chemical reaction to proceed in the reverse direction, thereby changing the product chemicals back into the original reactant chemicals. When removed from the charger, the battery once again contains enough of the original chemicals to react and generate electricity. Rechargeable batteries have long been used in car batteries and miner's lamps, and more recently in electronic equipment such as laptop computers and video cameras.

In the battery shown in Figure 3, lithium and sulfur are the reactants. When these two chemicals react, electrons

are released. This type of battery produces a large amount of energy for its small size and light weight. Lithium-sulfur batteries are commonly used to power cellular phones and other electronic devices. Batteries that can provide enough electricity to start or operate a car or other large device are often very heavy; although powerful lightweight batteries have been developed, they are very expensive. Scientists and engineers are searching for inexpensive, lightweight materials from which to make improved batteries.

Hydrogen Fuel Cells

Fuel cells are similar to batteries in that they use stored chemical energy to produce electricity. However, fuel cells generate electricity much more effectively than batteries. When hydrogen combines with oxygen in a combustion reaction, the resulting products are water and a lot of heat, but no electricity. A fuel cell is a device that uses catalysts to control the reaction of hydrogen with oxygen; the products of a fuel cell reaction are electricity and hot water that has a temperature of about 70°C.

Figure 4 is a diagram of a hydrogen fuel cell. The atoms of hydrogen gas fed into one side of the fuel cell are broken into protons and electrons. Only the protons can cross a membrane that separates the oxygen and hydrogen. The electrons flow through a circuit on their way to meet with the protons and oxygen atoms to form water molecules. This flow of electrons creates the electricity produced by the fuel cell. Because the reaction in a fuel cell proceeds at low temperatures, the nitrogen in the air does not react with oxygen to form NO_x pollutants. The only by-product of a hydrogen fuel cell is hot water.

Figure 4 Direct Hydrogen Fuel Cell

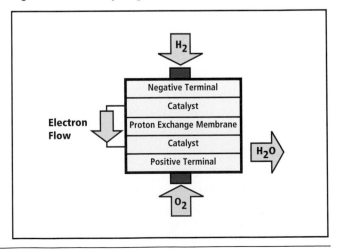

H_2
Negative Terminal
Catalyst
Electron Flow
Proton Exchange Membrane
Catalyst
H_2O
Positive Terminal
O_2

Fuel cells are being used in increasing numbers of office buildings around the country to provide electricity and hot water for heating, cooling, and dehumidifying. The U.S. Department of Energy is evaluating fuel cell buses in three California public transit districts. For example, AC Transit in the San Francisco East Bay area began a program with three hydrogen fuel cell buses in 2005. The program will gather information to compare hydrogen fuel cell buses to the diesel buses currently in use.

One of the problems with fuel cells is that pure hydrogen fuel is not naturally abundant. One of the challenges in fuel cell technology is that hydrogen gas is scarce, so it has to be produced. Natural gas, or methane (CH_4), is a source of hydrogen that is quite abundant: the methane can be "reformed" to produce hydrogen gas and CO_2. However, natural gas is a nonrenewable energy resource, and like oil, its consumption releases CO_2 as a by-product.

Another way of obtaining hydrogen is to use electricity to split water into hydrogen gas and oxygen gas in a process known as **electrolysis**. Electrolytic production of hydrogen has long been used on a large scale in industry. Hydrogen can be stored in pressurized containers until needed. Like hydrocarbon fuels, compressed hydrogen gas is flammable and must be handled with care. Environmental hazards associated with the production of hydrogen gas are related to the source of the energy used for reforming methane or for the electrolytic process.

Hydrogen is a fuel, but it is not a resource in the sense that fossil fuels or biomass are resources. The overall process of making hydrogen and then using it to react with air in a fuel cell consumes more energy than it yields. In the same way, more energy is required to generate electricity than is carried by the electricity. Hydrogen fuel, like electricity, "carries" energy in a form that is more useful than the energy inputs that are used to produce it. Some people imagine that someday there will be a device in every home that uses solar energy to make hydrogen by electrolysis. The hydrogen can then be used in fuel cells that provide electricity and heat for the home and as fuel for the car. Hydrogen fuel could also be available at solar-powered filling stations, or perhaps generated on board a vehicle from a fuel tank filled with water!

Analysis ▼?

Individual Analysis

1. Do you think that any of the energy storage technologies described in this reading could help us to develop a sustainable source of energy? Explain why or why not.

2. What additional information would you want to have about each type of storage technology to accurately evaluate its potential for widespread use in the future? Be as specific as you can.

33.4 Investigating the Decomposition of H₂O₂

Purpose ▶ **D**esign and carry out an investigation to determine the amount of energy transfer that takes place as hydrogen peroxide undergoes a decomposition reaction.

Introduction ▼

A significant transfer of energy accompanies the decomposition of hydrogen peroxide, which is shown below. Iron (II) chloride (FeCl₂) can be used to catalyze this reaction and make it easier to measure the temperature change associated with the energy transfer.

hydrogen peroxide → water + oxygen gas + heat

As you studied in Activity 26, "Catalysts, Enzymes, and Reaction Rates," catalysts like the FeCl₂ used in this activity decrease the energy required to start a chemical reaction. This often has the effect of speeding up the reaction. While a catalyst does participate in the reaction, it appears at the end of the reaction in the same form it had at the beginning. In this activity, use of the FeCl₂ catalyst will cause the decomposition of hydrogen peroxide to take place within a single class period rather than over several weeks.

Both cups contain equal amounts of H₂O₂. A catalyst for the decomposition of H₂O₂ has been added to the cup on the right.

33.4 Investigating the Decomposition of H$_2$O$_2$

Materials

 For each group of four students

    5.0 g iron (II) chloride (FeCl$_2$)

    250 mL hydrogen peroxide (H$_2$O$_2$)

    1   balance

    1   50-mL graduated cylinder

    1   immersion thermometer

    1   calorimeter

    1   styrofoam insert for calorimeter

■ **For each student**

    1   pair of safety glasses

Safety Note

 Wear safety glasses at all times during this activity. Hydrogen peroxide can bleach hair and clothes. Use care in all laboratory procedures and wash your hands at the end of the investigation.

Procedure

1. Work with the other members of your group to develop a step-by-step experimental procedure that will allow you to achieve the purpose of this activity. Keep in mind that your investigation is limited by the materials you have been given and the fact that you must complete the experiment within the allotted time.

2. Write out the experimental procedure clearly, so that someone else could follow it.

3. Check with your teacher to have your procedure approved.

4. Carry out your experiment, recording all data in clearly labeled data tables.

5. Prepare any appropriate graphs, equations, and calculations.

6. Describe, interpret, and explain your results. Make sure to relate your explanation to the laws of thermodynamics and the kinetic molecular theory that you were introduced to in Activities 3 and 4.

Note: The specific heat of the system can be approximated as 1.0 cal/g • °C

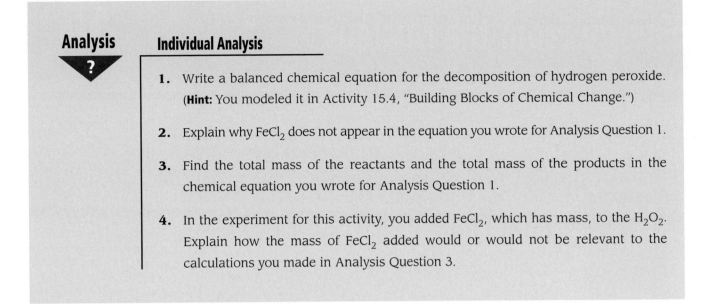

Analysis
?

Individual Analysis

1. Write a balanced chemical equation for the decomposition of hydrogen peroxide. (**Hint:** You modeled it in Activity 15.4, "Building Blocks of Chemical Change.")

2. Explain why FeCl$_2$ does not appear in the equation you wrote for Analysis Question 1.

3. Find the total mass of the reactants and the total mass of the products in the chemical equation you wrote for Analysis Question 1.

4. In the experiment for this activity, you added FeCl$_2$, which has mass, to the H$_2$O$_2$. Explain how the mass of FeCl$_2$ added would or would not be relevant to the calculations you made in Analysis Question 3.

Energy From the Nucleus

34.1   Cool Light

Purpose ▶ **I**nvestigate various types of radiation and explore the association between radiation and energy transfer.

Introduction

So far, you have observed energy in many forms and many places. Energy is constantly being transferred from one object to another, converted from one form to another, and transported from one place to another. Energy is transferred when heat flows from a hot object to a cold object or when an object is lifted. Energy changes from one form to another when a fuel is burned, when a person uses a bicycle to get to work, or when electricity is converted to light and heat. Energy is transported from one place to another through a wide variety of processes, including **radiation**. As you most likely have experienced firsthand, a glowing light bulb radiates both light and heat. Some other exam-

Satellite dishes, whether large or small, collect incoming radiation. The energy contained in the radiation can be used to reveal information about the radiation's source.

ples of radiation are radio, television, and cellular phone signals, sunlight, x-rays, nuclear radiation, and cosmic radiation. In this activity, you will observe and measure some of the effects of the radiant energy transfer associated with a glowing light bulb.

Materials

■■ For each group of four students

  1   slide containing thin plastic film
  1   rectangular piece of thick plastic

■■ For each team of two students

  1   9-V bulb
  1   9-V battery and harness
  1   bulb and filter stand
  1   immersion thermometer
     access to a clock with a second hand

Procedure

1. Without attaching the wires to the light bulb, set up the equipment as shown in Figure 1. Make sure there is a 1-cm gap between the top of the light bulb and the tip of the thermometer bulb. Make sure the slot in the stand is half-way between the light bulb and the thermometer bulb.

Figure 1   Correct Positioning of Thermometer

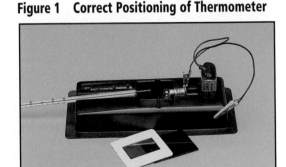

2. Read Steps 3 and 4 carefully and prepare an appropriate data table.

3. Record the initial temperature, then attach the wires so the light bulb glows. Note the time and take temperature readings every 30 seconds for 5 minutes. Record the data in your data table. After 5 minutes have elapsed, unclip one wire so that the bulb stops glowing.

4. Calculate the change in temperature (ΔT) caused by the glowing bulb. Add this information to your data table.

5. Carefully inspect and describe both the plastic film in the slide and the rectangular piece of plastic. As part of your description, include an estimate of the percentage of light that can pass through each sample of plastic.

6. Predict how much you would expect the temperature to change if you were to repeat Step 3 with

   a. the rectangular piece of plastic placed in the slot between the light bulb and the thermometer bulb.

   b. the slide with the plastic film placed in the slot between the light bulb and the thermometer bulb.

7. Do experiments to test each of the predictions you made in Step 6. Record all of your data in appropriately designed data table(s).

Energy From the Nucleus

34.1 Cool Light

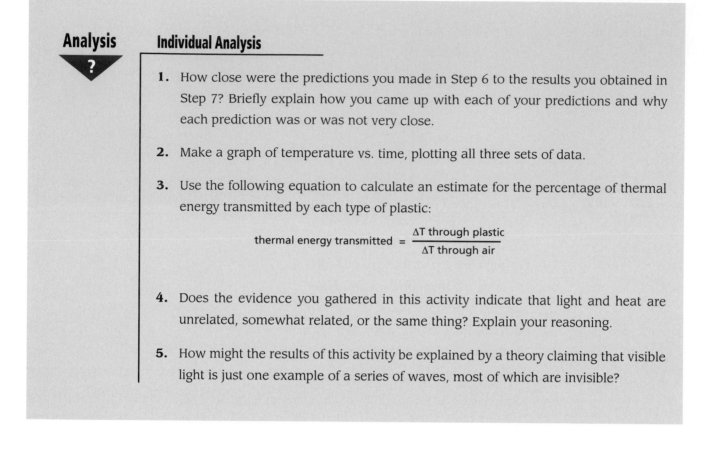

Analysis

?

Individual Analysis

1. How close were the predictions you made in Step 6 to the results you obtained in Step 7? Briefly explain how you came up with each of your predictions and why each prediction was or was not very close.

2. Make a graph of temperature vs. time, plotting all three sets of data.

3. Use the following equation to calculate an estimate for the percentage of thermal energy transmitted by each type of plastic:

$$\text{thermal energy transmitted} = \frac{\Delta T \text{ through plastic}}{\Delta T \text{ through air}}$$

4. Does the evidence you gathered in this activity indicate that light and heat are unrelated, somewhat related, or the same thing? Explain your reasoning.

5. How might the results of this activity be explained by a theory claiming that visible light is just one example of a series of waves, most of which are invisible?

34.2 Electromagnetic Wave Relationships

Purpose

> **D**erive some important mathematical relationships among the characteristics of electromagnetic waves.

Introduction

All waves have measurable characteristics by which they can be described. These characteristics include **wavelength**, **amplitude**, **frequency**, wave velocity, and energy. Two of these characteristics are illustrated in Figure 2. Frequency is the number of complete waves that pass in a second, wave velocity is the speed of the wave, and wave energy is the energy carried by the wave. Many of these characteristics are related to one another; the relationships can often be described by mathematical equations. These relationships allow you to use a few measured characteristics to accurately predict unmeasured characteristics and describe unobserved wave behavior. Wavelength (λ) and amplitude (A) are measured in meters; frequency (f) is measured in hertz (Hz), or wave cycles per second; wave velocity (v_w) is measured in meters per second.

Figure 2   Some Important Wave Characteristics

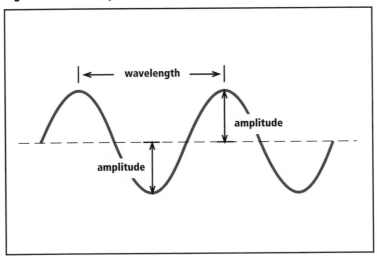

Procedure

1. Carefully study the data shown in Figure 3. Produce any graphs or calculations necessary for you to fully and clearly answer the Individual Analysis Questions.

   Note: $6.0 \times 10^{-3} = 0.006$; $6.0 \times 10^{-6} = 0.000006$

Figure 3   The Electromagnetic Spectrum

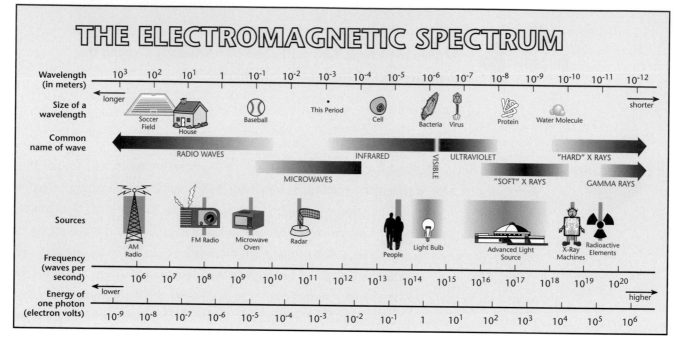

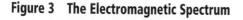

Analysis

?

Individual Analysis

1. Describe qualitatively the relationship between

   a. wavelength (λ) and frequency (f).

   b. energy (E) and frequency (f).

2. Explain why radios and microwave ovens can safely be used in the home, but special precautions and training are needed to operate an x-ray machine safely.

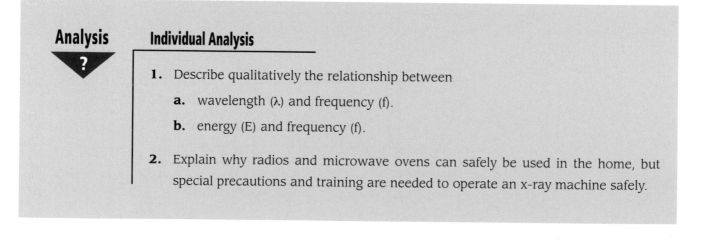

Extension   Use the information provided in Table 1 to derive the mathematical equation that shows quantitatively the relationship between

1. wavelength (λ) and frequency (f).

2. energy (E) and frequency (f).

Hint: Consider the units involved.

Table 1   Some Examples From the Electromagnetic Spectrum

Electromagnetic Radiation	Wavelength (meters)	Frequency (hertz)	Energy (joules)
Radio wave	6.0×10^{1}	5.0×10^{6}	33×10^{-28}
Microwave	6.0×10^{-3}	5.0×10^{10}	33×10^{-24}
Infrared ray	6.0×10^{-6}	5.0×10^{13}	33×10^{-21}
Visible light	6.0×10^{-7}	5.0×10^{14}	33×10^{-20}
Ultraviolet ray	6.0×10^{-8}	5.0×10^{15}	33×10^{-19}
X-ray	6.0×10^{-10}	5.0×10^{17}	33×10^{-17}
Gamma ray	6.0×10^{-12}	5.0×10^{19}	33×10^{-15}

34.3 Nuclear Radiation

Purpose ▶ **E**xplore high-energy electromagnetic radiation and its sources and uses.

Introduction ▼

The **electromagnetic spectrum** described in the last activity includes very high-energy **gamma radiation** (γ). Nuclear fusion reactions occurring in the sun and stars emit huge quantities of gamma radiation; gamma radiation is also emitted in fairly small amounts during the nuclear decay of naturally occurring radioactive elements, or radioisotopes, found on Earth. Low-level radiation emitted by radioisotopes can be used for many purposes, such as helping determine the age of old objects

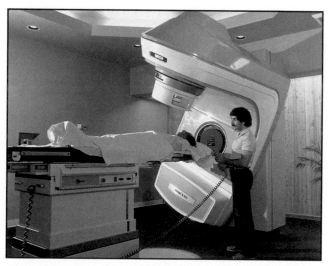

Electromagnetic radiation is used to help diagnose a variety of medical conditions.

and making medical diagnoses. When some of these radioisotopes are mined, refined, and concentrated, they produce higher levels of radiation that can be used to generate heat and electricity, preserve foods, and make weapons. You were introduced to the conversion of mass to energy during nuclear reactions in Activity 31.2, "Fuels for the Future." According to Einstein's equation, as shown below, the energy released when 1 kg of mass is destroyed during a nuclear reaction is 9×10^{16} J.

$$E = mc^2 = (1 \text{ kg}) \bullet (3 \times 10^8 \text{ }^m\!/_s)^2 = 9 \times 10^{16} \text{ kg} \bullet {}^{m^2}\!/_{s^2} = 9 \times 10^{16} \text{ joules}$$

9×10^{16} J is enough energy to raise the temperature of 2 billion liters (2 billion kg) of water from its freezing point at 0°C to its boiling point at 100°C.

Introduction
(cont.)

In addition to the observable macroscopic effects of the energy released during nuclear reactions—such as the destruction caused by nuclear weapons and the heat generated in a nuclear reactor—specialized equipment provides clues to what is actually happening to the atoms. Evidence indicates that during a nuclear reaction, radiant energy and parts of the nucleus are ejected outward from the nucleus at high velocities and in all directions. Some common nuclear reactions and the types of radiation they emit are shown in Figure 4.

Figure 4   Common Nuclear Reactions and Their Products

Hydrogen Fusion

| H | + | H | → | He | + | n | + | γ |

Hydrogen + Hydrogen → Helium + neutron + gamma ray

Uranium Decay

| U | → | Th | + | α | + | γ |

Uranium → Thorium + alpha particle + gamma ray

Krypton Decay

| Kr | → | Rb | + | β | + | ν | + | γ |

Krypton → Rubidium + beta particle + neutrino + gamma ray

Alpha particles, **beta particles**, and **neutrinos** are high-velocity particles emitted by a decaying nucleus. These particles carry away some of the energy generated by the reaction. Each of the elements that undergo radioactive decay has a specific and constant rate of decay. This rate is measured as the element's **half-life**, which is the time it takes for half of a sample of that element to decay. Elements with a shorter half-life decay more rapidly, and thus emit radiation at a faster rate, than elements with a longer half-life. Conversely, elements with a longer half-life remain radioactive for a much longer time than elements with a shorter half-life. In this activity, you will simulate the process by which a radioactive element decays.

Materials   ■■ **For each team of two students**

1   piece of paper
1   pair of scissors
1   marker

Procedure

1. Carefully fold your paper in half and cut along the fold. Make an obvious mark on one of the paper halves, then set aside the marked half. The marked piece of paper represents the half of the sample that degrades during the first half-life of a radioactive element.

2. Repeat Step 1 using the unmarked half piece of paper, setting aside the marked half in the same place as before.

3. Repeat Step 2 until you have 10 marked pieces of paper.

4. Make a table and a graph that shows the fraction of unmarked paper remaining after each cut you made. The graph shows the decay curve for your element.

Analysis
?

Group Analysis

1. What does the remaining unmarked piece of paper represent?

2. How many cuts would you have to make before all of your paper was marked and set aside? Explain.

3. Describe the shape of your decay curve.

4. If the element whose decay was simulated in this activity had a half-life of 10 years, how long would this experiment have taken?

5. Describe one difference between the nuclear decay reaction you modeled here and the chemical reactions you studied in earlier activities.

Analysis
(cont.)

?

Individual Analysis

6. There is concern about our use, and subsequent disposal, of radioisotopes. Using evidence from this activity, discuss why this concern exists.

7. The equation below provides a detailed and accurate description of radioactive decay in the element uranium. Examine it carefully. Is the equation balanced? What is your evidence?

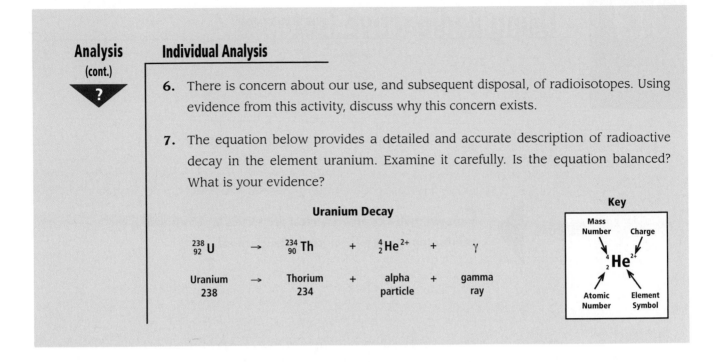

Uranium Decay

$$^{238}_{92}\text{U} \rightarrow ^{234}_{90}\text{Th} + ^{4}_{2}\text{He}^{2+} + \gamma$$

Uranium → Thorium + alpha + gamma
238   234   particle   ray

Key

Mass Number   Charge

$^{4}_{2}\text{He}^{2+}$

Atomic Number   Element Symbol

Using Radioactive Isotopes

Purpose ▶ **C**onsider half-life and chemical properties when choosing the appropriate radioisotope to use for a particular purpose.

Introduction ▼

Radioactive decay releases various forms of energy at a known, predictable rate. As you learned in Activity 31.2, "Fuels for the Future," nuclear power plants use this energy to create heat and generate electricity. In Activity 29, "Refrigeration Technology," you explored how this energy can be used to preserve foods. The energy emitted by radioactive decay can also be used to estimate age or locate the position of materials containing radioisotopes. Some of the most common radioisotopes used for these purposes are shown in Table 2.

Every isotope decays at its own unique rate, as measured by its half-life. We can determine the length of time the radioisotope has been decaying by comparing the amount of decay products to the amount of undecayed radioisotope remaining. This comparison enables us to estimate the age of objects that contain certain radioisotopes. **Tracers** are tags placed on moving objects that allow observers to track the object's position as it moves. Radioisotopes can be used as tracers because the energy given off by the nuclear decay can be detected and followed over time. If a radioisotope is added to a fluid, radioactivity will be detected in areas where the fluid flows. These characteristics allow observation of the movement of blood and other fluids in the body or the flow of oil or water in a pipeline. In this activity, you will explore how radioisotopes and other technologies are used in a variety of situations, and decide which technology you think is most appropriate for each situation.

Procedure

1. Consider the radioisotopes listed in Table 2 for use in each of the three scientific investigation scenarios described on the next page.

2. Record the radioisotope you would recommend for use in each situation.

Table 2   Some Common Useful Radioisotopes

	Half-Life	Decay Product	Major Radiation	Other Information
$^{14}_{6}\text{C}$ (Carbon)	5,568 yrs	$^{14}_{7}\text{N}$ (Nitrogen)	beta	used to date carbon containing samples less than 10 half-lives old
$^{41}_{20}\text{Ca}$ (Calcium)	103,000 yrs	$^{41}_{19}\text{K}$ (Potassium)	electron capture	more difficult to measure; longer half-life, which means that studies over longer time periods can be done
$^{47}_{20}\text{Ca}$ (Calcium)	4.5 days	$^{47}_{21}\text{Sc}$ (Scandium)	beta	smallest half-life, decreasing the exposure of the patient to radiation
$^{45}_{20}\text{Ca}$ (Calcium)	162.6 days	$^{45}_{21}\text{Sc}$ (Scandium)	beta	most researched of the calcium isotopes, and most readily available
$^{147}_{62}\text{Sm}$ (Samarium)	1.06×10^{11} yrs	$^{143}_{60}\text{Nd}$ (Neodymium)	alpha	used to date rocks containing samarium and parent or daughter compounds
$^{182}_{73}\text{Ta}$ (Tantalum)	114.7 days	$^{182}_{74}\text{W}$ (Tungsten)	beta	non-irritating and immune to body liquids

Scenarios

A. A crown rumored to be the Iron Crown of Charlemagne was recently found at an archaeological site. Charlemagne lived from A.D. 794–814, and ruled over western Europe and the Holy Roman Empire from 800–814. Historical records indicate that the crown found by the archaeologists must have been made sometime between the late Roman and Middle Ages, a spread of several hundred years. The main materials that make up the crown are gold, a strip of iron, a nail reputedly from the True Cross, and precious stones. During a more careful study of the crown, it was discovered that its precious stones were held in place by a mixture of beeswax and clay, a common adhesive of the times. This adhesive contains organic material rich in carbon. How would you gather data to decide whether the crown the archaeologists found is actually Charlemagne's Iron Crown?

B. In an effort to gather precise information about the bones of a living organism, an injection of a radioisotope is given. In a few hours, after the bone tissue has had enough time to absorb the injected radioisotope, the radiation emitted by the radioisotope can be used to make a very detailed image of the bone. Which radioisotope would you choose? Why?

C. Toads, like many animals, hibernate during certain parts of the year. To better protect the declining toad population, a local Fish and Wildlife unit has proposed tagging the toads before they go into hibernation. The tag would enable biologists to locate hibernating toads, determine the terrain they prefer, and find out how many toads are in a hibernation group. What kind of tag would you use?

Analysis ?

Group Analysis

1. Which factors did you find most important in deciding which isotope to use in each scenario? Explain.

2. What other information would be useful to you in making each decision?

3. Table 3 outlines non-radioactive methods that could be used for the purposes described in the three scenarios. If you were the chief scientist of the team investigating each of the topics described in the scenarios, would you recommend using any or all of these methods instead of the radioisotopes you chose? Explain.

4. Radioactive decay provides the energy used to generate a large portion of the electricity supply in some parts of the world. What characteristics of nuclear power make it desirable for this purpose? What characteristics make it undesirable? Explain.

Table 3   Three Non-Radioactive Diagnostic Procedures

Technique	How It Works	Other Information
Oxidizable Carbon Ratio	The two chemical analyses used in this procedure determine the total percentage of carbon and the percentage of readily oxidizable carbon in the sample. The results are converted to a ratio of total carbon to the readily oxidizable carbon in each sample, the Oxidizable Carbon Ratio ("OCR"). The OCR value is then factored into an environmentally based contextual formula and an estimate of age results.	Used for carbon containing charcoal samples from wood burned and then preserved in soil. Tested only for samples less than 8000 years old. Results are affected by soil porosity, soil moisture levels, and the length of time between cutting and subsequent burning of the wood.
Bone density (DEXA scan) with x-ray frequencies	Low-dose x-ray of two different energies are used to distinguish between bone and soft tissue, giving a very accurate measurement of bone density at these sites.	This type of scan only determines the bone density. It does not determine the rate of absorption of Ca into the bones or what happens to the bone over time.
Surgical implantation of a radio transmitter in the toad	Radio-transmitter implants can be used with toads. The implant is surgically implanted in the toad's stomach or colon and is excreted naturally after some time. The radio signal transmitted by the implant can be received remotely and the motion of the toad can be traced.	Implant packages should be small enough to allow amphibians to turn around in tight spaces. Implants should be coated with substances to keep the implants from being rejected by the animal's immune system.

Mechanical Energy

35.1 Energy to Move Mountains

Purpose ▶ **M**easure the amount of energy it takes to move an object and consider the trade-offs involved in using mechanical devices.

Introduction ▼

In previous activities you have seen how energy can be transformed, or converted from one form to another. Light can be converted into chemical energy, electricity, and heat; chemical energy can be converted into electricity and heat; nuclear energy can be converted into heat and electricity. **Mechanical energy**—the energy of moving objects—can also be converted into heat, as you observed in Activity 4.4, "Shaking the Shot." In fact, it is almost impossible to prevent mechanical energy from being at least partially converted to heat! In this activity, you will investigate mechanical energy in more detail as you explore some of the ways mechanical energy is used, measured, transferred, and transformed.

All motion in the universe occurs in response to a force. Forces are everywhere. For example, the force of gravity makes objects fall, keeps the gases of Earth's atmosphere from escaping into space, and keeps planets and satellites in their orbits. The exertion of a force involves energy. When a person exerts a force and uses energy to lift a chair, some of the energy is transferred to the chair, where it is stored as potential energy. The chair, because it is now in a higher position, has more potential energy: if it falls, it will fall farther and transfer more energy to the floor or any other object it falls onto. The amount of energy that is transferred to or from an object when its vertical position changes can be calculated using the formula in Equation 1.

Introduction

(cont.)

▼

Equation 1   Energy Transferred When the Vertical Position of an Object Changes

energy transferred   =   weight of object • vertical distance moved

joules (J) =   newtons (N)   •   meters (m)

Note: 1 J = 1 newton-meter (N•m)

1 N = 0.224 lbs

Equation 1 is similar to the equation used to find the amount of energy expended to move an object, which is given as Equation 2.

Equation 2   Energy Expended to Move an Object

energy expended   =   force exerted   •   distance over which force was exerted

(J)   =   (N)   •   (m)

People use mechanical devices to accomplish tasks more easily or to do tasks that would otherwise be impossible, such as moving very heavy loads. Mechanical devices like door knobs, scissors, and ramps decrease the amount of force needed to accomplish a task. There is a common misconception that using mechanical devices saves energy; this is not true. The amount of energy expended depends on both the force exerted and the distance the object is moved, as Equation 2 indicates. Although many mechanical devices decrease the force that is needed, they increase the distance over which the force must be exerted. So, the amount of energy expended to do a task using a mechanical device is actually greater than the amount of energy expended to do the same task without using a mechanical device! (For the purposes of this activity, electrical devices, such as telephones, stereos, and computers, are not considered mechanical devices.)

Whenever one moving object is in contact with another object, the **friction** between them always converts some of the expended energy into heat. In addition to friction, there are usually many other avenues by which energy is "lost" to the environment. The **efficiency** of a mechanical device is a measure of how much usable energy is lost in the process of using it. Given two identical devices, the one with less friction will be more efficient than the one with more friction. Efficiency, often reported as a percentage, can be calculated using Equation 3.

Equation 3   Calculating the Efficiency of a Mechanical Device

$$\text{efficiency} = \frac{\text{energy transferred}}{\text{energy expended}}$$

Materials

For each group of four students

1   spring scale
1   object with a hook
2   meter sticks
1   ring stand and clamp

Procedure

1. Hook the spring scale to your object and lift the object 30 cm from the table top. Record the force needed to do this. The force you measured is the weight of the object.

2. Use two meter sticks and a ring stand to set up a ramp 30 cm high, similar to the one shown in Figure 1.

Figure 1 Setting Up the Ramp

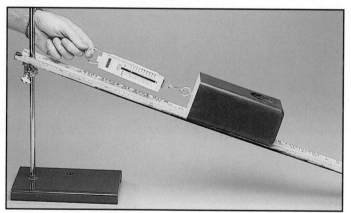

3. Place your object on the bottom of the ramp and use the spring scale to measure the force needed to drag it slowly but steadily to the top of the ramp. Record the force needed.

4. Use the weight of the object measured in Step 1 to calculate the energy transferred to the object when it is raised 30 cm.

5. Use Equation 2 and the force measured in Step 1 to calculate the energy expended in lifting your object 30 cm.

6. Use Equation 1 to calculate the energy transferred to the object when using the ramp.

Hint: Using the ramp does not change the weight of the object, and you raised the object only 30 cm!

Procedure
(cont.)

7. Using the force measured in Step 3, calculate the energy expended when using the ramp. Remember, you had to drag the object the full length of the meter stick to raise it 30 cm!

8. Calculate the efficiency of using the ramp to lift the object. Convert your result into a percentage.

Analysis

?

Group Analysis

1. In Procedure Steps 1 and 3, the object was raised a distance of 30 cm. Compare the force needed to do this with and without the help of a ramp. Describe any differences in the force required.

2. The mechanical advantage of a machine is a measure of how much less force is required when using that machine. Calculate the mechanical advantage of the ramp using the following formula:

$$\text{mechanical advantage} = \frac{\text{force used without machine}}{\text{force used with machine}}$$

3. Most machines are more complicated than a ramp. Would you expect more complex machines to be more or less efficient than a ramp? Explain your reasoning, using the concepts of force and mechanical advantage.

4. As you have seen, using a ramp involves a trade-off between the size of the force applied and the distance over which the force must be exerted. Describe this trade-off.

Individual Analysis

5. Does the efficiency of a machine affect its contribution to sustainable development? Explain.

6. Using a machine involves issues other than those discussed above. Describe one of these other issues and the trade-offs associated with each side of the issue.

35.2 Inertia

Purpose **S**how how inertia affects the force needed to change the motion of an object.

Introduction

To make something move or stop moving, inertia must be overcome. **Inertia** is the tendency of matter to stay in motion if it is already in motion, and to stay at rest if it is already at rest. In the late 1600s, Isaac Newton described inertia in his **First Law of Motion** and emphasized that inertia is always present in matter. Inertia wasn't "discovered" before the 17th century because it is often hard to observe in the day-to-day world. It is difficult to observe because friction can mask its effects. Today, you and your class will investigate a few instances in which inertia can be observed.

Procedure

1. Carefully observe the demonstration performed by your teacher.

2. Prepare a data table to record your observations and briefly explain how they relate to inertia.

3. Follow the instructions provided by your teacher to perform your own inertia activity.

4. Prepare a data table to record your observations and briefly explain how they relate to inertia.

Analysis

?

Group Analysis

1. Describe an example of inertia that was not demonstrated or discussed during this activity.

2. Describe an example of inertia that is difficult to observe in everyday life. Explain why it is difficult to observe.

35.3 Rambling Rates and Fumbling Forces

Purpose

Experiment with the amount of force needed to move objects of different mass a distance of one meter at various rates of acceleration.

Introduction

One way to observe the transfer of energy is to observe the acceleration of an object. **Acceleration** is the rate at which an object speeds up, slows down, or changes direction. Newton's **Second Law of Motion** describes the relationship between the force exerted on an object, the mass of the object, and the acceleration of the object that results from the force. In this activity, you will investigate this relationship by varying each of the three variables, one at a time.

One of the reasons that most tricycles do not move very fast is that children cannot exert a very large force on the pedals.

Materials   **For each group of four students**

1	balance
1	spring scale
2	objects of different mass
2	meter sticks
1	ring stand and clamp
1	stopwatch

Procedure

1. Prepare a data table similar to the one below.

Table 1   Forces Necessary to Move an Object at Rest

Situation	Mass of Object	Distance	Time	Force Applied			
				Trial 1	Trial 2	Trial 3	Average
A		1 meter	5 seconds				
B		1 meter	2 seconds				
C		1 meter	10 seconds				
D		0.5 meter	1 second				

2. Use a balance to determine the mass, in grams, of one of your objects.

3. Attach the spring scale to the object, as you did in Activity 35.1, "Energy to Move Mountains." Start with the object at rest, then pull on the spring scale to make the object move one meter in 5 seconds. Measure the force required to move the object. Record the force, in N, in your table as situation A, trial 1.

4. Repeat Step 3 two more times and record your results as Trials 2 and 3, then find the average.

5. Repeat Steps 3 and 4 for situations B, C, and D, varying the distance and time as shown in Table 1. Enter all results in your data table.

6. Repeat Steps 1–5 with the other object.

Analysis

?

Group Analysis

1. In situations A–C, the object was moved the same distance in a different amount of time. This results in the object speeding up, or accelerating, at a different rate in each situation. Describe the relationship between acceleration and applied force.

2. Look at your data from situation D. Explain why these data do or do not follow the relationship described in Analysis Question 1.

Individual Analysis

3. In which situation did you expend the most energy? Explain.

4. Compare the force applied to move the object with the higher mass 1 meter in 5 seconds with the force required to move the object with the lower mass 1 meter in 5 seconds. Describe the relationship between force and mass.

5. You have just conducted an investigation in which you measured the effect of mass and acceleration on force.

   a. Record your prediction for the relationship between force and acceleration.

   b. Design an investigation to determine the relationship between force and acceleration. Write a step-by-step procedure for your investigation.

   c. Perform your investigation at your teacher's direction.

A Multitude of Machines

Purpose ▶ **O**bserve the use of tools in countries from *Material World* and explore the trade-offs associated with using tools.

Introduction ▼

You have just investigated the use of machines and some of the natural laws that describe mechanical energy. Almost every tool and device people use is a machine. Now, you will consider the use of a variety of devices in specific circumstances. You will evaluate the use of each device based on the task to be accomplished, the availability of materials for making the device, and energy resources needed to operate the device. As you analyze the use of each device, keep in mind the following lessons from the last few activities:

- A force is required to cause an object to change its direction or its speed.

- In general, when a machine is used to do a task, a smaller force is required but more energy is expended.

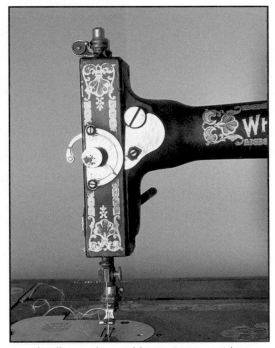

People all over the world sew. Some people use a needle and thread, some use manually powered machines like this one, and others use electric sewing machines.

Materials   ■■ **For each team of two students**

1   copy of *Material World*

Procedure

1. Working in teams of two, look through *Material World* to find examples of the devices listed below. Examine each of the following photographs:

   • wood-burning stove, page 45

   • yoke and oxen with plow, page 75

   • rickshaw, page 69

   • shopping cart, page 140

   • refrigerator, page 132

2. For each photograph, list the reasons why you think the people are using this device.

Analysis   **Group Analysis**

?

1. Consider the trade-offs involved in using each of the devices as pictured. Discuss whether each device could be replaced with some other technology. What are the trade-offs that would be involved in this replacement?

Individual Analysis

2. The woman on a bicycle pictured on page 60 of *Material World* could have used a car to carry her groceries. What trade-offs are associated with the introduction of widespread automobile use in less developed countries?

35.5 Can I Drop You Off?

Purpose ▶ **O**bserve and measure motion at constant velocity or at constant acceleration. Relate these data to energy transferred and energy expended.

Introduction ▼

You have observed that energy is expended when a force exerted on an object causes the object to accelerate. You have also observed the types of tools that are used in various countries for moving people and objects. This activity will focus on describing and measuring movement. Objects can have a position, a **velocity**, and an **acceleration**. This activity will help you understand the differences among these concepts.

Think about what happens when you are driving or riding in a car. If the car is stopped along the road, you could describe its position as "x" number of meters from the last intersection. If the car is moving steadily, you could describe its velocity by recording the speed of the car and the direction in which it is moving. If the driver of the car slows down for a yield sign, speeds up to pass, or swerves to avoid an accident, these changes in velocity can be described as acceleration or deceleration. All motion can be described by velocity and acceleration; the values of these characteristics can be calculated using Equations 4 and 5.

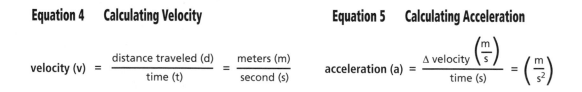

Equation 4   Calculating Velocity

$$\text{velocity (v)} = \frac{\text{distance traveled (d)}}{\text{time (t)}} = \frac{\text{meters (m)}}{\text{second (s)}}$$

Equation 5   Calculating Acceleration

$$\text{acceleration (a)} = \frac{\Delta \text{velocity} \left(\frac{m}{s}\right)}{\text{time (s)}} = \left(\frac{m}{s^2}\right)$$

You are probably familiar with one aspect of an object's velocity: its speed. In the U.S., speed is commonly measured in miles per hour, rather than meters per second. Velocity measures not only speed, but also the direction of movement. Because of this, acceleration—which you may think of as speeding up or slowing down—can also be a change in direction without a change in speed.

Option A   Accelerating an Automobile

Materials

▦ **For the class**

1	car
4	bathroom scales
4	towels or small foam pads
10	droppable, non-bouncing objects (e.g. small sand bags or traffic cones)
2	stopwatches
4	meter (or 2-meter) sticks

■ **For each student**

1	piece of graph paper or graphing calculator or computer graphing program

Procedure

1. Your teacher will assign you a job for the portion of this activity that takes place outdoors. There will be two rounds, so you may have a different job during each round. The jobs include those listed in the table below.

Table 2   Job Assignments

Inside the Car	Outside the Car
1 driver	4 people pushing
1 dropper	4 people measuring the distance
1 timer	4 people recording the distance
	4 people recording the scale readings
	1 timer checker

Safety Note

The car's engine should not be running during the trials and the transmission should be in neutral. The driver should not use brakes or the steering wheel during the trial. However, the driver should pay attention and apply brakes as necessary for safety. When driving the car, make sure all students have moved a safe distance away from the car.

Procedure
(cont.)

2. During the first round, each of the four car pushers should hold a bathroom scale and place it flat up against the back of the car, holding it in place and using a towel to protect the finish of the car. They should push on the scale until the car starts rolling, then continue walking at a constant speed. The constant speed can be accomplished by counting aloud as you take each step.

3. The dropper in the car should drop one droppable object before the car begins to move, and one every 4 seconds thereafter. The timer in the car should start the stop watch when the car begins to move and alert the dropper at every 4-second interval to drop another object. The timer checker should stand outside of the car and use a stopwatch to check the timing of the objects being dropped.

4. When the car has traveled the distance assigned by your teacher, the driver should use the brake to stop the car. The measurers should measure the distance between cones and make sure the recorders report the data for every 4 seconds. The measurers should use a data table similar to the one below.

Round 1: Motion at a Constant Velocity

Time	Distance	Change in Distance	Velocity	Change in Velocity	Acceleration
Start	0	0	0	0	0
4					

The number of rows needed will vary depending on the velocity of the object and the distance it travels.

Note: You will not fill in the velocity or acceleration columns until you do Group Analysis Questions 1 and 2.

5. Have the driver and car pushers carefully return the car to the starting position.

6. During the second round, instead of a constant velocity, the car will be pushed with a constant force. The same procedure for dropping objects every 4 seconds should be followed by the timers and the dropper. However, this time the car pushers will push so that all their bathroom scales always register the same force. They should decide on a force between 5 and 10 pounds, depending on how fast they want to run and how far the car has to travel. The recorders should prepare and complete a second data table similar to the one on the next page and record the force in the table's title.

Procedure

(cont.)

Round 2: Motion Produced by the Exertion of Constant Force (_____ lbs.)

Time	Distance	Change in Distance	Velocity	Change in Velocity	Acceleration
Start	0	0	0	0	0
4					

The number of rows needed will vary depending on the velocity of the object and the distance it travels.

Note: You will not fill in the velocity or acceleration columns until you do Group Analysis Questions 1 and 2.

7. Return the car to a safe parking space.

8. Return to the classroom and record all the data collected during the activity.

Analysis

?

Group Analysis

1. Calculate the velocity for each time interval and add this data (in the appropriate units) to your data table. Remember that each time interval is 4 seconds.

2. Calculate the acceleration for each time interval and add this data (in the appropriate units) to your data table. Again, remember that each time interval is 4 seconds.

3. Prepare a graph for each trial showing how distance, velocity, and acceleration changed over time. Use different colors or symbols to identify each curve.

4. Describe in words the curves for each trial, using numerical values from your data.
   Example: In the case of constant velocity, the increase in distance was linear, the velocity was constant at 2 m/s, and the acceleration was 0 m/s^2 throughout the trial.

5. The same amount of energy is transferred when a car drives at a constant speed from one city to the next as when it speeds up and slows down repeatedly on the way. Do both cases require the same force to be exerted? How does this relate to the volume of gasoline consumed?

Individual Analysis

6. Relate the curves that you prepared in Analysis Question 3 to the concept of energy transfer. In which trial was more energy transferred to the car? Did the car expend any energy? Explain.

Option B   Accelerating Anything

Materials

■■ **For each team of two students**

1	ticker tape
1	ticker timer
1	object (e.g. brick, block, or cart)
1	spring scale
1	string, 10–15 cm in length (optional)
1	meter stick
1	roll of masking tape

■ **For each student**

1	sheet of graph paper or graphing calculator or computer graphing program

Procedure

1. Attach the spring scale to the object, using the string if necessary.

2. On the flat surface where you will conduct your investigation, use a small piece of masking tape to mark the starting point and each 10 cm for a total of 50 cm.

3. Set up the ticker timer as directed by your teacher. Record the rate at which the timer marks the paper (for example, every $\frac{1}{60}$ second).

4. Pull the spring scale and attached object so that the object always travels the distance between consecutive marks in the same amount of time. Record the force applied at the 10-cm, 30-cm, and 50-cm mark and save the ticker tape.

5. Measure the distance between 20 consecutive dots found near the middle of your ticker tape and record the distances between each dot in a data table similar to the one below.

Round 1: Motion at a Constant Velocity

Time	Distance	Change in Distance	Velocity	Change in Velocity	Acceleration
0	0	0	0	0	0

The number of rows needed will vary depending on the velocity of the object and the distance it travels.

Note: You will not fill in the velocity or acceleration columns until you do Group Analysis Questions 1 and 2.

6. Pull the spring scale and attached object so that the spring scale always measures the same force. Note the force and save the ticker tape.

Procedure
(cont.)

7. Measure the distance between 20 consecutive dots found near the middle of your ticker tape and record the distance between each dot in a data table similar to the one below. Record the force in your table's title.

Round 2: Motion Produced by the Exertion of Constant Force (_____ lbs.)

Time	Distance	Change in Distance	Velocity	Change in Velocity	Acceleration
0	0	0	0	0	0

The number of rows needed will vary depending on the velocity of the object and the distance it travels.

Note: You will not fill in the velocity or acceleration columns until you do Group Analysis Questions 1 and 2.

Analysis

Group Analysis

1. Calculate the velocity for each time interval and add these data (in the appropriate units) to your data table. Recall that you recorded the time interval between each dot on the ticker tape in Procedure Step 3.

2. Calculate the acceleration for each time interval and add these data (in the appropriate units) to your data table.

3. Prepare a graph for each trial showing how distance, velocity, and acceleration changed over time. Use different colors or symbols to identify each curve.

4. Describe in words the curves for each trial, using numerical values from your data. **Example:** In the case of constant velocity, the increase in distance was linear, the velocity was constant at 2 m/s, and the acceleration was 0 m/s^2 throughout the trial.

5. The same amount of energy is transferred when a car drives at a constant speed from one city to the next as when it speeds up and slows down repeatedly on the way. Do both cases require the same force to be exerted? How does this relate to the volume of gasoline consumed?

Individual Analysis

6. Relate the curves that you prepared in Analysis Question 3 to the concept of energy transfer. In which trial was more energy transferred to the object? Did the object expend any energy? Explain.

Trade-offs of Energy Use

36.1   By-Products of Combustion

Purpose ▶ **C**ompare the potentially polluting by-products formed during the combustion of two types of fuel.

Introduction ▼

You have been studying how energy resources are converted to forms that can be used by humans, and how humans use that energy. In this activity, you will consider a number of the trade-offs associated with the use of two of the world's major energy resources—fossil fuels and nuclear fuels.

You have seen that energy—especially energy produced by combustion—plays an important role in powering the machinery on which our society has come to depend. In Activity 31, "Fueling Trade-offs," you saw examples of energy produced by combustion. Now you will examine some of the by-products of combustion.

Combustion, which provides much of the energy we use for transportation, industry, and in the home, also produces vast quantities of pollution.

Materials

For each group of four students

1 glass fuel burner containing ethanol

1 glass fuel burner containing kerosene

1 30-mL dropper bottle of bromthymol blue (BTB)

1 30-mL dropper bottle of 0.05 M sodium hydroxide (NaOH)

access to water

■■ For each team of two students

1 piece of aluminum foil

1 180-mL plastic bottle with cap

2 30-mL graduated cups

■ For each student

1 pair of safety glasses

Safety Note

Wear safety glasses. Burners can be hot after they are turned off and no longer appear to be hot. Sodium hydroxide is a strong base and can damage skin and clothing. Avoid direct contact with it. If any sodium hydroxide spills, rinse repeatedly with water and notify your teacher.

Procedure

1. Adjust the wick of each burner so that it barely projects above the metal casing. Light the wicks and observe the two flames. Make a table to compare the similarities and differences you observe between the flames.

2. Decide which team of two will investigate which fuel first. (You will switch fuels in Step 11). Carefully hold one piece of aluminum foil about 5 cm above the top of the flame for 15 seconds. Describe any changes in the foil.

3. Exchange your piece of foil with the other team in your group and compare the appearance of the two pieces.

4. Place 10 mL of water in each of the two graduated cups, add two drops bromthymol blue (BTB) to each one, and swirl each cup to thoroughly mix the liquids.

Procedure
(cont.)

5. Carefully place the open mouth of your team's 180-mL bottle over the lit wick of your burner and let it rest on the metal casing of the burner as shown in Figure 1. When the flame is extinguished, keep the bottle in an upside-down and vertical position and quickly but carefully lift it up just enough to screw on the cap.

6. As quickly as possible, turn the bottle right-side up, uncap it, pour in the contents of one of your graduated cups of BTB solution, and recap the bottle.

7. Shake the bottle vigorously 20 times so that the BTB solution mixes thoroughly with the gaseous products of combustion. Pour the BTB solution back into the empty graduated cup and describe the appearance of the liquid. Use the other cup containing BTB solution as a standard for comparison.

8. Add 0.05 M NaOH one drop at a time to the graduated cup containing the BTB solution that was shaken in the bottle, swirling after each drop, until the original blue color returns.

Figure 1   Placing an Inverted Bottle Over a Lit Burner

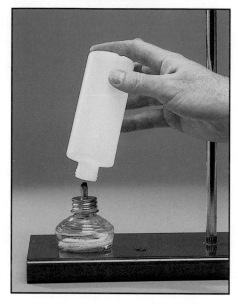

9. Rinse out your used lab equipment and repeat Steps 4–8 two more times so that you have data for three trials.

10. Prepare a data table to display the results of each trial as well as any appropriate averages.

11. Switch fuel burners with the other team in your group and repeat Steps 4–10.

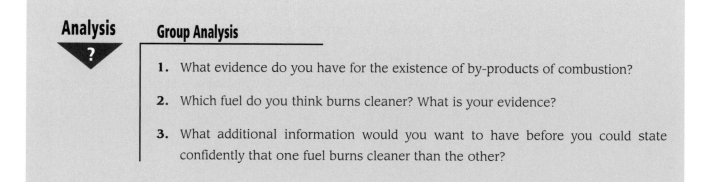

Analysis
?

Group Analysis

1. What evidence do you have for the existence of by-products of combustion?

2. Which fuel do you think burns cleaner? What is your evidence?

3. What additional information would you want to have before you could state confidently that one fuel burns cleaner than the other?

36.2 How Air Quality Affects You

Purpose ▶ **E**xplore the health effects of air pollution and discover which section of your community has the highest levels of particulate pollution.

Introduction ▼

Pollutants released into the atmosphere are spread by wind and can often affect communities far from the source.

There are a variety of pollutants that come from a number of sources. A **point-source pollutant** can be tracked back to a specific location, such as a smokestack. Non-point source pollution, such as smog, is typically created by different sources at different locations, such as cars on a highway or backyard barbecues scattered throughout a community. Air pollution does not necessarily spread out equally in all directions from a source. In fact, in some cases areas farther from a source may have higher pollutant levels than areas nearer to the source. Local and regional atmospheric conditions, such as wind, can affect the spread of air pollutants, and different pollutants interact with the environment in different ways. Various methods are used to detect, identify, and determine the amount of pollutants present in the air.

Many states have a department of health services that uses an air quality reporting method called the **Pollutant Standards Index** (PSI). The PSI provides a simple, uniform way to report daily levels of air pollution and to advise the public when the air quality may be harmful to health. A value of 100 on the PSI index represents the maximum acceptable concentration for each pollutant as set by the U.S. Environmental Protection Agency (EPA). The PSI index is divided into six air-quality categories: Good (0–50), Acceptable (51–100), Poor (101–200), Extremely Poor (201–300), Dangerous (301–400), and Extremely Dangerous (401–500).

Part A Pollutant Standards Index

Prediction

Do you think your community has an air pollution problem?

Procedure

1. Collect the PSI ratings for your community from your local newspaper or television news station each day for a week.

2. Use this information and the data in Tables 1 and 2 to analyze the danger of air pollution in your community.

Table 1 Some Major Pollutants and Their Effects

Pollutant	Primary Ambient Air Quality Standards (100 on the PSI)	Health Effects When Standard Exceeded
Carbon Monoxide CO	9 parts per million averaged over an 8-hour period	Interferes with oxygen carrying capacity of the blood. Weakens contractions of the heart. Individuals with anemia, emphysema, or other lung disease, as well as those living at high altitudes, are more susceptible to the effects of CO. Even at relatively low concentrations, CO can affect mental function, visual acuity, and alertness.
Ozone O_3	0.12 parts per million averaged over a 1-hour period	Irritates eyes and nose. Increases the aging process and susceptibility to infection. Impairs normal functioning of the lungs.
Particulates	150 micrograms per cubic meter of air averaged over a 24-hour period	Aggravates acute respiratory illness. Small suspended particles can be inhaled deeply into the lung where they are hard to dislodge and may carry other pollutants deeper into the lung.
Sulfur Dioxide SO_2	0.14 parts per million averaged over a 24-hour period	Increases the risk of acute and chronic respiratory disease.
Nitrogen Oxides NO_x	0.25 ppm over 1 hour period (California standard)	Irritates the lungs and lowers resistance to respiratory diseases (influenza). Important contributor to O_3 formation and acid rain.

Table 2   Pollutant Standards Index (PSI)

Index Value	Air Quality	General Health Effects	What You Should Do	Other Precautions Taken
0–50	Good	None	Have a great day.	
51–100	Acceptable	None	Have a good day.	
101–200	Poor	Mild aggravation of symptoms in susceptible people. Symptoms of irritation in the healthy population.	People with existing heart or respiratory ailments should reduce physical exertion and outdoor activity.	
201–300	Extremely Poor	Significant aggravation of symptoms and decreased exercise tolerance in people with heart or lung disease. Widespread symptoms among the healthy population.	The elderly and people with heart or lung disease should stay indoors and reduce physical activity. Vigorous outdoor activity should be curtailed. Normally healthy children are not affected for brief periods, e.g., during school recesses.	If an air pollution ALERT is issued: • A public advisory will be issued to the mass media. • Industry voluntarily cuts back operations that pollute. • Citizens are asked to voluntarily curtail all non-essential motor vehicle use.
301–400	Dangerous	Premature onset of certain heart and respiratory diseases, as well as significant aggravation of their symptoms. Decreased exercise tolerance in healthy people.	The elderly and people with existing heart and respiratory diseases should stay indoors and avoid physical exertion. The general population should avoid outdoor activity.	If an air pollution WARNING is issued: • A public advisory will be issued to the mass media. • Mandatory cutback takes place of industrial operations that pollute. • Citizens are asked to reduce motor vehicle use.
401–500	Extremely Dangerous	Premature death of ill and elderly. Healthy people will experience adverse symptoms that affect their normal activity.	Everyone should remain indoors, keeping windows and doors closed. Everyone should minimize physical exertion and avoid traffic.	If an air pollution EMERGENCY is issued: • A public advisory will be issued to the mass media. • Mandatory curtailment to the national holiday levels will take place in all industries, places of employment, schools, colleges, and libraries. • Citizens are required to curtail motor vehicle use.

Analysis

Group Analysis

1. Do you think the pollutant concentration levels that are used to define the six categories of the PSI are appropriate? Explain why or why not.

2. Which pollutant(s) do you think are the most dangerous? Why?

3. Think about your answer to the Prediction question. Now do you think your community has an air pollution problem? What is your evidence?

Part B Particulate Pollution in the Community

Prediction

What part of your community do you think will have the greatest amount of particulate pollution? Explain your reasoning.

Procedure

1. With the other members of your group, create an experimental procedure that each student in the class could do at home to test your prediction.

2. Discuss your ideas with the class, and agree on one procedure.

3. Write down the procedure agreed upon by the class.

4. Collect your data.

5. On the due date, report your data to the class.

Analysis

?

Individual Analysis

1. Using the results from the entire class, write a report following the guidelines given to you by your teacher.

2. Explain why you agree or disagree with the following quote from Michael Deland, an EPA administrator: "Now the problem (of air pollution) is us—you, me, our cars, our woodstoves."

36.3 Measuring Particulate Pollution

Purpose ▶ **A**nalyze a set of information to identify any relationships among particulate concentration, concentrations of other pollutants, and atmospheric conditions.

Introduction ▼

Levels of airborne particulate pollution are often determined by using a scale that measures the blackness of a filter used to collect particulates. A measure of the amount of light that can pass through the filter is called the **attenuation value**. Attenuation (ATN) values can be estimated by sight or analyzed more precisely using a specially calibrated light source and light detector. The higher the concentration of particulates, the blacker the filter becomes. The blacker the filter, the less light that can pass through it, resulting in a higher ATN value.

In this activity you will estimate the ATN values for a series of filters used to collect airborne particulates throughout a four-day air pollution episode in southern California. In addition to data on the amount of airborne particulates, you will analyze data on other atmospheric conditions collected during the same time period.

One of the other data sets contains measurements of the concentration of ozone in the lower atmosphere. As you learned in Activity 29.4, "Unforeseen Consequences," ozone is naturally distributed in the upper atmosphere, where it has the beneficial effect of shielding surface ecosystems—including human beings—from the sun's damaging ultraviolet rays. Ozone also forms in the lower atmosphere. When it is concentrated in urban smog, ozone is a harmful pollutant.

Part A   Estimating ATN Values

Materials   **For each team of two students**

1   light attenuation scale
1   magnifier

Procedure

1. Below are several simulated particulate filters. Rank them from lightest to darkest.

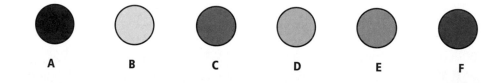

A   B   C   D   E   F

2. Check to see how your results compare with those of the other team in your group. Describe any differences you notice.

3. Use the magnifier and light attenuation scale to determine the ATN value for each of the simulated filters. Record these values.

4. Record the values determined by the other team in your group and calculate the average for your group. Record this average and report it to your teacher, who will calculate the class average for each filter.

Part B Investigating Variations in Particulate Accumulation

Materials

For each group of four students

1   copy of Transparency 36.4, "Variations in Air Temperature and Surface Ozone Levels"

1   copy of Transparency 36.5, "Variations in Wind Speed and Direction"

For each team of two students

1   Particulate Accumulation Data Sheet

1   light attenuation scale

1   magnifier

Procedure

1. Examine each of the 14 filters on the Particulate Accumulation Data Sheet and use the light attenuation scale to determine the ATN value of each.

2. Calculate your group's average ATN value for each filter.

3. Make a line graph showing how the ATN value varies over the sampling period. Make your graph the same size as the ones on the transparencies.

4. Compare the variation in ATN value with that of the four other atmospheric measurements shown on the transparencies.

Analysis

?

Individual Analysis

1. Is there a relationship between the amount of ozone in the air and the ATN value of the filters? Explain why or why not.

2. Which, if any, of the atmospheric measurements seem to have a relationship to the amount of particulates in the filter?

3. Which of the variables seems to have the strongest relationship to the amount of particulates in the filter? Explain.

4. Are you confident that the variable you described in Analysis Question 3 is responsible for the variation in ATN values? Explain why or why not.

36.4 Effects of Radiation on DNA

Purpose ▶ **S**imulate the effects of radiation on the DNA replication process.

Introduction

Unlike combustion, nuclear fuels do not produce harmful gaseous or particulate pollutants. However, there are risks associated with nuclear energy—exposure to the radiation produced by nuclear fuel or the wastes from nuclear reactions can be dangerous. Also, because the amount of energy that can be produced by a nuclear reaction is so huge, there is potential for catastrophic accidents.

Radiation is potentially harmful because it can break bonds in the structure of the DNA molecule. These breaks may disrupt the normal functioning of the DNA. Damaged DNA can result in the death of cells and organisms, and can also cause changes, or **mutations**, in a gene. Mutations may cause abnormal cell function or result in birth defects and other inherited genetic abnormalities.

Natural radiation from sunlight, cosmic rays, and decay of radioactive elements—as well as radiation emitted from devices such as x-ray machines or nuclear reactors—carries enough energy to harm DNA molecules. In Activity 19.1, "Genes, Chromosomes, and DNA," you built a model of a DNA molecule and used it to model DNA replication. In this activity, you will investigate the effects of high-energy radiation on a molecule of DNA.

Materials   ■■ **For each team of two students**

    1   DNA model set

Procedure   **Figure 2   Radiation Striking a DNA Segment**   **Figure 3   DNA Damage Resulting From Radiation**

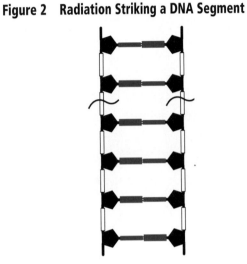
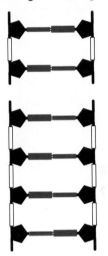

1. Build a model of a DNA strand as shown in Figure 2. Use whatever base pairs you would like, then make a labeled sketch of your strand.

2. The squiggly lines in Figure 2 indicate high-energy radiation striking the bonds that hold the nucleotides together. In your model, break the two nucleotide bonds that are struck by the radiation to produce two DNA fragments similar to those shown in Figure 3. This type of radiation damage to a DNA molecule is known as a double-stranded break (DSB).

3. Often, as the cell attempts to repair the damage created by a DSB, a DNA fragment will reconnect with a strand other than the one from which it originally broke. To simulate this, connect one of your fragments with a fragment produced by the other team in your group.

4. Make a labeled sketch of the DNA strand formed when you reconnected the fragments.

Analysis

Group Analysis

1. It is not unusual for a DSB to be repaired incorrectly.

   a. Describe two possible consequences for a cell in which a DSB has been repaired incorrectly.

   b. Explain which consequence you think is more likely, and why.

Individual Analysis

2. Discuss two environmental factors other than radiation that could introduce an error in the genetic code of a cell.

3. Describe two of the trade-offs between the benefits of high-frequency radiation and the risks involved in genetic mutation.

36.5 Not In My Back Yard

Purpose ▶ **E**xamine trade-offs involved in the disposal of waste produced by nuclear power plants.

Introduction

Over the past 20 years, some Asian and European countries have built new nuclear power plants for generation of electricity. In that same time span, the U.S. has built no new nuclear power plants and has taken several older ones out of service. When deciding whether or not to use nuclear power, the many risks associated with the use of nuclear energy technology must be considered. For example, there is the possibility of serious accidental explosions, the potential that the technology and materials could be used to build nuclear weapons, the possibility of sabotage, and the need to prevent the release of high-frequency radiation throughout the **nuclear fuel cycle**—including ore mining, fuel fabrication, transportation, storage, reactor operation, reprocessing, and disposal of radioactive waste. As you modeled in Activity 36.4, "Effects of Radiation on DNA," radiation from nuclear reactions can cause cell mutations. It can also damage the molecular structure of many materials.

This activity focuses on the disposal of **nuclear wastes**, which are non-useful radioactive substances that are produced along with the useful energy. The radiation emitted from these wastes can remain dangerous for tens to hundreds of thousands of years. Of particular concern is high-level nuclear waste created during fission reactions. Fission is a special kind of radioactive decay that occurs when a neutron hits the nucleus of a heavy element and causes the nucleus to split into two roughly equal parts. Splitting of the nucleus releases energy and energetic particles and creates two different, lighter elements. In a nuclear power plant, the energy released during nuclear fission is used to boil water and create steam. This steam is used to turn the turbines of electricity generators. The majority of high-level nuclear waste comes from nuclear power plants. This reading outlines some of the issues surrounding the storage and disposal of those wastes.

Disposing of Radioactive Waste

Radioactive wastes come primarily from nuclear power plants, military activities, research laboratories, industrial processes, and medical facilities. Some nuclear wastes are classified as low-level wastes, which can be disposed of in concrete canisters at specialized sites. Although low-level wastes remain radioactive for long periods of time, they are not nearly as concentrated as the high-level waste from nuclear power plants. It is the disposal of high-level wastes that creates the biggest problem and causes the most concern.

The radioactive decay of a nuclear fuel never completely stops; there are always some atoms that continue to emit radiation. So, even when the fuel is "spent"—no longer providing enough energy to be useful—it is still radioactive. Spent fuel becomes part of the high-level waste from a nuclear power plant. Another component of nuclear waste is the material that is made radioactive by exposure to the nuclear reaction. As you have learned, all radioactive material has a half-life. The long half-lives of the isotopes contained in many of the wastes from nuclear technologies cause these wastes to remain potentially dangerous far into the future. For this reason, special precautions must be taken to store and dispose of these wastes. Any radiation that escapes into the outside environment could endanger the health and safety of people and the environment, not only in this generation but also in future generations.

In a nuclear power plant, high-level nuclear waste is periodically removed from the heavily shielded reactor. Wastes are usually stored temporarily at the reactor site before being transported to a permanent storage facility. At temporary storage sites, nuclear wastes are placed in small steel containers that are usually kept above ground in tanks or, occasionally, placed in underground pools of water. These containers are rigorously tested for chemical and physical strength, not only to make sure they are adequate for on-site storage but also to make sure they can withstand the rigors of transportation by train or truck.

Once the wastes have been transported to the permanent storage facility, they are sealed with glass and concrete and stored deep underground or in caverns deep inside mountains. Sites for permanent storage facilities are chosen based on both technical and political considerations. Potential locations for these facilities should be unpopulated, seismically inactive, and relatively dry.

Hanford, Washington, is the only permanent storage facility in the U.S. that currently accepts high-level nuclear waste. The Hanford site is experiencing difficulties with leaking underground storage tanks and possible groundwater contamination, and its storage capacity has nearly been reached. Research is in progress to locate another suitable site. Much political and technical controversy surrounds the selection of new waste-disposal sites because of the hundreds of thousands of years during which the nuclear wastes can remain radioactive and potentially dangerous. It is not possible to ensure that any waste storage facility will be safe and secure for that long.

Many people living in states such as Nevada and New Mexico, where prospective sites are located, are opposed to locating long-term waste storage facilities near their communities. Understandably, they are reluctant to assume the risks associated with the storage of nuclear wastes without having received the benefits associated with the use of nuclear energy. Similar resistance exists in communities bordering the highways or railways along which nuclear wastes would have to be transported to the disposal site. The U.S. government has proposed to "compensate" the recipient states for the risks they are asked to assume, but so far the proposed compensation has not overcome the political resistance. To date, Congress has been reluctant to force unwanted facilities on less populous and less wealthy states, in part because such an action poses a threat to political values that are embodied in the U.S. Constitution and in over 200 years of political practice.

Nuclear waste is stored in much the same way in many other countries. However, in France and the United Kingdom spent fuel is reprocessed to extract plutonium, a decay product of the fission of uranium. Plutonium, the preferred material for making nuclear bombs, can also be used as a fuel for nuclear reactors—it actually yields more energy than did the initial fission of uranium!

Nuclear power generated using the "plutonium cycle" is for practical purposes "inexhaustible" and cuts down on the volume of waste generated. However, because of the risks associated with plutonium—its explosiveness, toxicity, and links to the proliferation of nuclear weapons—a political decision was made in the U.S. not to use the "plutonium cycle" for nuclear energy.

Analysis
▼
?

Group Analysis

Recall what you have learned about fossil fuels and alternative energy sources as you respond to the following questions.

1. What are some of the trade-offs involved in the disposal of nuclear wastes from power plants?

2. Considering the trade-offs associated with other possible sources of electricity, do you think that the risks of using nuclear power to generate electricity outweigh the benefits? Explain.

3. When analyzing the trade-offs of resource use, should the distribution of risks and benefits be considered? Have the beneficiaries of yesterday's (and today's) resource use paid the full cost, or are these costs being paid (or will be paid) by other people in other places and times who do not share in the benefits?

Global Perspectives on Sustainability 37

37.1 Energy Use and Sustainability

Purpose ▶ **A**nalyze levels of energy production and consumption in various countries throughout the world in terms of sustainability.

Introduction ▼

Human society is, in part, based on our ability to use and produce energy. Energy use varies widely from individual to individual and from society to society. Energy production and resources also vary widely from region to region. Access to energy is very important in maintaining or increasing a country's economic output and its residents' standard of living. Throughout the history of the human race, the amount of energy used per capita has increased steadily. One measure of the potential for future access to energy is the size of the known energy reserve—that is, current estimates of the amount of energy that can be supplied using the natural resources that have been discovered and can be exploited using today's technologies. However, it must be kept in mind that reserve estimates can change dramatically from one year to the next.

Throughout the world, humans rely on many different sources of energy. There are advantages and disadvantages associated with the use of each energy source.

Materials

■■ **For each team of two students**

1   copy of *Material World*

■ **For each student**

supply of graph paper

Procedure

1. Tables 1 and 2, on the next page, include many statistics related to energy resources that can be used to compare and categorize different countries. Choose one or two statistics, or some combination of statistics, that you think will best allow you to distinguish those countries that use energy more sustainably from those that use energy less sustainably.

2. Use your chosen statistic(s) to create one or two bar graphs that illustrate which countries use energy more sustainably and which use energy less sustainably.

    Note: 1 exajoule = 1 x 10^{18} J
    1 gigajoule = 1 x 10^{9} J

3. Calculate the number of years that the world's reserves of fossil fuels will last, based on the 1993 estimate of the reserves and the 1999 annual production.

Table 1   1999 Energy Production

Country	Total (exajoules)	Per Capita (gigajoules)	Energy Source (%)			
			Fossil Fuels	Hydroelectric	Nuclear	Other**
World	411.2	69	79.6	2.3	6.7	11.2
United States	70.6	252	81.2	1.5	11.9	4.5
*Brazil	5.6	33	48.4	18.7	0.8	32.1
*China	45.7	36	78.5	1.6	0.4	19.5
*Cuba	0.1	20	46.2	0.2	0	53.6
*Ethiopia	0.7	12	0	0.8	0	99.2
*Guatemala	0.2	20	24.0	3.3	0	72.7
*India	9.9	17	49.4	1.7	0.8	48.1
*Thailand	0.9	26	63.3	0.8	0	35.9
Iceland	0.1	344	0	22.9	0	77.1
Israel	0.03	4	11.4	0.5	0	88.3
Italy	1.2	20	70.2	14.1	0	14.5
Japan	4.4	34	4.7	7.1	79.1	8.0
Kuwait	4.4	2,360	100	0	0	0
Mexico	9.34	96	91.8	1.3	1.2	5.8
Russia	39.8	272	94.4	1.5	3.4	0.5
United Kingdom	11.8	199	90.3	0.2	8.8	0.6

*Less developed country

**Geothermal, solar, wind and traditional (wood, charcoal, dung, etc.)

Table 2   1999 Energy Consumption and Estimated 1993 Reserves

	1999 Energy Consumption			Estimated 1993 Reserves		
	Total Energy Use (exajoules)	% Change Since 1989	Per Capita Total Energy Use (gigajoules)	Fossil Fuels (exajoules)	Uranium (exajoules)	Hydroelectric (exajoules/yr)
World	406.2	13	68	41,000	37,000	n/a**
United States	95.0	16	339	7,400	61,000	9.4
*Brazil	7.5	31	45	110	27,000	34
*China	45.6	29	36	3,500	0	67
*Cuba	0.5	(26)	47	0.6	0	n/a
*Ethiopia	0.8	25	12	0.8	0	5.1
*Guatemala	0.3	47	23	2.5	0	1.4
*India	20.1	38	20	2,100	0	5.8
*Thailand	3.0	85	48	36	0	0.2
Iceland	0.1	57	478	0	0	2.0
Israel	0.8	57	131	0.1	0	0.1
Italy	7.1	12	123	12	800	1.5
Japan	21.6	24	170	25	1,100	3.6
Kuwait	0.7	1	392	560	0	0
Mexico	6.2	23	64	370	1,900	2.3
Russia	25.2	n/a	173	2,000	50,000	n/a
United Kingdom	9.6	9	163	64	0	0.1

*Less developed country

**n/a = not available

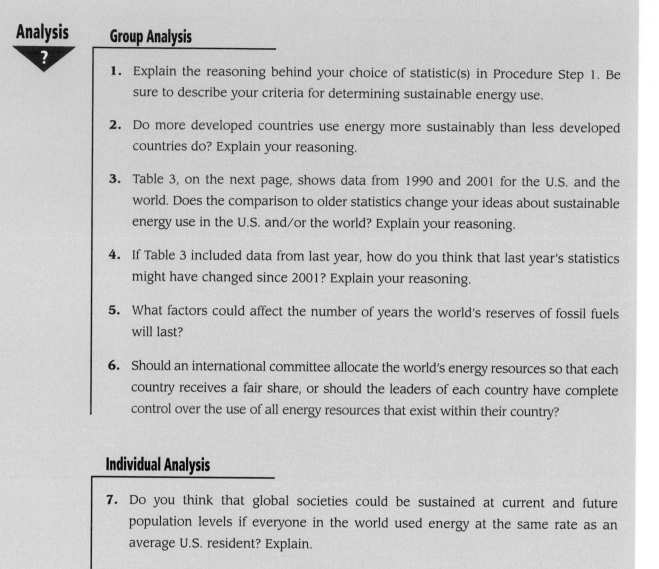

Analysis ?

Group Analysis

1. Explain the reasoning behind your choice of statistic(s) in Procedure Step 1. Be sure to describe your criteria for determining sustainable energy use.

2. Do more developed countries use energy more sustainably than less developed countries do? Explain your reasoning.

3. Table 3, on the next page, shows data from 1990 and 2001 for the U.S. and the world. Does the comparison to older statistics change your ideas about sustainable energy use in the U.S. and/or the world? Explain your reasoning.

4. If Table 3 included data from last year, how do you think that last year's statistics might have changed since 2001? Explain your reasoning.

5. What factors could affect the number of years the world's reserves of fossil fuels will last?

6. Should an international committee allocate the world's energy resources so that each country receives a fair share, or should the leaders of each country have complete control over the use of all energy resources that exist within their country?

Individual Analysis

7. Do you think that global societies could be sustained at current and future population levels if everyone in the world used energy at the same rate as an average U.S. resident? Explain.

8. How do you propose—on an individual level, a national level, and a global level—that humans prepare to meet future energy demands? What trade-offs are involved in your proposals?

Table 3   1990 and 2001 Energy Consumption in Exajoules

	1990		2001	
	World	**United States**	**World**	**United States**
Total	378.0	87.9	419.9	95.5
Fossil Fuels	321.2 (85.0%)	78.6 (89.4%)	333.8 (79.5%)	82.3 (86.2%)
Hydroelectric	8.5 (2.3%)	1.1 (1.3%)	9.2 (2.2%)	0.8 (0.8%)
Nuclear	23.2 (6.1%)	7.0 (8.0%)	29.0 (6.9%)	8.8 (9.2%)
Renewables (Geothermal, Solar, Wind, and Traditional)	25.1 (6.6%)	1.2 (1.4%)	47.9 (11.4%)	3.6 (3.8%)

37.2 Global Shuffle!

Purpose ▶ **C**onsider the impact of present-day decisions on the future of a community.

Introduction

Leaders of all countries—whether they are royalty, democratically elected officials, military leaders, or have obtained their leadership role through other means—often make decisions that affect the livelihood, environment, and quality of life of the people who live within their borders. The leaders of different countries can choose to enact policies that promote strategies with widely varying short- and long-term effects. Some countries try to maximize profits; some countries try to increase the quality of life of the residents; some countries try to conserve natural resources; some countries try to preserve the environment.

In this activity, you will be the leader of your community. You will make decisions about the development of your community and see some of the consequences of your decisions.

Materials

  For each group of four students

4 playing pieces
1 game board
1 set of "Something's Happening" cards
1 set of "You Make the Call" cards
1 die

■ **For each student**

1 copy of Student Sheet 37.1

Overview

In this game, each player assumes the role of leader of the community in which he or she lives. As leader, you must have a clear vision of your goals for your community. As you roll the die to move through the year, you will pick cards, most of which will ask you to make a decision that will affect your community's wealth, energy or material resources, air or water quality, and population size. At the end of three years, you will analyze your performance in terms of how well your decisions helped achieve your goals and how your decisions affected the sustainability of your community.

Procedure

1. Have each person in your group choose a playing piece and place it on the "Start" area of the game board.

2. Briefly describe your goals for your community on your copy of Student Sheet 37.1. Write down your strategy for how you will make decisions to accomplish your goals.

3. Shuffle the "Something's Happening" cards, then put them on the appropriate place on the game board. Do the same with the "You Make the Call" cards.

4. When all your group members have completed Step 2, determine who will go first. The order of play should be clockwise.

5. To begin play, roll the die. Move your playing piece that number of months, beginning with January. Depending on which month you land on, pick the topmost "Something's Happening" card or the topmost "You Make the Call" card. Record the type of card you chose on Student Sheet 37.1.

6. Read the card carefully. If your card provides you with two or more options, study the short- and long-term consequences, make your choice based on the goals you have for your community, and then record the point values of your decision on Student Sheet 37.1. If your card does not give you options, simply record the point values indicated on the card.

Procedure
(cont.)

7. When you move from December to January on the game board and cross the "Happy New Year" point, you must take a census to determine if the population has increased or decreased before you pick your first card of the new year.

    If population is unchanged, enter zero in all columns in the row labeled "Year-End Adjustment."

    If population has increased, enter –5 points in both the "Air and Water" category and the "Resources" category in the row labeled "Year-End Adjustment."

    If population has decreased, enter +5 points in both the "Air and Water" category and the "Resources" category in the row labeled "Year-End Adjustment."

8. Continue playing until each player has completed three full years.

9. Determine your cumulative totals for your three-year term of office.

10. Discuss the Group Analysis Questions with other members of your group, then complete the Individual Analysis Questions on your own.

Analysis
?

Group Analysis

1. What were your goals for your community and why did you choose these goals?

2. Describe the short- and long-term consequences of your three-year term of office.

3. Do you think that the leader of an actual community or country would be wise to have the goals you had for your community? Explain.

4. What goals do you think are appropriate for the U.S. to have? Discuss the trade-offs that need to be considered when pursuing these goals.

Individual Analysis

5. Make up a new "You Make the Call" card that differs in a significant way from any of the existing cards. Explain why this is an important development issue that is facing today's leaders and how you determined the point values.

Glossary

acceleration The rate of change in speed or direction of a moving object; expressed in m/s^2.

acetylsalicylic acid The chemical name for aspirin.

acid rain Rain contaminated by air pollutants that make it low in pH; may contain sulfuric, nitric, carbonic, and chloric acids.

activation energy The energy required to break the bonds in reactant molecules to start a chemical reaction.

active transport A process in which energy is expended to move a substance across a cellular membrane from a region of lower concentration to a region of higher concentration, opposite the direction of normal diffusion; generally carried out by proteins in the cell membrane.

air pollutant Any of a broad class of compounds that can cause environmental damage when released into the air. Major classes of air pollutants include chlorofluorocarbons, which can cause destruction of the ozone layer; sulfur and nitrogen oxides, which contribute to acid rain and smog; and particulates, which can cause asthma and other respiratory problems.

allele Any of the alternative forms of a gene that occupy the same place, or locus, on a specific chromosome and that help determine a specific trait.

alloy A solid mixture of two or more elements, at least one of which is metallic.

alpha particle A small particle given off during some forms of nuclear decay; identical to a helium nucleus, having two protons, two neutrons, and a mass number of 4.

alternative energy Energy obtained for human use through methods other than the combustion of fossil fuels. Examples include solar energy, wind energy, and nuclear energy.

arable land Land that is suitable for agricultural use.

atom The smallest particle of an element; made of protons, electrons, and (in most cases) neutrons.

attenuation value (ATN) A measure of how much light can pass through a filter; used to measure the amount of particulates in the air.

beta particle A small particle given off during some forms of nuclear decay; an electron.

bioaccumulation The accumulation of toxic material at higher trophic levels of food webs.

biodegradable Capable of being broken down by the activity of microorganisms or other living things.

biofuel Any biological material that can be burned directly or turned into a product that burns.

biomass The total mass of biological material in a defined area.

biotechnology The application of biological science, especially through genetic engineering and recombinant DNA technology, in order to produce products or outcomes considered beneficial to humans.

birth rate The ratio between births and individuals in a specific population at a particular time. Typically reported as the number of births per 1,000 individuals per year.

boiling point The temperature at which a specific substance changes from a liquid to a gas and vice versa; different for every substance.

bond See **chemical bond**.

Boyle's law For a constant number of molecules of a gas at constant temperature, the product of pressure and volume is constant; $P_1V_1 = P_2V_2$. Stated qualitatively, as pressure goes up, volume goes down.

by-product Any material produced by a chemical reaction in addition to the principal product.

Calorie The unit most often used to report the energy content of food products. 1 Calorie = 1000 calories = 1 kilocalorie.

calorie A unit used in the measurement of energy, equal to 4.18 joules; the amount of energy required to raise the temperature of 1 gram of water 1°C.

calorimeter An insulated vessel designed to measure the energy released or absorbed during chemical and physical changes.

carbon A chemical element, each atom of which has six protons. The most common isotope of carbon, carbon-12, has six neutrons and an atomic mass of 12 amu. Carbon atoms almost always form four bonds with other atoms, creating a wide variety of carbon-containing compounds. Carbon is found in all organic substances, such as proteins, fats, carbohydrates, and DNA, and commonly appears in petroleum products and in biological reactions.

carrying capacity The maximum population that an ecosystem can support without undergoing deterioration.

catalyst A substance that lowers the energy necessary to start a chemical reaction. Catalysts take part in the chemical reaction, but retain their original form when the reaction is complete.

cell membrane The physical boundary around a cell that separates the chemicals in the cell from those outside the cell.

cell wall A rigid structure, produced by plant cells, that surrounds the cell membrane and provides structural support for the plant.

cellular respiration A chemical process occurring in an organism which provides energy necessary for that organism's survival. In animals and many other organisms, oxygen and sugars react to release stored chemical energy, also producing carbon dioxide and water.

change of state The change from one phase of matter to another. The change between solid and liquid (melting/freezing) and between liquid and gas (boiling/condensing) are changes of state.

Charles' law For a constant number of molecules of a gas at constant pressure, the ratio of volume and temperature is constant; $V_1/T_1 = V_2/T_2$. Stated qualitatively, as temperature goes up, volume goes up.

chemical bond The attractive force that holds atoms together. Different types of bonds involve different amounts of energy. The strongest, highest-energy bonds are those occurring between atoms in a molecule.

chemical equation A mathematical representation of a chemical reaction. Reactants appear on the left side of the equation; products appear on the right.

chemical family A column of elements on the Periodic Table of the Elements, having similar chemical and physical properties.

chlorofluorocarbon (CFC) Any compound containing carbon, chlorine, and fluorine. CFCs in the upper atmosphere have damaged the protective ozone layer. CFCs were used in aerosols until the early 1990s and continue to be used as solvents and refrigerants.

chlorophyll The compound in green plants and algae that absorbs energy from the sun, enabling photosynthesis and giving these organisms their green color.

chloroplast The organelle in plant cells that contains chlorophyll.

chromosome Cell structure in which the genes are located; made up of molecules of deoxyribonucleic acid (DNA) that encode information for the growth, maintenance, and development of cells; found in the nucleus of eukaryotic cells.

clone A living organism which has genetic material identical to that of its parent; in nature, offspring resulting from asexual reproduction. Humans have also found ways to genetically engineer clones of organisms that reproduce sexually.

cloning The process that results in offspring having genetic material identical to that of the parent organism.

combustion A chemical reaction in which a fuel reacts with oxygen to form carbon dioxide, water, and energy; other products may also result when combustion is incomplete or the reactants contain impurities.

compound A pure substance made up of atoms of two or more elements bonded together in definite ratios.

computer model A computer-generated representation of a system that enables manipulation of variables and observation of the consequences of changing those variables; a form of mathematical model.

conceptual model A diagram or other pictorial representation of the components of a system that enables observation and clarification of the relationships among those components.

conductivity The relative ability of a material to transmit heat or electricity.

consumer An organism, often represented as part of a food web, that derives energy from ingesting other organisms.

criteria pollutant Any chemical whose presence in the environment is used as an indicator of the general level of pollution.

cross-linked polymer A web-like structure created when crosswise connections form between polymers. Cross-linking can result from the addition of substance like sodium borate, in the case of Slime, or by hydrogen bonding of the original polymer molecule to itself and others, as in the case of Kevlar.

crude oil A liquid mixture of many different hydrocarbons that have formed naturally beneath Earth's surface; also called petroleum. Crude oil is commonly obtained by drilling a hole to the depth of the subsurface oil and pumping the oil to the surface.

cytoplasm The mixture of water and other chemicals enclosed by the cell membrane, excluding genetic material.

daughter cell A new cell produced as a result of cell division.

death rate The ratio between deaths and individuals in a specific population at a particular time. Typically reported as the number of deaths per 1,000 individuals per year.

decomposer An organism that uses dead organic material as its source of energy. By feeding on and breaking down this material, decomposers release inorganic nutrients needed by many plants and other producer organisms.

decomposition The process by which complex materials are broken down into simpler chemical forms through chemical reactions. Decomposition can result from the action of microorganisms, chemicals, heat, or light.

degradation A process similar to decomposition, in which larger or more complex materials are broken down into smaller or less complex materials, but not necessarily through chemical reactions.

demographic transition A theory of population change according to which human birth and death rates change over time as a result of socioeconomic changes rather than environmental pressures.

demography The study of human populations, especially population size, density, distribution, and vital statistics.

density The mass of a material per unit of volume;

$$d = \frac{m}{v}$$

deoxyribonucleic acid (DNA) The chemical name for the molecule found in chromosomes that carries the genetic information in cells. DNA molecules are made of ribose sugar rings, phosphate groups, and nitrogenous ring structures called bases.

diagnostic properties Properties of a material that distinguish it from all others and thus can be used to identify it.

diatomic element An element made up of atoms bonded in pairs.

diffusion The random spontaneous movement of substances from areas of high concentration to areas of low concentration, often through permeable membranes; this process tends to result in equal concentrations on either side of the membrane.

disease resistance The ability of a living organism to avoid infection by disease-causing organisms.

dominant allele An allele that determines the phenotype whenever present in a gene.

ductility The relative ability of a material to be shaped into a wire.

ecological impact The effect that a specific process or action performed by humans or other species has on the environment.

ecology A branch of the life sciences that focuses on the interactions among organisms and between organisms and their surroundings.

ecosystem A group of living and non-living things which, with the addition of energy (typically from the sun), form a sustainable system.

efficiency In any machine, the ratio of the amount of useful energy transferred to the total amount of energy expended.

element A substance made of only one kind of atom; cannot be chemically broken down into different, simpler substances.

endothermic A term describing any process that requires energy from its surroundings; often observed by a decrease in temperature.

energy The capacity for movement or the ability to cause movement; found in many different forms.

energy consumption The use of energy produced from any source; generally referring to the use of non-renewable energy sources.

energy flow diagram A graphical representation of the energy transfers that occur during a chemical or physical change.

energy production The generation of usable energy from any renewable or non-renewable energy source.

energy reserves The projected amount of energy that could be economically produced from known sources, using known technology.

energy source Any material or organism from which energy is transferred during a chemical or physical change.

energy transfer The process through which energy is exchanged; typically associated with a chemical or physical change.

enzyme A biological catalyst; often made of one or more proteins.

equilibrium A balanced state; in chemistry, the condition that exists in a solution when a chemical reaction is proceeding at the same rate forwards and backwards, resulting in almost constant concentrations for all components.

eukaryotic cell A cell that has a nucleus.

eutrophication The process by which a body of fresh water fills up with organic material. This process can be accelerated by the introduction of nutrients from agricultural runoff or wastewater treatment that promote plant growth.

evaporation The change of state from liquid to gas at a temperature below the boiling point.

evaporative cooling The decrease in temperature brought about when a liquid absorbs energy from its surroundings as it evaporates.

exothermic A term describing any process that produces energy and releases it to the surroundings; often observed by an increase in temperature.

expended Used or given up to do a job.

experimental error A value used to express the degree of similarity between an experimental value and an ideal or expected value; generally calculated using the equation

$$\text{error} = \frac{\text{experimental value} - \text{expected value}}{\text{expected value}}$$

exponential growth Expansion that increases by a greater amount during each time interval; follows the mathematical relationship $y = x^n$, where n is a non-negative, real number. Typically seen in populations where resources are plentiful and space is not a constraint.

extrapolate To work from a set of known data in order to make inferences outside of the range of the known data.

false-color image An image, often a map, in which colors do not represent the actual appearance of the objects depicted, but are instead used to enhance or distinguish characteristics of objects in the image.

feedback loop A situation in which information transferred between interacting objects continually alters the nature of their interaction.

fermentation A chemical reaction through which organisms break down an energy-rich compound, typically sugars, without consuming oxygen and often resulting in the formation of alcohol; anaerobic respiration.

fertility rate The ratio between births and adult females in a specific population at a particular time. Typically reported as the number of births per 1,000 adult females per year.

fertilizer Substances rich in plant nutrients.

First Law of Thermodynamics All heat entering a system adds that amount of energy to the system; a subset of the Law of Conservation of Energy.

first-generation offspring The progeny of the organism being studied; the children.

fixation The process of stabilizing potentially mobile substances, usually contaminants or nutrients, to prevent leaching.

food chain A sequential ordering of organisms in an ecosystem, in which each uses the previous organism as a food source.

food web A representation of the relationship among all interacting food chains within an ecosystem, showing how each organism derives energy required for survival.

force Any influence that can cause acceleration.

fossil fuel A non-renewable energy source derived from originally living matter and typically extracted from below Earth's surface; any combustible product derived from coal, petroleum, and natural gas.

fractional distillation A process by which a mixture of liquids can be separated into its various components based on differences in boiling point. When the mixture is heated, the component with the lowest boiling point vaporizes and these gases are collected, cooled, and condensed, forming the first fraction. With continued heating, a series of fractions is collected, each containing a relatively pure sample of one of the substances that made up the original mixture.

frequency (a) The number of repetitions of an event within a given time. (b) In the study of waves, the number of wavelengths that pass in a second; expressed in Hertz.

friction Resistance to movement resulting from contact between two surfaces.

fuel Any chemical substance that reacts to produce energy; many common fuels combust when reacting with oxygen.

gamma rays High-energy electromagnetic radiation produced from nuclear reactions. Low-dose exposure occurs naturally, but exposure to high doses can cause significant genetic damage and, as a result, health problems.

gene A sequence of DNA base pairs that codes for one hereditary trait; found on a chromosome.

genetic engineering See **biotechnology**.

genetics The study of heredity and gene behavior.

genome The collective set of genetic information that describes a species.

genotype The genes possessed by an individual organism, including dominant and recessive alleles.

geothermal energy Energy harnessed from the heat stored in Earth's interior.

gray (Gy) A unit for measuring radiation dosage.

growth rate The relative change in the value of a variable. Often calculated using the equation

$$\text{growth rate} = \frac{\text{final value} - \text{initial value}}{\text{initial value}}$$

half-life The time required for half of the atoms in a sample of a radioactive isotope to decay.

heat capacity The maximum amount of thermal energy a sample of material can store.

heat Energy transferred between substances of different temperatures.

heat flow The transfer of thermal energy.

heat of combustion The amount of energy generated from a combustion reaction; different for each fuel.

heat of fusion The amount of heat that must be transferred to freeze or melt a 1-gram sample of a substance.

heat of vaporization The amount of heat that must be transferred to boil or condense a 1-gram sample of a substance.

heavy metals Metal elements with a large atomic mass; components of many toxic substances.

heterozygous Having one dominant allele and one recessive allele of a given gene.

hidden costs Effects of a process or action that do not involve obvious financial costs in the short term, but that do have less apparent negative consequences, such as environmental damage and/or resource depletion, in both the short and long term.

homeostasis A relatively stable state of equilibrium; the maintenance of a balance of energy and material flow into and out of living cells.

homeotherm Any animal species that maintains a relatively constant body temperature, regardless of surroundings; examples include mammals and birds.

homozygous Having either two dominant alleles or two recessive alleles of a given gene.

humus Soil that is very rich in organic matter.

hybrid (a) The offspring of a cross between two unlike parents (of the same or different species). (b) A trait that mixes the characteristics of two genes.

hydrocarbon A compound containing only hydrogen and carbon; often found in petroleum, natural gas, and coal.

hydrochlorofluorocarbon (HCFC) A compound containing carbon, chlorine, fluorine, and hydrogen; currently used in refrigerants as a replacement for chlorofluorocarbons (CFCs).

hydroelectric energy Energy harnessed from water in the form of electricity; generally accomplished by allowing a river or other waterway to flow over a water wheel, causing it to turn. This turning motion is used to power an electric generator.

hypertonic solution A solution that has a higher chemical concentration than another solution, to which it is being compared; often, a solution that has a higher concentration than the cytoplasm of a cell that it surrounds.

hypotonic solution A solution that has a lower chemical concentration than another solution, to which it is being compared; often, a solution that has a lower concentration than the cytoplasm of a cell that it surrounds.

incomplete combustion Any reaction of a fuel with oxygen in which the products are not limited to carbon dioxide and water. Products typically include carbon monoxide, compounds containing impurities found in the fuel and air, and non-combusted fuel that can persist in the atmosphere as pollutants.

independent assortment The process by which two unlinked genes are inherited independently of each other; generally the case with genes found on different chromosomes.

inertia A property of matter by which it remains at rest or in uniform motion unless acted upon by some external force.

infant mortality rate The ratio between the number of infants that die before age 1 and the total number of births in a specific population at a particular time. Typically reported as the number of infant deaths per 1,000 births per year.

inference A conclusion reached by considering available evidence and making judgements based upon it.

infrared radiation Electromagnetic radiation that has a frequency just above that of light visible to the human eye; sensed as heat. Infrared radiation is visible to some insects.

insulation The prevention of transmission of heat or other types of energy through a material; a material which prohibits such transfer.

interpolate To estimate the value of an unmeasured variable that falls between two or more known values.

ionizing radiation Electromagnetic radiation containing sufficient energy to disrupt electron configurations and cause the formation of ions in exposed materials.

isotonic solution A solution that has the same chemical concentration as another solution, to which it is being compared; often, a solution that has the same concentration as the cytoplasm of a cell that it surrounds.

J-curve The type of curve that appears in a graph describing exponential growth.

joule The metric unit of energy equal to 0.239 calories.

kinetic energy Energy associated with motion.

lactase The enzyme needed to convert lactose to glucose so that it can be metabolized.

lactose The sugar found in milk.

Law of Conservation of Energy Energy cannot be created or destroyed.

leachate Liquid that results from the movement of aqueous material through contaminated soil or landfill, carrying contaminants to new areas.

less developed country A country in which the average resident does not have access to many economic, material, and energy resources; a country that uses less than the energy equivalent of 1,200 kg of oil per person per year.

limiting factor A component in a system that prevents further change in a particular variable.

limiting resource A material needed for a reaction that is in short supply; in biology, materials needed for continued cell or population growth.

line of best fit A straight line drawn through a plotted set of data points used to define the relationship between two variables.

linear growth Expansion that increases by the same amount during each time interval; follows the mathematical relationship y = mx + b, where m and b are non-negative, real numbers.

luster The reflectivity of a material's surface; shininess.

macronutrient Any nutrient of which the human body needs to receive at least 12 mg/day.

malachite ore A common ore of copper containing a variety of copper-bearing minerals.

malleability The relative ability of a material to be bent.

mathematical model A quantitative description of the relationship between two or more variables, often expressed as an equation.

mechanical energy Energy of motion; commonly associated with moving machines and devices.

meiosis A process of cell division that results in the production of sex (sperm and egg) cells. Sex cells contain half the number of chromosomes found in the parent cell.

melting point The temperature at which a solid turns into a liquid and vice versa.

metabolism The processes by which organisms use energy.

micronutrient Any nutrient needed to be received by the human body in quantities greater than 0 but less than 12 mg/day.

mineral (a) Any chemical element required by the human body. (b) Any naturally occurring crystalline substance.

mitochondrion A cell organelle associated with respiration and energy production.

mitosis A process of cell division that results in two daughter cells identical to the parent cell; asexual reproduction.

model Any representation of a system or its components.

molecule Two or more atoms tightly bonded together to form a neutral particle.

monomer The smallest individual repeating unit of a polymer.

more developed country A country in which the average resident has access to many economic, material, and energy resources; a country that uses more than the energy equivalent of 1200 kg of oil per person per year.

mutations Changes in the chemical structure (specifically, the sequence of base pairs) of the DNA molecule; can occur spontaneously or as the result of environmental factors.

Newton's first law of motion Objects in motion tend to stay in motion; objects at rest tend to stay at rest unless acted on by an unbalanced force.

Newton's second law of motion Any unbalanced force exerted on an object will cause a change in its motion, or acceleration; often expressed as the equation F=ma.

Newton's third law of motion For every action, there is an equal and opposite reaction.

nitrogen base One component of the DNA molecule. The four nitrogen bases found in DNA are adenine, guanine, cytosine, and thymine. The sequence of these bases in DNA codes for the production of specific proteins.

non-point source pollutant Any contaminant that cannot be traced to a single place of origin. For example, smog is a mixture of pollutants derived from a variety of sources.

non-renewable resource A resource that cannot be replenished within a single human lifetime; often used to refer to fossil fuels and mineral resources.

nuclear decay The process by which the nuclei of some atoms break into pieces, forming new atom(s) and releasing energy.

nuclear energy Energy recovered from the exothermic fission (splitting apart) of heavy atoms, usually of uranium or plutonium, or fusion (joining) of light atoms, usually of hydrogen. Fission is used in nuclear power plants and nuclear weapons; fusion is the process that powers the sun.

nucleotide Subunit of DNA consisting of one deoxyribose molecule, one phosphate group, and one nitrogen base.

nucleus The inner portion. (a) In a eukaryotic cell, the organelle in which chromosomes are found. (b) In an atom, the region where protons and neutrons are found.

nutrient cycle The movement of a nutrient through an ecosystem.

optimum The value, or set of conditions, which results in the most favorable outcome.

ore Any rock containing a sufficiently high concentration of elements for use by industry.

organelle A specialized, membrane-bound component of a cell, analogous to an organ. Examples include the nucleus, chloroplasts, and mitochondria.

osmosis Movement of a water or another solvent across a membrane into an area of higher chemical concentration.

oxygenated fuel Gasoline to which an additive has been added in order to promote complete combustion; common additives include ethanol and methyl tert-butyl ether (MTBE).

ozone A molecule consisting of three oxygen atoms having the chemical formula O_3. Ozone is considered an air pollutant when found near Earth, because it is toxic when breathed and contributes to the "greenhouse effect." However, ozone is a vital component of the stratosphere because it reduces the amount of UV radiation that reaches Earth's surface.

ozone layer A part of the stratosphere characterized by a high concentration of ozone molecules; prevents much of the UV radiation that enters Earth's atmosphere from reaching the planet's surface.

particulate Any small solid carried in the air or water.

per capita Per person (in Latin, *per capita* means "by head").

Periodic Table of the Elements A systematic table used to organize chemical elements based on physical and chemical properties. The organization of the modern periodic table was developed by Dmitri Mendeleev.

permeability The quality of a substance that allows molecules to move through it.

petroleum A natural resource formed from decaying plant and animal material over millions of years.; used primarily in the manufacture of plastics, pharmaceuticals, and fuels.

phenotype The characteristics which appear in an individual as determined by both genotype and environment.

photodegradable Able to be broken down by the action of light.

photosynthesis The process through which plants and phytoplankton use energy from sunlight to convert carbon dioxide and water into oxygen and sugars. Organisms that photosynthesize are the primary producers of most of the energy in the food web.

physical model A three-dimensional representation of a variable or system, constructed to show physical relationships between components.

physical system A group of objects and/or organisms which are related or connected in some way; such systems are defined by an outside observer for the purposes of investigation.

phytoplankton See **plankton**.

plankton Microscopic aquatic organisms, typically found at or near the bottom of aquatic food webs. Phytoplankton derive energy from the sun through photosynthesis; zooplankton derive energy from what they eat.

plant cell The structural unit of a plant, consisting of cytoplasm, organelles, cell membrane, and cell wall.

poikilotherm Any animal species whose internal body temperature changes based on the temperature of its surroundings; examples include reptiles and insects.

point-source pollutant Any contaminant that can be traced to a single place of origin. For example, some pollutants can be traced back to a specific smokestack or industrial plant.

Pollutant Standards Index (PSI) A standard used to quantify the concentration levels of criteria pollutants based upon their effects on human health.

polymer A long chemical compound made of many repeating structural units.

population The total number of organisms of a particular species occupying a given ecosystem at a given time.

population crash A sharp decrease in the number of organisms of one or more species, generally due to a natural disaster or environmental crisis.

population curve A graph showing the number of organisms in an ecosystem plotted against time; typically used to represent the combined effects of birth rate, death rate, and migration.

population density The total number of organisms inhabiting a given area at a given time.

population doubling time The time span during which the number of organisms of a particular species in a defined ecosystem has, or is projected to have, increased by a factor of 2.

population dynamics The study of changes in birth rate, death rate, and migration that affect the total number of organisms in a given ecosystem.

porosity The percentage of open space in a solid material.

potential energy Stored energy; often used to refer to energy resulting from an object's position relative to Earth's surface or to the energy stored in the bonds of molecules, as in food.

predator-prey relationship A relationship between two species of organisms in which one is a food source for the other.

producer An organism that converts energy from the environment into chemical energy, typically through photosynthesis; an organism at the first trophic level.

product A substance created as a result of a chemical reaction. The number and type of elements found in the products of a chemical reaction are identical to those found in the reactants.

prokaryotic cells Cells that do not have a nucleus. Members of the kingdom Monera have prokaryotic cells.

properties Chemical or physical characteristics of a substance or material.

protein A biological molecule made of amino acids; used in the body for maintenance and repair of tissues.

prototype A physical model designed to show features of a manufactured object; generally produced to test possible designs before actual production.

qualitative analysis A non-numerical manner of reporting and collecting data; in science, often used as the first stage of investigation.

quality of life A comparative measure used to assess the relative well-being of an individual or group, based on qualitative indicators such as health, happiness, sense of self-worth, self-confidence, and peace.

quantitative analysis A numerical manner of reporting and collecting data.

radioactivity The property possessed by some elements of spontaneously emitting high-energy particles while undergoing nuclear decay.

raw material Any unprocessed material that can be used to make products.

reactant A substance consumed during a chemical reaction. The elements found in the reactants of a chemical reaction are used to form the products.

reaction pathway A set of elementary reactions whose overall effect is a net chemical reaction; the mechanism for a reaction.

recessive allele An allele that determines the phenotype only when paired with an identical allele to form a homozygous recessive gene.

recombinant DNA DNA molecules that have been cut apart and then put back together again in a new way. Creation of recombinant DNA allows the transfer of genetic material from one species or individual to another.

refining The process by which ores or other raw materials are used to make metals or other products.

refrigeration cycle The process by which a continuously circulating refrigerant transfers heat from one area to another.

remote sensing Any of a number of methods for the retrieval of data using technologies that reach beyond the human body; often used to refer specifically to satellite technologies.

renewable resource A resource that can be replenished within a short time span, such as food crops or solar energy.

reservoir A stored supply of a given resource; often used in reference to stored water.

resource allocation The decision-making process involved in determining which people or industries will have access to limited supplies, such as food, materials, or energy.

runoff Water that does not soak into the ground but instead travels along, or very near to, Earth's surface, picking up and transporting suspended particulates and dissolved substances.

satellite An object that orbits a celestial body; often used to refer to manufactured objects placed in orbit around Earth for use in global communication and remote sensing.

saturated solution A solution in which as much solute has been dissolved as possible; if more solute were added, that amount would remain undissolved.

scatterplot A type of graph in which x and y data values are used to create a set of points that describe the relationship between the x and y variables. The set of plotted points can be used to determine the curve, and its associated mathematical equation, that best describes the relationship between the variables x and y.

scientific law An observation that has been made repeatedly by many individuals and that does not conflict with other scientific laws, according to the best current knowledge.

S-curve The type of curve that appears in a graph describing a period of exponential growth followed by a period of zero growth, as in a population that has reached carrying capacity.

Second Law of Thermodynamics Heat will always flow from an area of higher temperature to an area of lower temperature.

second-generation offspring The progeny of the progeny of the original organism being studied; the grandchildren.

selective breeding The practice of changing the reproductive behaviors or processes of plants and animals in order to create offspring with specific, desired traits.

semi-permeable Term describing any material, often a membrane, that permits the passage of some materials and prohibits the passage of others.

smog The brownish haze caused by the action of solar UV radiation on hydrocarbons and nitrogen oxides in the atmosphere from automobile exhaust and industrial emissions; typically found in urban areas.

soil The mixture of organic and inorganic materials in which most plants grow; can be more or less fertile depending on the environment.

specific heat The amount of energy required to increase the temperature of a 1-gram sample of matter 1°C.

standard of living A comparative measure used to assess the relative economic level of an individual or group of people, based on quantitative indicators such as average income and the monetary value of their possessions.

stored energy See potential energy.

stratosphere The part of the atmosphere 10-50 km from Earth's surface. The ozone layer is located in the stratosphere.

structural formula A diagram of a molecule showing the geometrical arrangement of atoms and the bonds between them.

subsoil The layer of soil found beneath the topsoil; generally less rich in nutrients and water content than the topsoil layer.

sulfuric acid A strong acid with the chemical formula H_2SO_4; used in many industries, including the refining of copper ore.

support energy Energy that is indirectly required to produce a product. For example, the energy required to produce fertilizer used in agriculture is support energy for the production of the food crop.

survival The continuation of life or existence.

sustainability Characteristic of a process or habit which allows this process or habit to continue indefinitely; involves minimizing energy and resource loss.

sustainable development The achievement of industrial and economic growth in ways that do not cause environmental degradation; development that results in increased standard of living and quality of life in the present without decreasing the ability of future generations to attain an equal or greater standard of living and quality of life.

technology Tools that make tasks easier; often applied specifically to computers and other electrical devices.

temperature A measure of the average kinetic energy of the molecules of a substance; units of measurement include degrees Celsius (°C), degrees Fahrenheit (°F), and Kelvins (K).

test cross The practice of breeding an organism whose genotype for a specific trait is unknown with an organism that is known to be homozygous recessive, in order to determine the genotype of the first organism.

theory A scientifically acceptable principle or group of principles proposed to explain observations and to extend current scientific understandings of the natural world.

thermal conductor A material that permits heat to be transmitted easily.

thermal energy Energy of atomic or molecular motion.

thermal insulator A material that does not permit heat to be transmitted easily.

thermodynamics The study of heat and its relationships to other forms of energy.

titration A quantitative method used to determine the concentration of a chemical in solution through the use of indicators; typically used for acids and bases.

topsoil The nutrient-rich layer of soil where plants grow most easily.

toxic waste Any by-products that are harmful to humans and/or the environment; usually, such waste must be disposed of according to governmental regulations.

trade-off A balancing of risks and benefits associated with choosing a course of action; giving up one thing in favor of another.

traditional fuels Renewable energy sources such as wood, dung, or charcoal.

transferred Conveyed to another person or object.

transgenic organism An organism that contains genes from a different organism.

trophic level Any of the hierarchical levels of a food web, characterized by organisms that are the same number of steps removed from the producers. Producers are at the first trophic level; organisms at the second trophic level, called primary consumers, eat producers; and organisms at higher trophic levels, called secondary or tertiary consumers, eat other consumers.

ultraviolet (UV) light Electromagnetic radiation that has a frequency slightly higher than visible light; the lowest-energy ionizing radiation.

variable An element of an experiment or investigation which can be changed independently of all other elements.

velocity The speed and direction of motion.

viscosity The property of resistance to flow in a fluid or semifluid. Highly viscous fluids flow poorly; liquids with low viscosity flow very easily.

wastewater Water and water-borne wastes originating from municipal, industrial, and commercial processes; sewage.

wastewater treatment The use of physical, chemical, or biological methods to reduce the toxicity of wastewater before it is released into natural water bodies.

water retention The absorption and retainment of moisture by a soil or other substance.

wavelength A measure of wave size; typically measured as the distance between consecutive crests (or troughs) of a wave train.

x-ray A type of high-frequency, high-energy electromagnetic radiation; often used to obtain images of bones within the body.

yeast Single-celled fungi commonly found on plants, on skin, in intestines, and in soil and water. Yeast obtain energy through fermentation and are often used in the process of brewing beer and making bread. Some varieties also cause infections in humans.

zooplankton See **plankton**.

Index

Credits

PART 1

Activity 1 11 (Fig. 2), 12 (Fig. 3): Tessera* **14 (photo):** Reprinted courtesy of the Cotsen Institute of Archaeology, University of California, Los Angeles. Photo by George Lee.

Activity 3 35 (Fig. 2): After Knut Schmidt-Nielsen, *Animal Physiology*, 2nd ed. (Cambridge: Cambridge University Press, 1979), p. 185.

Activity 4 45 (Fig. 1), 47 (photo), 54 (Fig. 4), 55 (Fig. 5): Tessera* **58–59:** Text and figure from *Collected Works of Count Rumford, Volume 1, The Nature of Heat*, edited by Sanborn C. Brown. Copyright ©1968 by the President and Fellows of Harvard College. Reprinted by permission of The Belknap Press of Harvard University Press.

Activity 5 60 (photo): Tessera*

Activity 6 68 (Table 1): Data source for schooling and percent income spent on food: *Material World* (Sierra Club Books), 248–249; for area and population: Population Reference Bureau, *2004 World Population Data Sheet*; for energy use, World Bank, *World Development Report 2001*, **71 (Table 2):** Committee on Environment and Natural Resources, National Science and Technology Council, **73 (photo):** Tessera*

Activity 7 77: Text adapted from "What Is a Model Anyway?" in "What Are Computer Models and What Can They Tell Us About Waquoit Bay?" *Waquoit Bay National Estuarine Research Reserve Science and Policy Bulletin*, series 1, p. 1. **79–82:** The deer population models used in this activity were developed by Michael Monroe and Raul Zaritsky of the Education Division of the National Computational Science Alliance and are included by kind permission of NCSA. ©NCSA. All rights reserved. This work was funded by the North Central Regional Technology Educational Consortium. For additional information, visit their website at http://www.ncsa.uiuc.edu/edu/. STELLA® is a registered trademark of High Performance Systems, Inc. For more information on HPS programs, visit their website at http://www.hps-inc.com/. **84 (Fig. 4):** Data source: United Nations Department of Economic and Social Information and Policy Analysis, Population Division, "World Population Growth From Year 0 To Stabilization." (New York: United Nations Development Program, 6/7/94). Posted on the Internet at gopher://gopher.undp.org:70/00/ungophers/popin/wdtrends/histor/. **84 (Fig. 5):** Data source: United Nations Department of International Economic and Social Affairs. *Long-Range World Population Projections: Two Centuries of Population Growth, 1950–2150* (New York: United Nations, 1992), p. 14.

Activity 8 92 (Fig. 5): Drawing by Sue Boudreau. **98–103:** See note for Activity 7, pp. 79–82, above.

Activity 9 108 (Table 4): Data source: John Weeks, *Population*, 5th ed. (Belmont, CA: Wadsworth Publishing Company, 1992), p. 47 (data for 1650 and 1950); "World Population Data Sheet," Population Reference Bureau, Washington, D.C., 1996 (data for 1996 and 2025).

PART 2

Activity 11 136 (Table 1): 2004 *World Population Data Sheet*, Population Reference Bureau, **145 (Table 3), 146 (Table 4):** Data source: FAOSTAT (Statistical database of the Food and Agriculture Organization of the United Nations, http://apps.fao.org); **(Table 4):** Data source: Lester Brown and Hal Kane, *Full House: Reassessing the Earth's Population Carrying Capacity* (New York: W.W. Norton & Co., 1994), pp. 66–67; **147 (table):** Data source: Brown and Kane, *Full House*, p. 64, table 4–1.

Activity 13 170 (Fig. 5), 172 (Fig. 6): Tessera*

Activity 14 179 (Fig. 2), 180 (Fig. 3), 181 (Fig. 4a, b, c): Tessera*

Activity 15 193 (photo): Tessera*

Activity 16 204: Quoted passages from *Experiments Upon Vegetables*, by Jan Ingenhousz (London, 1779). Reprinted in "Jan Ingenhousz: Plant Physiologist," by Howard S. Reed, *Chronica Botanica* 11 (1947–48): 309–83. **206 (Fig. 2):** From *Biology* by Helena Curtis. ©1968, 1975, 1979, 1983 by Worth Publishers. Used with permission.

Activity 17 212 (photo): Laura Baumgartner* **215 (photo):** © Lab-Aids, Inc.®

Activity 18 221 (photo): Laura Baumgartner*

Activity 19 240 (Fig. 11): ©1996 Monsanto Company. May be reproduced for educational purposes without permission.

* Photographed for SEPUP